An Introduction to

Differential Geometry and Topology in Mathematical Physics

An Introduction to Differential Geometry and Topology in Mathematical Physics

Wang Rong & Chen Yue

Zhejiang University, P R China

World Scientific

Singapore • New Jersey • London • Hong Kong

Published by

World Scientific Publishing Co. Pte. Ltd.
P O Box 128, Farrer Road, Singapore 912805
USA office: Suite 1B, 1060 Main Street, River Edge, NJ 07661
UK office: 57 Shelton Street, Covent Garden, London WC2H 9HE

British Library Cataloguing-in-Publication Data
A catalogue record for this book is available from the British Library.

AN INTRODUCTION TO DIFFERENTIAL GEOMETRY AND TOPOLOGY IN MATHEMATICAL PHYSICS

ISBN 981-02-3559-3

This book is printed on acid-free paper.

Printed in Singapore by Uto-Print

Acknowledgments

I wish to express my gratitude to Professors D. Y. Gan, B. Y. Hou, and C. L. Shen for several very helpful conversations and suggestions, and to Professor Yue Chen for preparing the English version of my manuscripts.

WANG, RONG

Preface

The purpose of writing this book is to give an outline of the classical results and the most recent developments of differential geometry and topology, especially those important developments which are closely related to the new discoveries in the field of theoretical physics. The contents of this book consist of three parts:

In Part I, we introduce the fundamental knowledge of differential manifolds, including exterior differential, tangent space, cotangent space and their transformation properties. Differential operation of tensors on manifolds, Riemann manifold together with its conformal transformation, and complex manifolds are also introduced.

In Part II, the global topological properties of differential manifolds are discussed in details, including the theories of homotopy, homology, de Rham cohomology, fibre bundles and their characteristic classes which were first introduced by S. S. Chern.

In Part III, we emphasize the index theorem and the properties of simply connected 4-manifolds. Atiyah-Singer's index theorem and its applications, such as the derivation of the dimension of instantons' moduli space, are demonstrated. This part also provides an important theorem of Donaldson on intersection forms of 4-manifolds, followed by a very exciting conclusion drawn by M. H. Freedman and C. Taubes: " There is an uncountable family of open subsets of R^4_{st} which are all homeomorphic to R^4, but no two are diffeomorphic." This is a very peculiar phenomenon not occurring in any R^n with $n \neq 4$, and it happens that the dimension of our universe is $n = 4$.

WANG, RONG

Contents

Part I. Differential Manifolds

Chapter 1. Preliminary Knowledge and Definitions

§ 1.1 What is a Manifold

The concept of manifold came from the Euclidean space. An n-dimensional real manifold can be considered as the result of gluing the blocks of R^n (the real n-dimensional Euclidean space) together. The most essential character of an n-dimensional real manifold is that there is an n-dimensional system of local coordinates in an open neighborhood of each point of the manifold.

Figure 1.1.1 shows two examples of 1-dimensional real manifolds, S^1 (circle) and an open interval $a<x<b$ in R^1.

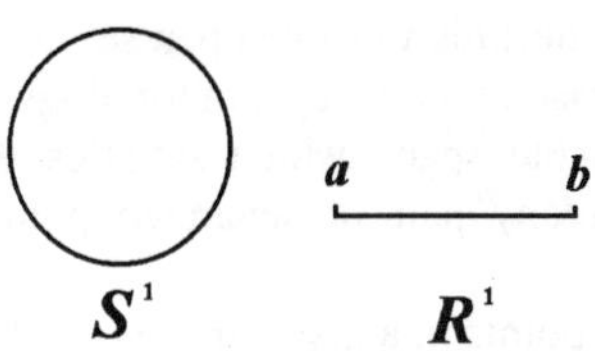

Figure 1.1.1

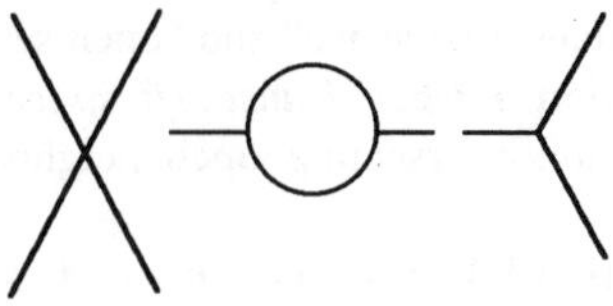

Figure 1.1.2

The 1-dimensional graphs given in Figure 1.1.2 are *not* 1-dimensional real manifolds, since their intersection points have neighborhoods not homeomorphic to R^1.

Figure 1.1.3 illustrates two 2-dimensional real manifolds, S^2 (sphere) and an open interval $0<x<a$, $0<y<b$ in R^2.

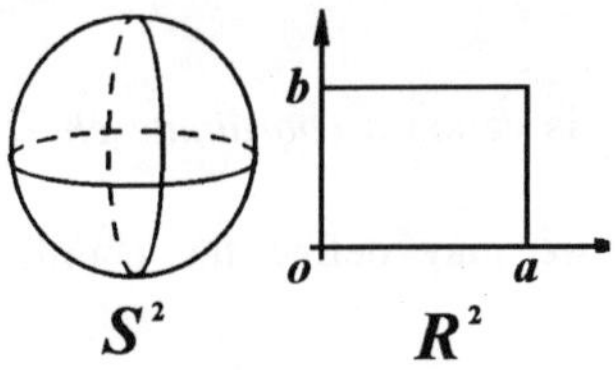

Figure 1.1.3

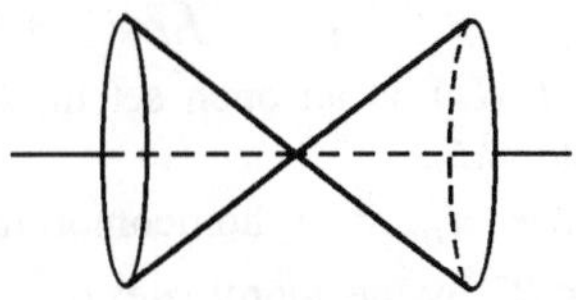

Figure 1.1.4

The 2-dimensional graph shown in Figure 1.1.4 is *not* a 2-dimensional real manifold, since the intersection point of the two cones has a neighborhood not homeomorphic to R^2.

The meaning of "homeomorphism" is explained in the following four definitions:

Definition 1.1.1: Suppose X and Y are two spaces, and $f(x)=y$ maps from X to Y. f: $X \to Y$ is *surjective* or *onto* if and only if for each $y \in Y$, there exists $x \in X$ (not necessarily unique) such that $f(x)=y$.

Definition 1.1.2: A mapping f: $X \to Y$ is *injective* or *1-1* if and only if for any two distinct points x_1, $x_2 \in X$, $y_1 = f(x_1)$ and $y_2 = f(x_2)$ are also distinct in Y.

Definition 1.1.3: A mapping f is called a *bijection* if f is both injective and surjective.

Definition 1.1.4: If there exists a bijection f between X and Y, and if both f and f^{-1} are continuous functions, then we say that *there exists a homeomorphic mapping between* X *and* Y, or simply that X is *homeomorphic* to Y.

The properties invariant under a homeomorphic mapping are called *topological properties*, e.g. compactness, separation properties, connexity, open sets, dimension, etc.. Topological property is also called *topological invariant property*.

Base on the definition of "homeomorphism" and the fact that a real manifold is separable (obtained by gluing the blocks of R^n together), we can define the n-dimensional real manifold as the following:

Definition 1.1.5: An *n-dimensional real manifold* M is a Hausdorff space, each point of which has an open neighborhood homeomorphic to some open set in R^n.

"Hausdorff space" and "open set" are both essential concepts in topology.

Definition 1.1.6: *Hausdorff space* is a separable space which satisfies that there exists non-intersecting open neighborhoods for any pair of separated points in the space.

Definition 1.1.7: An *open set* A is a subset of points in a given space (e.g. Hausdorff space), with each of its points having a neighborhood completely contained in A.[1]

§ 1.2 Introducing Coordinates and Differential Structures into Manifolds

Let f_U be the homeomorphic mapping defined in 1.1.4, and U be some open set in a manifold M. Then we have

$$f_U : U \to f_U(U) \qquad (1.2.1)$$

where $f_U(U)$ is an open set in R^n. (U, f_U) is called a *coordinate chart* for the manifold M.

Since f_U is a homeomorphic mapping, we may define the coordinates of $f_U(z) \in R^n$ by the coordinates of z in M:

$$u^i = (f_U(z))^i, \quad z \in U, \quad i = 1, 2, \cdots, n \qquad (1.2.2)$$

where u^i's $(1 \le i \le n)$ are called the *local coordinates* of z $(z \in U)$.

Suppose V is another open set in M, then again there is a homeomorphic mapping f_V such that

$$f_V : V \to f_V(V) \qquad (1.2.3)$$

---

1 The definition of an open set A does not depend on the shape of the "neighborhoods" and the choice of the distance function.

where $f_V(V)$ is another open set in R^n. Both (U, f_U) and (V, f_V) are coordinate charts of M. In general, we can not expect the whole manifold M to be homeomorphic to R^n, therefore a family of open sets $\{U_\alpha\}$ is required to cover M. That is,

$$\bigcup_\alpha U_\alpha = M \tag{1.2.4}$$

Then we have a sequence of coordinate charts (U_1, f_{U_1}), (U_2, f_{U_2}), $\cdots$ The collection of all the coordinate charts is called an *atlas* A:

$$A = \left\{ \left(U_1, f_{U_1}\right), \left(U_2, f_{U_2}\right), \cdots \right\} \tag{1.2.5}$$

Suppose $U \cap V \neq \phi$, then $f_U(U \cap V)$ and $f_V(U \cap V)$ must be two non-empty open sets in R^n, with the following relationship:

$$\begin{aligned} f_V \cdot f_U^{-1}:\ & f_U(U \cap V) \to f_V(U \cap V) \\ f_U \cdot f_V^{-1}:\ & f_V(U \cap V) \to f_U(U \cap V) \end{aligned} \tag{1.2.6}$$

Similar to Eq.1.2.2, we may define the coordinates of $f_V(z) \in R^n$ by the coordinates of z in M:

$$v^j = (f_V(z))^j, \qquad z \in U, \qquad j = 1, 2, \cdots, n \tag{1.2.2$_1$}$$

Hence Eq.1.2.6 is in fact a transformation of the two kinds of coordinates u^i and v^j in the open set $U \cap V$ of M. $f_U \cdot f_V^{-1}$ and $f_V \cdot f_U^{-1}$ are inverse transformations to each other.

Since $f_U(U \cap V)$ and $f_V(U \cap V)$ in Eq.1.2.6 are both points in R^n, their coordinates must be continuous real functions, therefore the elements of the matrix of the coordinate transformation $f_U \cdot f_V^{-1}$ and $f_V \cdot f_U^{-1}$ must also be continuous real functions.

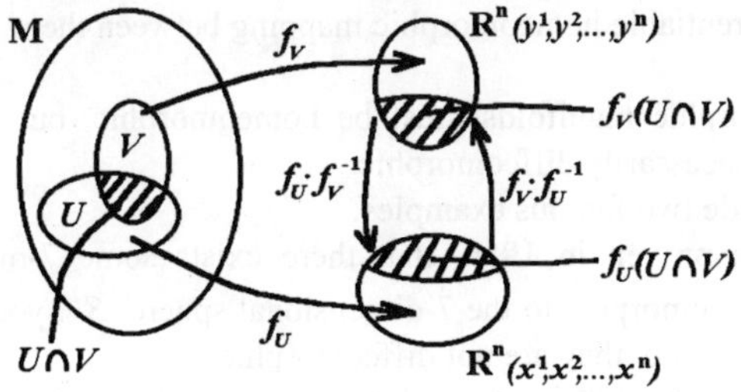

Figure 1.2.1

The continuous transformation between x^i and y^i shown in Figure 1.2.1 can be written as

$$y^i = \left(f_V\, f_U^{-1}\left(x^1, \cdots, x^m\right)\right)^i = \varphi^i\left(x^1, \cdots, x^m\right)$$

$$x^i = \left(f_U\, f_V^{-1}\left(y^1,\cdots,y^m\right)\right)^i = \psi^i\left(y^1,\cdots,y^m\right) \tag{1.2.7}$$

where φ^i and ψ^i are continuous functions satisfying

$$\varphi^i\left(\psi^1\left(y^1,\cdots,y^m\right),\cdots,\psi^m\left(y^1,\cdots,y^m\right)\right) = y^i$$
$$\psi^i\left(\varphi^1\left(x^1,\cdots,x^m\right),\cdots,\varphi^m\left(x^1,\cdots,x^m\right)\right) = x^i \tag{1.2.8}$$

If $U \cap V = \phi$, or if the coordinate transformation functions $\varphi^i\left(x^1,\cdots,x^m\right)$ and $\psi^j\left(y^1,\cdots,y^m\right)$ given in Eq.1.2.7 are in C^r when $U \cap V \neq \phi$, then the two coordinate charts (U, f_U) and (V, f_V) are called to be C^r-consistent. Here "C^r-" means "continuously differentiable up to the rth order".

Definition 1.2.1: Given an atlas $A = \left\{ \left(U_1, f_{U_1}\right), \left(U_2, f_{U_2}\right), \cdots \right\}$ on an n-manifold M, satisfying the following 3 conditions:

1. $\left(U_1, U_2, \cdots\right)$ is an open cover of M (i.e. $\bigcup U_i = M$);
2. every two coordinate charts in A are C^r-consistent;
3. A is the largest atlas, that is, if there is any coordinate chart $\left(\tilde{U}, f_{\tilde{U}}\right)$ for M which is C^r-consistent with every coordinate charts in A, then $\left(\tilde{U}, f_{\tilde{U}}\right)$ must belong to A.

Then A is a C^r *differential structure of* M.

Definition 1.2.2: If there is a C^r differential structure of M, then M is a C^r *differential manifold.*

Definition 1.2.3: If there is a C^∞ differential structure of M, then M is a *smooth manifold*, or *differential manifold.*

Definition 1.2.4: If the manifolds X and Y are not only homeomorphic, but also having a C^∞ differentiable homeomorphic mapping between them, then X and Y are *diffeomorphic*.

Two diffeomorphic manifolds must be homeomorphic, but two homeomorphic manifolds are not necessarily diffeomorphic.

Here we provide two famous examples:

1. Milnor has proven in 1956, that there exists some 7-manifolds which are (topologically) homeomorphic to the 7-dimensional sphere S^7, yet they have different differential structures (i.e. they are not diffeomorphic).

2. In the 80's of the 20th century, due to a series of research done by Donaldson, Freedman, Kirby, and Taubes, it is discovered that besides the proper differential structures, there are uncountably many exotic differential structures in the 4-dimensional Euclidean space R^4. In other words, there exists uncountably many R^4's which are not diffeomorphic, but are (topologically) homeomorphic to each other.

The examples of differential manifolds given below are quite typical:

Example 1: The n-dimensional sphere S^n.

$$S^n = \left\{ \left(x^1, x^2, \cdots, x^{n+1}\right) \in R^{n+1} \,\middle|\, \sum_{i=1}^{n+1} \left(x^i\right)^2 = 1 \right\}$$

can be covered by two open sets U_+ and U_- (Note: $1, 2, \cdots, n+1$ at the upper right corner of each x are superscripts.) where $U_+ = S^n - (0,0,\cdots,-1)$ and $U_- = S^n - (0,0,\cdots,1)$. The corresponding maps f_+ and f_- are:

$$f_+ : \left(x^1, x^2, \cdots, x^{n+1}\right) \to \left(\frac{x^1}{1+x^{n+1}}, \frac{x^2}{1+x^{n+1}}, \cdots, \frac{x^n}{1+x^{n+1}} \right)$$

$$f_- : \left(x^1, x^2, \cdots, x^{n+1}\right) \to \left(\frac{x^1}{1-x^{n+1}}, \frac{x^2}{1-x^{n+1}}, \cdots, \frac{x^n}{1-x^{n+1}} \right)$$

The maps f_+ and f_- are stereographic projections from the south and north poles respectively.

It can be verified that they are C^∞-consistent, therefore S^n is a smooth manifold. So far it is known that besides the differential structures mentioned above, there are other differential structures for S^7, S^9 and S^{11} , which are (topologically) homeomorphic, but not diffeomorphic.

Example 2: The real Grassmann manifold *Gr(N, k)*.

Gr(N, k) is a set of *k*-dimensional linear subspaces through the origin in an *N*-dimensional vector space. A *k*-dimensional linear subspace has its dimension *k*, and its degree of freedom in all directions in a *N*-dimensional space is $(N - k)$, therefore the total dimension of *Gr(N, k)* is $k(N - k)$. *Gr(N, k)* has C^∞ differential structures, therefore is a $k(N - k)$ dimensional differential manifold.

Example 3: *RP(n)*, the real projective space of dimension *n*.

RP (n) is a real Grassmann manifold *Gr(N, k)* with $N = n+1$ and $k = 1$, thus has the dimension $k(N - k)=n$. *RP (n)* has C^∞ differential structures, and thus is an *n*-dimensional differential manifold.

Example 4: The complex Grassmann manifold *Gr(N, k, c)*.

Gr(N, k, c) is a set of *k*-dimensional complex linear spaces through the origin in a complex vector space C^N (*c* represents the complex numbers). A *k*-dimensional complex linear subspace has its dimension *k*, and its degree of freedom in all directions in an *N*-dimensional complex vector space is $(N - k)$. Therefore the dimension of *Gr(N, k, c)* is $k(N - k)$ (i.e. the dimension of complex numbers).

Example 5: *CP(n)*, the complex projective space of dimension *n*.

CP(n) is a complex Grassmann manifold *Gr(N, k, c)* with $N = n+1$ and $k = 1$, thus has the dimension $k(N - k)=n$. *CP(n)* has C^∞ differential structures, and thus is an *n*-dimensional complex differential manifold.

Now let us take a look at the properties of *CP*(1).

CP(1) is just *Gr*(N=2, k=1, *c*). It has two complex coordinates, denoted as

$z=(z_1, z_2)$ where z_1 and z_2 are not zeros simultaneously. Take two open covers:

1. $U_{z_1\neq 0}$ with coordinate $\zeta=\dfrac{z_2}{z_1}$, and
2. $U_{z_2\neq 0}$ with coordinate $\zeta'=\dfrac{z_1}{z_2}$.

In the area of intersection (where z_1 and z_2 are both non-zero), take

$$\zeta=\frac{z_2}{z_1}=u+iv\,,\quad \zeta'=\zeta^{-1}=\frac{z_1}{z_2}=\frac{u-iv}{u^2+v^2}\,,\ \text{and}\ \ \zeta\cdot\zeta'=1\,.$$

There exists a mapping $f:\ U_{z_1}\to U_{z_2},\ f:\ \zeta\to\zeta'$. That is,

$$f:\ (u,v)\to\left(\frac{u}{u^2+v^2},\,-\frac{v}{u^2+v^2}\right).$$

This is a complex analytic mapping, hence $CP(1)$ is a complex analytic manifold. If we make a change of the variables: $u=\dfrac{x}{1+z}$, and $v=\dfrac{-y}{1+z}$ (since $x^2+y^2+z^2=1$, there are only two independent variables), then

$$\frac{u}{u^2+v^2}=\frac{x/(1+z)}{(1-z^2)/(1+z)^2}=\frac{x}{1-z}\,,\qquad \frac{-v}{u^2+v^2}=\frac{y/(1+z)}{(1-z^2)/(1+z)^2}=\frac{y}{1-z}$$

Thus we can see that $f:\ U_{z_1}\to U_{z_2}$ of the mapping (*) is exactly

$$f:\ \left(\frac{x}{1+z},\,\frac{-y}{1+z}\right)\to\left(\frac{x}{1-z},\,\frac{y}{1-z}\right)$$

Compare with the case of S^2, the sphere of dimension 2 in the Example 1 (n=2), the mapping from U_+ to U_- is

$$f':\ U_+\to U_-,\quad f':\ \left(\frac{x}{1+z},\,\frac{y}{1+z}\right)\to\left(\frac{x}{1-z},\,\frac{y}{1-z}\right)$$

It is clear that the mapping between the two open covers U_{z_1} and U_{z_2} of $CP(1)$, and the mapping between the two open covers of S^2 are only differed by a reflection ($\dfrac{-y}{1+z}\to\dfrac{y}{1-z}$ and $\dfrac{y}{1+z}\to\dfrac{y}{1-z}$ respectively). Therefore the two manifolds $CP(1)$ and S^2 are not only homeomorphisms, but also diffeomorphisms (homeomorphism and diffeomorphism are invariant under reflections).

Definition 1.2.5: Let $A=\{U_i\}$ be a set of open sets $U_1, U_2, \cdots$ (each U_i has a finite n-dimensional volume) on an n-manifold M. If A satisfies $\bigcup U_i=M$, then $A=\{U_i\}$ is called an *open cover* of the manifold M.

Definition 1.2.6: If A' is a subset of the open cover A, and is itself an open cover of M, then A' is called a *subcover* of A.

Definition 1.2.7: If the number of U_i's in an open cover $A=\{U_i\}$ is finite, then A

is called a *finite cover*.

Definition 1.2.8: If every open cover of a manifold M has a finite subcover, then M is said to be *compact*, otherwise it is said to be *incompact*.

We can deduce from the above definitions that R^n is incompact, since it is impossible to cover an infinite *n*-dimensional space by any finite subcover. On the other hand, S^n, *RP*(*n*), *CP*(*n*) and $S^1 \times S^1 = T^2$ (a 2-dimensional ring surface) are all compact manifolds.

Compactness is an invariant property of topology, and it is invariant under homeomorphic mappings.

Definition 1.2.9: The *product manifold* $M \times N$ of two differential manifolds M and N is still a differential manifold. If $\{U_i, f_i\}$ and $\{V_j, g_j\}$ are atlases of M and N respectively, and if there exists smooth projections

$$\Pi_1 : \ M \times N \to M \quad \text{and} \quad \Pi_2 : \ M \times N \to N,$$

then the atlas of $M \times N$ is $\{U_i \times V_j ; f_i \cdot \Pi_1 \otimes g_j \cdot \Pi_2\}$. The differential structures of the product manifold can therefore be determined: dimension of M + dimension of N = dimension of $M \times N$.

§ 1.3 Tangent Spaces and Cotangent Spaces

Suppose $y = f(x)$ is a function of x, and $x = x(t)$ is a function of t. Then in a neighborhood of $x_p = x(t_p)$ we may obtain the following linear approximation:

$$\begin{aligned} y &= f(x_p) + (t - t_p)\left.\frac{dx}{dt}\right|_{t=t_p} \left.\frac{d}{dx} f(x)\right|_{x=x_p} \\ &= f(x_p) + (t - t_p) v_p \left.\frac{d}{dx} f(x)\right|_p \end{aligned} \tag{1.3.1}$$

Generalizing to the *n*-manifold M, we suppose that $y = f(x^1, \cdots, x^n)$ is a function of $x^1, \cdots, x^n$, and $x^i = x^i(t)$ is a function of t. Then in a neighborhood of $x^i = x^i_p = x^i(t_p)$, we have the following linear approximation:

$$\begin{aligned} y &= f(x^1_p, x^2_p, \cdots, x^n_p) + (t - t_p) \left.\frac{d}{dt} f(x^1, \cdots, x^n)\right|_p \\ &= f(x^1_p, x^2_p, \cdots, x^n_p) + (t - t_p)\, v^i_p \left.\frac{\partial}{\partial x^i} f(x^1, \cdots, x^n)\right|_p \end{aligned} \tag{1.3.2}$$

where $v^i_p = \left.\dfrac{dx^i}{dt}\right|_p$ is the "velocity" in the direction of x^i.

The changes of $f(x)$ and $f\left(x^1,\cdots,x^n\right)$ with respect to t at p are given by Eq.1.3.1 and Eq.1.3.2. We may hence define X_p as the following $\left(\left.\frac{dx^i}{dt}\right|_p = v_p^i\right)$:

$$X_p f(x) = \left.\frac{d}{dt}f(x)\right|_p = \left.\frac{dx}{dt}\right|_p \left.\frac{d}{dx}f(x)\right|_p = v_p \left.\frac{d}{dx}f(x)\right|_p \tag{1.3.1$_1$}$$

$$\begin{aligned} X_p f\left(x^1,\cdots,x^n\right) &= \left.\frac{d}{dt}f\left(x^1,\cdots,x^n\right)\right|_p \\ &= \left.\frac{dx^i}{dt}\right|_p \left.\frac{\partial}{\partial x^i}f\left(x^1,\cdots,x^n\right)\right|_p = v_p^i \left.\frac{\partial}{\partial x^i}f\left(x^1,\cdots,x^n\right)\right|_p \end{aligned} \tag{1.3.2$_1$}$$

The functions $f(x)$ and $f\left(x^1,\cdots,x^n\right)$ in Eq.1.3.1$_1$ and Eq.1.3.2$_1$ are arbitrary. Only that

$$X_p = \left.\frac{d}{dt}\right|_{t=0} = v_p \left.\frac{d}{dx}\right|_p \tag{1.3.1$_2$}$$

in Eq.1.3.1$_1$ and

$$X_p = \left.\frac{d}{dt}\right|_{t=0} = v_p^i \left.\frac{\partial}{\partial x^i}\right|_p \tag{1.3.2$_2$}$$

in Eq.1.3.2$_1$ do not depend on $f(x)$, $f\left(x^1,\cdots,x^n\right)$, and the choice of the system of coordinates. Therefore we may define $X_p = \left.\frac{d}{dt}\right|_{t=0}$ in Eq.1.3.1$_2$ and Eq.1.3.2$_2$ as the *tangent vector*. For example, if there is a curve in an *n*-manifold M that passes through the point p when $t = 0$,

$$x^i = x^i(p) + \eta^i t + O\left(t^2\right),$$

Then the vector which is tangent to this curve at the point $t = 0$ can be obtained as

$$Y_p = \left.\frac{d}{dt}\right|_{t=0} = \eta^i \left.\frac{\partial}{\partial x^i}\right|_p \tag{1.3.3}$$

which is consistent to Eq.1.3.2$_2$.

Definition 1.3.1: The collection of all the tangent vectors through the point p forms the *tangent space* $T_p(M)$ of a manifold M to the point p. If there are *n* partial derivatives $\frac{\partial}{\partial x^1},\cdots,\frac{\partial}{\partial x^n}$, then $T_p(M)$ is a *tangent space of dimension n* to the point p.

Generalize the tangent vector X_p such that it is well-defined at every point of the manifold M, thus X_p becomes $X(x)$, and Eq.1.3.2$_2$ is extended to

$$X(x) = \sum_j \xi^j(x) \frac{\partial}{\partial x^j} \tag{1.3.4}$$

where $X(x)$ is a function of x with the properties of a field, and is hence named a *field of tangent vectors*.

Suppose that there are two fields of tangent vectors

$$X(x) = \sum_i \xi^i(x) \frac{\partial}{\partial x^i} \quad \text{and} \quad Y(x) = \sum_j \eta^j(x) \frac{\partial}{\partial x^j} \tag{1.3.5}$$

Their *Lie bracket* $[X,Y]$ is defined as the following:

$$\begin{aligned}[X,Y]f &= X(Yf) - Y(Xf) \\ &= \left(\xi^j(x) \frac{\partial \eta^k(x)}{\partial x^j} - \eta^j(x) \frac{\partial \xi^k}{\partial x^j} \right) \frac{\partial}{\partial x^k} f\end{aligned} \tag{1.3.6}$$

We are going to encounter the Lie brackets when we discuss the Lie derivatives later.

Now introducing the *cotangent space*: The basis vectors of a cotangent space are dx^i's ($i = 1, 2, \cdots, n$), while the basis vectors of a tangent space are $\frac{\partial}{\partial x^i}$'s. Here n is the dimension of the manifold M.

To explain the dual relationship between tangent space and cotangent space, we introduce the symbol $<,>$ with the rule

$$< df, \frac{\partial}{\partial x^i} > = \frac{\partial}{\partial x^i} f \tag{1.3.7}$$

Hence

$$< dx^j, \frac{\partial}{\partial x^i} > = \frac{\partial x^j}{\partial x^i} = \delta^j_i \tag{1.3.8}$$

This is the dual relationship between the basis $\frac{\partial}{\partial x^i}$ of a tangent vector and the basis dx^j of a cotangent vector. Thus the tangent vector space $T_p(M)$ with basis $\frac{\partial}{\partial x^i}$ ($i = 1, 2, \cdots, n$) and the cotangent vector space $T^*_p(M)$ with basis dx^j ($j = 1, 2, \cdots, n$) are also dual to each other (here p represents an arbitrary point on M).

Definition 1.3.2: The union of all tangent vector spaces to points p of a manifold M

$$T(M) = \bigcup_p T_p(M) \tag{1.3.9}$$

is called the *tangent bundle* of M.

Definition 1.3.3: The union of all cotangent vector spaces to points p of a manifold M

$$T^*(M) = \bigcup_p T^*_p(M) \tag{1.3.10}$$

is called the *cotangent bundle* of M.

$T_p(M)$ and $T_p^*(M)$ are both 2*n*-manifolds, with local fields being direct product manifolds (the direct product of an *n*-dimensional coordinate space and an *n*-dimensional vector space).

Since dx^j is a basis of the cotangent vector, the cotangent vector at the point p should be written as:

$$\sigma(p) = \sigma_j(p)\, dx^j$$

Generalizing from the point p to an arbitrary point $x\left(x^1,\cdots,x^n\right)$ on M, we now form the *field of cotangent vectors*

$$\sigma(x) = \sigma_j(x)\, dx^j \tag{1.3.11}$$

Further more, assume that $\sigma_j(x)$ is differentiable, then $\sigma(x)$ in Eq.1.3.11 is called a *differentiable field of cotangent vectors*, or *1-form* (since dx^j is contained). Take $\sigma_j(x) = \delta_{ij}$, then $\sigma(x) = dx^i$, and thus dx^i is itself a 1-form.

§ 1.4 Differential Forms and Exterior Differentials

Definition 1.4.1: The *exterior product* $dx^i \wedge dx^j$ of 1-form dx^i and 1-form dx^j is

$$dx^i \wedge dx^j = \left(dx^i \otimes dx^j - dx^j \otimes dx^i\right) = \delta^{ij}_{kl}\, dx^k \otimes dx^l \tag{1.4.1}$$

$$(\delta^{ij}_{kl} = \delta^i_k \delta^j_l - \delta^i_l \delta^j_k)$$

$dx^i \wedge dx^j$ is a 2-form (the product of two 1-forms), and is a basis of a set of tensors with two indices *i* and *j*.

A tensor field can be constructed on a manifold M, for example:

$$K_2(x) = \frac{1}{2}\sum_{i\neq j} f_{ij}(x)\, dx^i \wedge dx^j \tag{1.4.2}$$

If $f_{ij}(x)$ is differentiable, then $K_2(x)$ is a differentiable tensor field, and a 2-form as well. The definitions can be extended to a higher ordered *p*-form:

$$dx^{i_1} \wedge\cdots\wedge dx^{i_p} = \delta^{i_1\cdots i_p}_{k_1\cdots k_p}\, dx^{k_1} \otimes dx^{k_2} \otimes\cdots\otimes dx^{k_p} \tag{1.4.3}$$

$$\delta^{i_1\cdots i_p}_{k_1\cdots k_p} = \begin{vmatrix} \delta^{i_1}_{k_1} & \cdots & \delta^{i_1}_{k_p} \\ \vdots & \ddots & \vdots \\ \delta^{i_p}_{k_1} & \cdots & \delta^{i_p}_{k_p} \end{vmatrix} \qquad (i,k = 1,\cdots,n)$$

$$= \begin{cases} 1 & \text{if the subscript is an even permutation of the superscript} \\ -1 & \text{if the subscript is an odd permutation of the superscript} \\ 0 & \text{otherwise} \end{cases} \tag{1.4.3$_1$}$$

The basis of an exterior differential tensor with p indices and a tensor field of p-form can therefore be constructed.

For examples:

$$p\text{-form:}\quad \alpha_p(x)=\frac{1}{p!}f_{i_1\cdots i_p}(x)\,dx^{i_1}\wedge\cdots\wedge dx^{i_p}$$

$$q\text{-form:}\quad \beta_q(x)=\frac{1}{q!}g_{j_1\cdots j_q}(x)\,dx^{j_1}\wedge\cdots\wedge dx^{j_q} \tag{1.4.4}$$

Another example:

$$\begin{aligned}\alpha_p(x)\wedge\beta_q(x)&=\frac{1}{p!q!}f_{i_1\cdots i_p}\,g_{j_1\cdots j_p}\,dx^{i_1}\wedge\cdots\wedge dx^{i_p}\wedge dx^{j_1}\wedge\cdots\wedge dx^{j_q}\\&=\frac{(p+q)!}{p!q!}\frac{1}{(p+q)!}f_{i_1\cdots i_p}\,g_{j_1\cdots j_p}\,dx^{i_1}\wedge\cdots\wedge dx^{i_p}\wedge dx^{j_1}\wedge\cdots\wedge dx^{j_q}\\&=\frac{(p+q)!}{p!q!}\gamma_{p+q}(x)\end{aligned} \tag{1.4.5}$$

where

$$\gamma_{p+q}(x)=\frac{1}{(p+q)!}f_{i_1\cdots i_p}\,g_{j_1\cdots j_p}\,dx^{i_1}\wedge\cdots\wedge dx^{i_p}\wedge dx^{j_1}\wedge\cdots\wedge dx^{j_q} \tag{1.4.6}$$

Notice that $\gamma_{p+q}(x)$ in Eq.1.4.6 has the same form as those of $\alpha_p(x)$ and $\beta_q(x)$ in Eq.1.4.4, and that the functions $f_{i_1\cdots i_p}(x)$ and $g_{j_1\cdots j_q}(x)$ are completely anti-symmetric with respect to their subscripts. If M is n-dimensional, the it is required that $p+q\le n$, otherwise $\alpha_p(x)\wedge\beta_q(x)=0$, since there must be some repetitions among the indices $i_1\cdots i_p j_1\cdots j_q$ in Eq.1.4.6.

The exterior product satisfies the laws of association and distribution:

$$(\alpha\wedge\beta)\wedge\gamma=\alpha\wedge(\beta\wedge\gamma) \tag{1.4.7}$$

$$(\alpha+\beta)\wedge\gamma=\alpha\wedge\gamma+\beta\wedge\gamma \tag{1.4.8}$$

The p-form $\alpha_p(x)$ and the q-form $\beta_q(x)$ also have the following law of commutation:

$$\alpha_p(x)\wedge\beta_q(x)=(-1)^{pq}\beta_q(x)\wedge\alpha_p(x) \tag{1.4.9}$$

which can be obtained from Eq.1.4.3 and the property of $\delta^{i_1\cdots i_p}_{k_1\cdots k_p}$ Eq.1.4.3$_1$. So far we have known that

(a) the Lie brackets of the tangent vector fields X, Y, Z, $\cdots$ have the following properties:

I. $[\lambda_1 X+\lambda_2 Y,Z]=\lambda_1[X,Z]+\lambda_2[Y,Z]$ (1.4.10)

II. $[X,Y]=-[Y,X]$ (1.4.11)

III. $[[X,Y],Z]+[[Y,Z],X]+[[Z,X],Y]=0$ (1.4.12)

which means that the set of tangent vector fields X, Y, Z, $\cdots$ forms a Lie algebra.

(b) the cotangent vectors $(dx^1),(dx^2),(dx^3),\cdots,(dx^n)$ also have an important property, which is explained by the following example: Let us consider $\underline{dx}^i = (dx^i)$ as a vector, which has n components on an n-manifold, e.g. the orthogonal coordinate system of a 2-manifold

$$\underline{dx}^1 = \begin{cases} (\underline{dx}^1)^1 = dx^1 \\ (\underline{dx}^1)^2 = 0 \end{cases}, \qquad \underline{dx}^2 = \begin{cases} (\underline{dx}^2)^1 = 0 \\ (\underline{dx}^2)^2 = dx^2 \end{cases}$$

gives $\underline{dx}^1 \wedge \underline{dx}^2 = \delta^{12}_{ij}(\underline{dx}^1)^i(\underline{dx}^2)^j = dx^1 \cdot dx^2$; the orthogonal coordinate system of a 3-manifold

$$\underline{dx}^1 = \begin{cases} (\underline{dx}^1)^1 = dx^1 \\ (\underline{dx}^1)^2 = (\underline{dx}^1)^3 = 0 \end{cases}, \qquad \underline{dx}^2 = \begin{cases} (\underline{dx}^2)^2 = dx^2 \\ (\underline{dx}^2)^1 = (\underline{dx}^2)^3 = 0 \end{cases},$$

$$\underline{dx}^3 = \begin{cases} (\underline{dx}^3)^3 = dx^3 \\ (\underline{dx}^3)^1 = (\underline{dx}^3)^2 = 0 \end{cases}$$

gives $\underline{dx}^1 \wedge \underline{dx}^2 \wedge \underline{dx}^3 = \delta^{123}_{ijk}(\underline{dx}^1)^i(\underline{dx}^2)^j(\underline{dx}^3)^k = dx^1 \cdot dx^2 \cdot dx^3$; and so on and so forth. We then obtain for the orthogonal coordinate system of dimension n:

$$\underline{dx}^1 \wedge \underline{dx}^2 \wedge\cdots\wedge \underline{dx}^n = dx^1 \cdot dx^2 \cdots\cdots dx^n = \text{volume element} \tag{1.4.13}$$

If the coordinate system is rotated, e.g. take $dx^{1\prime} = dx^1 + \alpha\, dx^2$, still we have

$$\underline{dx}^1 \wedge \underline{dx}^2 \wedge\cdots\wedge \underline{dx}^n = dx^{1\prime} \wedge dx^2 \wedge\cdots\wedge dx^n = \text{volume element}$$

Therefore

$$dx^{1\prime} \wedge dx^{2\prime} \wedge\cdots\wedge dx^{n\prime} = \text{volume element} \tag{1.4.14}$$

is always true no matter how nonorthogonalized the coordinate axes are.

The differential operations for the differential forms include:

d — exterior differential operation;

D — covariant differential operation (will discuss in Chapter 3);

L_X — Lie derivative operation (will discuss in Chapter 2), etc.

Here we present the definition of the exterior differential d.

Suppose we have a p-form:

$$\alpha_p = \frac{1}{p!} f_{i_1\cdots i_p}\, dx^{i_1} \wedge\cdots\wedge dx^{i_p} \tag{1.4.15}$$

The corresponding exterior differential is:

$$\begin{aligned} d\alpha_p &= \frac{1}{p!} d f_{i_1\cdots i_p}\, dx^{i_1} \wedge\cdots\wedge dx^{i_p} \\ &= \frac{1}{p!} f_{i_1\cdots i_p, j}\, dx^j \wedge dx^{i_1} \wedge\cdots\wedge dx^{i_p} \end{aligned}$$

$$=\frac{1}{p!}\frac{1}{(p+1)!}\delta^{j,i_1\cdots i_p}_{k_1\cdots k_{p+1}} f_{i_1\cdots i_p,j}\, dx^{k_1}\wedge\cdots\wedge dx^{k_{p+1}} \tag{1.4.16}$$

Taking $(p+1)!$ as the denominator is because that there are $(p+1)!$ permutations for $k_1\cdots k_{p+1}$, which contribute equally.

Applying the exterior differential one more time, we have:

$d(df)=d(f_{,i}\,dx^i)=f_{,ij}\,dx^j\wedge dx^i=0$ (since $f_{,ij}=f_{,ji}$)

$d(d\alpha_p)=\dfrac{1}{p!}f_{i_1\cdots i_p,jk}\,dx^k\wedge dx^j\wedge dx^{i_1}\wedge\cdots\wedge dx^{i_p}=0$ (since j k in $f_{i_1\cdots i_p,jk}$ is symmetric, and $dx^k\wedge dx^j$ is anti-symmetric)

Therefore *d* has the following character

$$d^2=0 \tag{1.4.17}$$

If a differential form α satisfies

$$d\alpha=0 \tag{1.4.18}$$

Then α is a closed form. If a differential form α can be written as

$$\alpha=d\beta \tag{1.4.19}$$

Then α is an exact form. An exact form must be closed, but a closed form is not necessarily exact.

§ 1.5 Orientation of Manifolds and Integration of Differential Forms

In an open cover $\{U_\alpha\}$ of a differentiable manifold M, any open set U_α may have more than one coordinate systems. Taking any two systems (x^i) and (y^j), we may have two cases.

Case 1: the Jacobian $\dfrac{\partial(x^1,\cdots,x^n)}{\partial(y^1,\cdots,y^n)}>0$, then the coordinate systems (x^i) and (y^j) are said to have the same orientations.

Case 2: the Jacobian $\dfrac{\partial(x^1,\cdots,x^n)}{\partial(y^1,\cdots,y^n)}<0$, the coordinate systems (x^i) and (y^j) are said to have opposite orientations.

In an open cover $\{U_\alpha\}$ of a differential manifold M, the coordinate systems of any two open sets U_α and U_β with non-empty intersections, (x^α) and (y^β) respectively, have a non-zero Jacobian in the region $U_\alpha\cap U_\beta\ (\neq\phi)$. Hence there are also two possible cases.

Case 1: the coordinate systems (x^α) and (y^β) of any pair of U_α and U_β, respectively, in $\{U_\alpha\}$ has positive Jacobian in $U_\alpha\cap U_\beta\ (\neq\phi)$. In this case, the differential manifold is just an oriented manifold.

Case 2: not every pair of U_α and U_β in $\{U_\alpha\}$ has their coordinate systems (x^α) and (y^β) having positive Jacobian in $U_\alpha \cap U_\beta \ (\neq \phi)$. In this case, the differential manifold is unoriented.

Take 2-dimensional surface as an example. On the torus T^2 (see Figure 1.5.1), the coordinate frame can return to the original position where it started after moving along a complete circle. Thus we always have $\dfrac{\partial(x^1,\cdots,x^n)}{\partial(y^1,\cdots,y^n)} > 0$ for any pair of U_α and U_β with $U_\alpha \cap U_\beta \ (\neq \phi)$, which means that T^2 is an oriented 2-manifold.

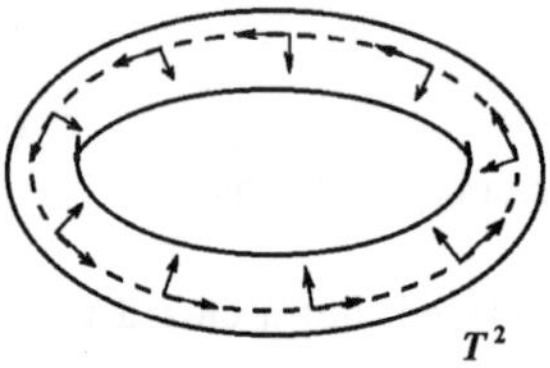

Figure 1.5.1

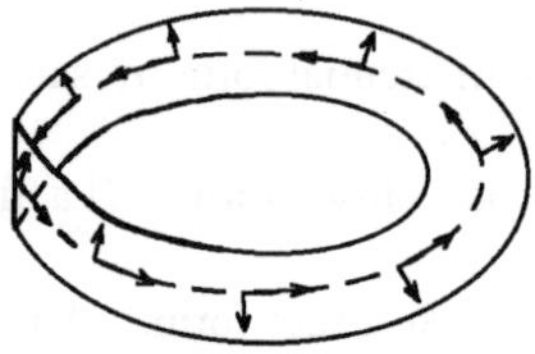

Figure 1.5.2 Möbius Strip

On a Möbius strip, however, one of the coordinate axes will change its direction after the coordinate frame completes a full circle (see Figure 1.5.2). Thus the Jacobian of the original coordinates and the coordinates after completing a full circle is

$$\begin{vmatrix} \dfrac{\partial x'}{\partial x} & 0 \\ 0 & \dfrac{\partial y'}{\partial y} \end{vmatrix} \sim \begin{vmatrix} 1 & 0 \\ 0 & -1 \end{vmatrix} < 0$$

which means that the Möbius strip is an unoriented 2-manifold.

Another difference between T^2 and the Möbius strip is that T^2 has two side faces, the inner side and the outer side, while the Möbius strip has only a single side face, which is the result of gluing the inner and the outer side faces together. It is the gluing of the inner and the outer side faces that turns one arrow up side down ($y' = -y$), while the other arrow's direction is unchanged ($x' = x$) after the coordinate frame completes a full circle.

Theorem 1.5.1: If there exists a continuous *n*-form ω which is non-zero everywhere on an *n*-manifold M, then M is orientable, and vice versa.

Proof: Suppose that ω is non-zero everywhere, then ω must be positive (or negative) everywhere. Therefore $\omega^U_{1,\cdots,n}$ and $\omega^V_{1,\cdots,n}$ must have the same sign in $U \cap V$, i.e. the Jacobian is everywhere positive. Thus M is orientable.

On the other hand, if a manifold M is orientable, then the Jacobian is positive in every $U \cap V$. Hence $\omega^U_{1,\cdots,n}$ and $\omega^V_{1,\cdots,n}$ must have the same sign, and there must be a continuous n-form ω which is non-zero (either >0 or <0) everywhere. □

When a manifold has only one coordinate chart (i.e. can be covered by a single open set), it is said to be automatically orientable.

About the integral on an n-manifold, we notice that the coordinate transformation of $dx^1 \wedge dx^2 \wedge \cdots \wedge dx^n$ happens to be

$$dy^1 \wedge dy^2 \wedge \cdots \wedge dy^n = \frac{\partial(y^1, y^2, \cdots, y^n)}{\partial(x^1, x^2, \cdots, x^n)} dx^1 \wedge dx^2 \wedge \cdots \wedge dx^n$$

which is consistent with the transformation of the volume element of the integrand. Thus when integrating on an n-manifold M, we may choose $\tau(x)$ of an n-form

$$\tau(x) = dx^1 \wedge dx^2 \wedge \cdots \wedge dx^n \tag{1.5.1}$$

to be the volume element of dimension n. Hence we may proceed integral operations of $f(x)$ on M after the open cover $\{U_\alpha\}$ of the n-dimensional differentiable manifold M is determined. The integral can be written as

$$\int_M f(x)\, \tau(x) = \int_M f(x)\, dx^1 \wedge \cdots \wedge dx^n \tag{1.5.2}$$

If the orthogonal coordinate system of R^n is chosen to be the $x^1, \cdots, x^n$, then the above equation can be simplified as

$$\int_M f(x)\, \tau(x) = \int_M f(x)\, dx^1 \cdots dx^n \tag{1.5.3}$$

If there are more than one coordinate charts U_α, U_β, $\cdots$ in the open cover $\{U_\alpha\}$ of M, the integral will have to be decomposed first and be summed up afterwards.

Stokes' Theorem: Suppose that M is an oriented smooth n-manifold with boundaries. ω_{n-1} is an $(n-1)$-form in M with compact support (the support of f is denoted as *supp* f, which is the smallest closed set containing the region where $f \neq 0$. $f \equiv 0$ outside its support. A support only contains but not necessarily equals the set $\{x, f(x) = 0\}$). ∂M is the non-zero boundary of M. Then we have the following theorem, namely the Stokes' Theorem:

$$\int_{\partial M} \omega_{n-1} = \int_M d\,\omega_{n-1} \tag{1.5.4}$$

Proof: Without loss of generality, we take ω_{n-1} in the open set

$$U \subset R^n_+ = \left(x = (x^1, \cdots, x^n) \in R^n,\, x_n \geq 0\right) \tag{1.5.5}$$

and assume that ω_{n-1} has a compact support (i.e. the region where $\omega_{n-1} \neq 0$ does not extend to infinities). In general ω_{n-1} may have the form

$$\omega_{n-1} = \sum_{i=1}^{n} (-1)^{i-1} \alpha_i\, dx^1 \wedge \cdots \wedge \widehat{dx^i} \wedge \cdots \wedge dx^n \tag{1.5.6}$$

where $\widehat{dx^i}$ means erasing the term dx^i. Then we have

$$d\omega_{n-1}=\sum_{i=1}^{n}\frac{\partial\alpha_i}{\partial x^i}dx^1\wedge\cdots\wedge dx^i\wedge\cdots\wedge dx^n \quad (1.5.7)$$

Thus

$$\begin{aligned}\int_U d\omega_{n-1}&=\sum_{i=1}^{n}\int_{R_+^n}\frac{\partial\alpha_i}{\partial x^i}dx^1\wedge\cdots\wedge dx^n\\&=\sum_{i=1}^{n}\int_{R_+^n}\frac{\partial\alpha_i}{\partial x^i}dx^1\cdots dx^n\end{aligned} \quad (1.5.8)$$

Here the symbol of exterior product $\wedge$ is omitted since the orthogonal coordinates of R_+^n have been chosen.

Now decompose R_+^n into R_+ and R^{n-1}, the right hand side of Eq.1.5.8 consists of two parts:

$$\begin{aligned}\int_U d\omega_{n-1}&=\sum_{i=1}^{n}\int_0^\infty dx^n\left(\int_{R^{n-1}}\frac{\partial\alpha_i}{\partial x^i}dx^i\right)dx^1\cdots\widehat{dx^i}\cdots dx^{n-1}\\&+\int_{R^{n-1}}\left(\int_0^\infty\frac{\partial\alpha_n}{\partial x^n}dx^n\right)dx^1\cdots dx^{n-1}\end{aligned} \quad (1.5.9)$$

Since the support of each α_i is compact, such that

$$\int_{-\infty}^{\infty}\frac{\partial\alpha_i}{\partial x_i}dx^i=\alpha_i(x^1\cdots x^i\cdots x^n)\Big|_{x^i=-\infty}^{x^i=+\infty}=0 \qquad (i=1,2,\cdots,n-1)$$

therefore the right hand side of Eq.1.5.9 only has the term involves α_n in the second part remained.

$$\int_U d\omega_{n-1}=-\int_{R^{n-1}}\alpha_n(x^1,x^2,\cdots,x^{n-1},0)\,dx^1\cdots dx^{n-1} \quad (1.5.10)$$

On the other hand,

$$\int_{\partial U}\omega_{n-1}=\int_{\partial R_+^n}\omega_{n-1}=\int_{\partial R_+^n}\sum_{i=1}^{n}(-1)^{i-1}\alpha_i(x^1,\cdots,x^n)\,dx^1\wedge\cdots\wedge\widehat{dx^i}\wedge\cdots\wedge dx^n$$

and $x^n=0$, $dx^n=0$ in ∂R_+^n. Therefore the sum has only one term, the one without dx^n, left:

$$\int_{\partial U}\omega_{n-1}=\int_{\partial R_+^n}(-1)^{n-1}\alpha_n(x^1,\cdots,x^{n-1},0)\;dx^1\wedge dx^2\wedge\cdots\wedge dx^{n-1} \quad (1.5.11)$$

Note: Although the manifolds ∂R_+^n and R^{n-1} are similar in their forms, they have different coordinate systems and orientations. The outer normal direction vector of R_+^n is $-e_n=(0,\cdots,0,-1)$, therefore the ordering of the orientation basis of R_+^n must be $(-e_n,e_1,\cdots,e_{n-1})$ with the normal basis $(-e_n)$ being the first component.

By contrast, the ordering of the orientation basis of R^n is $(e_1,e_2,\cdots,e_{n-1},e_n)$. Hence if the coordinates in R^n are taken as $(x^1,x^2,\cdots,x^{n-1},x^n)$, the corresponding coordinates in R_+^n must be taken as $(y^1=-x^n,y^2=x^1,\cdots,y^n=x^{n-1})$.

The Jacobian of the coordinate transformation is

$$\frac{\partial\left(x^1,x^2,\cdots,x^n\right)}{\partial\left(y^1,y^2,\cdots,y^n\right)}=(-1)\cdot(-1)^{n-1}=(-1)^n \tag{1.5.12}$$

which is consistent with $(-dx^n)\wedge dx^1\wedge\cdots\wedge dx^{n-1}=(-1)^n\,dx^1\wedge\cdots\wedge dx^{n-1}\wedge dx^n$.

Thus changing $\int_{\partial R^n_+}$ in Eq.1.5.11 into $\int_{R^{n-1}}$, we obtain by the coordinate transformation

$$\begin{aligned}\int_{\partial U}\omega_{n-1}&=\int_{R^{n-1}}(-1)^{n-1}\cdot(-1)^n\alpha_n(x^1,\cdots,x^{n-1},0)\,dx^1\wedge dx^2\wedge\cdots\wedge dx^{n-1}\\&=-\int_{R^{n-1}}\alpha_n(x^1,\cdots,x^{n-1},0)\,dx^1\wedge dx^2\wedge\cdots\wedge dx^{n-1}\end{aligned} \tag{1.5.13}$$

which is consistent with the right hand side of Eq.1.5.10. Therefore

$$\int_U d\omega_{n-1}=\int_{\partial U}\omega_{n-1} \tag{1.5.14}$$

which proves the Stokes' Theorem. □

Example: Suppose that D is a closed bounded region in R^3, and ∂D is the 2-dimensional boundary of D. Taking a vector $\vec{E}$ and the area element $d\vec{s}$, we have

$$\begin{aligned}&\int_D d(\vec{E}\cdot d\vec{s})\\&=\int_D d\left(\frac{1}{2}\varepsilon_{ijk}E_i\,dx^j\wedge dx^k\right)\\&=\int_D \frac{1}{2}\varepsilon_{ijk}\frac{\partial E_i}{\partial x^l}dx^l\wedge dx^j\wedge dx^k\\&=\int_D\frac{\partial E_1}{\partial x^1}dx^1\wedge dx^2\wedge dx^3+\int_D\frac{\partial E_2}{\partial x^2}dx^1\wedge dx^2\wedge dx^3+\int_D\frac{\partial E_3}{\partial x^3}dx^1\wedge dx^2\wedge dx^3\\&=\int_D\frac{\partial E_1}{\partial x^1}dx^1\wedge dx^2\wedge dx^3-\int_D\frac{\partial E_2}{\partial x^2}dx^2\wedge dx^1\wedge dx^3+\int_D\frac{\partial E_3}{\partial x^3}dx^3\wedge dx^1\wedge dx^2\\&=\int_{(\partial D)_1}E_1\,dx^2\wedge dx^3+\int_{(\partial D)_2}E_2\,dx^3\wedge dx^1+\int_{(\partial D)_3}E_3\,dx^1\wedge dx^2\\&=\int_{\partial D}E\cdot d\vec{s}\end{aligned}$$

In order to transform $\int_D$ into $\int_{\partial D}$, we have to shift dx^i which is perpendicular to $d\vec{s}$ to the most left position. In other words, dx^i of the normal vector which is orthogonal to $d\sigma_i$ (i.e. $dx^1\wedge dx^2$, $dx^2\wedge dx^3$, and $dx^3\wedge dx^1$) must be shifted to the most left position. That is the very reason why we should take the coordinates of R^n_+ as $(y^1=-x^n, y^2=x^1,\cdots,y^n=x^{n-1})$ when we were proving the Stokes' Theorem (see Eq.1.5.12).

Chapter 2. Properties and Operations of Tangent Vectors and Cotangent Vectors

§ 2.1 Mapping Transformation of Tangent Vector Fields and Cotangent Vector Fields

Suppose that there are two differential manifolds *M* and *N*, with a smooth mapping f and an inverse mapping f^{-1}:

$$\begin{aligned} f:\ & M \to N \\ & p \to q = f(p) \end{aligned}, \qquad \begin{aligned} f^{-1}:\ & N \to M \\ & q \to p = f^{-1}(q) \end{aligned} \tag{2.1.1}$$

(i) When f is acting on a *tangent vector field* of *M* (and is therefore called a *tangent mapping*), we have

$$X = \eta^i(x) \frac{\partial}{\partial x^i}\bigg|_p \tag{2.1.2}$$

$$f:\ X \to Y = \eta^i(x) \frac{\partial y^j}{\partial x^i} \frac{\partial}{\partial y^j}\bigg|_{q=f(p)} \tag{2.1.3}$$

Hence for a tangent vector field the mapping f appears to be $x \to y = f(x)$. f is also called a "*forward mapping* ($x \to y$)".

(ii) When f^{-1} is acting on a *cotangent vector field* of *N* (called a *cotangent mapping*), we have:

$$\omega_1(y) = \varphi_i(y) dy^i \tag{2.1.4}$$

$$f^{-1}:\ \omega_1(y) \to \omega_1(x) = \varphi_i(y) \frac{\partial y^i}{\partial x^j} dx^j \bigg|_{p=f^{-1}(q)} \tag{2.1.5}$$

Hence for a cotangent vector field the mapping f^{-1} appears to be $y \to x = f^{-1}(y)$. f^{-1} is also called a "*backward mapping* ($x \leftarrow y$)".

Generalize the backward mapping to the cotangent tensor of order *r*, we have:

$$\omega_r(y) = \frac{1}{r!} \varphi_{i_1 \cdots i_r}(y) dy^{i_1} \wedge \cdots \wedge dy^{i_r} \tag{2.1.6}$$

($\varphi_{i_1 \cdots i_r}$ is completely anti-symmetric with respect to $i_1 \cdots i_r$)

$$f^{-1}: \omega_r(y) \to \omega_r(x) = \frac{1}{r!} \varphi_{i_1 \cdots i_r}(y(x)) \frac{\partial(y^{i_1}, \cdots, y^{i_r})}{\partial(x^{k_1}, \cdots, x^{k_r})} dx^{k_1} \wedge \cdots \wedge dx^{k_r} \tag{2.1.7}$$

f and f^{-1} in Eq.2.1.1 are mappings of the manifolds $M \xrightarrow{f} N$ and $N \xrightarrow{f^{-1}} M$ respectively, while f in Eq.2.1.3 is a tangent mapping derived from $f: M \to N$, and f^{-1} in Eq.2.1.5 is a cotangent mapping derived from $f^{-1}: N \to M$.

To avoid confusion, we denote the derived mappings by f_* and f^*, and hence rewrite Eq.2.1.3 and Eq.2.1.5 as the following:

$$f_*: \quad X \to Y = \eta^i(x)\frac{\partial y^j}{\partial x^i}\frac{\partial}{\partial y^j}\bigg|_{q=f(p)} \tag{2.1.3$_1$}$$

where f_* is the "forward mapping";

$$f^*: \quad \omega_1(y) \to \omega_1(x) = \varphi_i(y(x))\frac{\partial y^i}{\partial x^j}dx^j\bigg|_{p=f^{-1}(q)} \tag{2.1.5$_1$}$$

where f^* is the "backward mapping".

Eq.2.1.7 can also be rewritten as

$$f^*: \omega_r(y) \to \omega_r(x) = \frac{1}{r!}\varphi_{i_1\cdots i_r}\frac{\partial(y^{i_1},\cdots,y^{i_r})}{\partial(x^{k_1},\cdots,x^{k_r})}dx^{k_1}\wedge\cdots\wedge dx^{k_r} \tag{2.1.7$_1$}$$

Applying the "forward mapping" successively twice gives us

$$\left(\frac{\partial}{\partial x}\xrightarrow{f_{1*}}\frac{\partial}{\partial y}\xrightarrow{f_{2*}}\frac{\partial}{\partial z}\right): \quad \frac{\partial}{\partial x^i} = \frac{\partial y^j}{\partial x^i}\frac{\partial z^k}{\partial y^j}\frac{\partial}{\partial z^k}$$

$$(f_2\cdot f_1)_* = f_{2*}\cdot f_{1*}, \qquad \left(\frac{\partial}{\partial x}\xrightarrow{(f_2\cdot f_1)_*}\frac{\partial}{\partial z}\right)\ ^{1} \tag{2.1.8}$$

Applying the "backward mapping" successively twice gives us

$$\left(dx \xleftarrow{f_1^*} dy \xleftarrow{f_2^*} dz\right): \quad \frac{\partial z^k}{\partial y^j}\frac{\partial y^j}{\partial x^i}dx^i = dz^k$$

$$(f_2\cdot f_1)^* = f_1^*\cdot f_2^*, \qquad \left(dx \xleftarrow{(f_2\cdot f_1)^*} dz\right)\ ^{2} \tag{2.1.9}$$

$(f_2\cdot f_1)_*$ in Eq.2.1.8 and $(f_2\cdot f_1)^*$ in Eq.2.1.9 are dual linear mappings.

f_* and f^* are also dual linear mappings with the following property:

$$< f^*\omega_1, X >_p = < \omega_1, f_*X >_{q=f(p)} \tag{2.1.10}$$

Proof: Suppose $\omega_1 = \omega_1(y) = \varphi_j(y)dy^j$, $X = X(x) = \varepsilon^i(x)\frac{\partial}{\partial x^i}$. Then

$$f_*X = \varepsilon^i(x)\frac{\partial y^j}{\partial x^i}\frac{\partial}{\partial y^j} \qquad \text{(according to Eq.2.1.3$_1$)}$$

$$f^*\omega_1 = f^*\omega_1(y) = \varphi_j(y)\frac{\partial y^j}{\partial x^i}dx^i \qquad \text{(according to Eq.2.1.5$_1$)}$$

Therefore

---

1 $(f_2\cdot f_1)_* = f_{2*}\cdot f_{1*}$ is similar to $(f_2\cdot f_1) = f_2\cdot f_1$.

2 $(f_2\cdot f_1)^* = f_1^*\cdot f_2^*$ is similar to $(f_2\cdot f_1)^{-1} = f_1^{-1}\cdot f_2^{-1}$.

$$< f^*\omega_1, X >_p = < \varphi_j(y)\frac{\partial y^j}{\partial x^i}dx^i, \varepsilon^k(x)\frac{\partial}{\partial x^k} >_p$$

$$= \varphi_j(y)\frac{\partial y^j}{\partial x^i}\delta^i_k \varepsilon^k(x) = \varphi_j(y)\frac{\partial y^j}{\partial x^i}\varepsilon^i(x)$$

$$< \omega_1, f_* X >_{q=f(p)} = < \varphi_j(y)dy^j, \varepsilon^i(x)\frac{\partial y^k}{\partial x^i}\frac{\partial}{\partial y^k} >_{q=f(p)}$$

$$= \varphi_j(y)\delta^j_k \varepsilon^i(x)\frac{\partial y^k}{\partial x^i} = \varphi_j(y)\frac{\partial y^j}{\partial x^i}\varepsilon^i(x)$$

which proves Eq.2.1.10. □

Note: The left hand side of Eq.2.1.10 is a function on the manifold *M*; *X* is a tangent vector, and $f^*\omega_1$ is a 1-form pulled from the manifold *N* backward to *M*. The right hand side of Eq.2.1.10 is a function on the manifold *N*; ω_1 is a 1-form of *N* and $f_* X$ is a tangent vector pushed forward from *M* to *N*. The equal sign in Eq.2.1.10 means that the two sides of the equation have the same value on the corresponding points (points *p* and *q=f(p)*).

§ 2.2 *Submanifolds and Foliation*

In an *n*-manifold *M*, a submanifold $S^{(m)}$ of dimension *m* (*m*<*n*) is a smooth subset of *M*. The neighborhood of any point on $S^{(m)}$ has a system of coordinates $(x^1, x^2, \cdots, x^m)$, while the other $(n-m)$ coordinates (in *M*) are constants $(a^1, a^2, \cdots, a^{n-m})$. That is, the coordinates of any point of the submanifold S^m are $(x^1, \cdots, x^m; a^1, \cdots, a^{n-m})$ in the system of coordinates of the *n*-manifold *M*.

Now by analyzing tangent vector fields, we investigate submanifolds, beginning with 1-dimensional submanifolds.

The 1-dimensional submanifold of *M* is a curve, of which each point has a tangent vector. If reversely, given a field of tangent vector, $X(x) = \xi^i(x)\frac{\partial}{\partial x^i}$, on *M*, the question is: through an arbitrary point of the manifold, can one find a curve of which the tangent line is exactly the vector field of that point? The answer is that there are two cases. In case 1, if $X_P = 0$ at $P \in M$, then *P* is called a critical point of the field of tangent vectors. The tangent vector field near a critical point is related to the properties of global topology, and is rather complicated, so we will not discuss here. In case 2, near the non-critical points (or called *generic points*), one can indeed find one (and only one, in the direction of X_P) smooth curve through the point *P*, of which the tangent line coincides with the direction of the vector field X_P (and this direction is the same as v_p^i in Eq.1.3.2). Such a curve is called the *integral curve* of the vector field *X*. Since there is only one direction for vector field of each point, there can be only one integral curve through any single point, and thus the integral curves do not intersect. Each integral curve is a 1-dimensional submanifold $S^{(1)}$ of *M*.

Generalize to the 2-dimensional submanifolds $S^{(2)}$ (see Figure 2.2.1 where *M* is a 3-manifold). On each $S^{(2)}$, x^1 and x^2 are independent variables, while $a^1\ (=x^3)$ is a constant. Varying the constant a^1 gives different submanifolds $S^{(2)}$. A foliation is hence formed by piling the submanifolds $S^{(2)}$ up (see Figure 2.2.1).

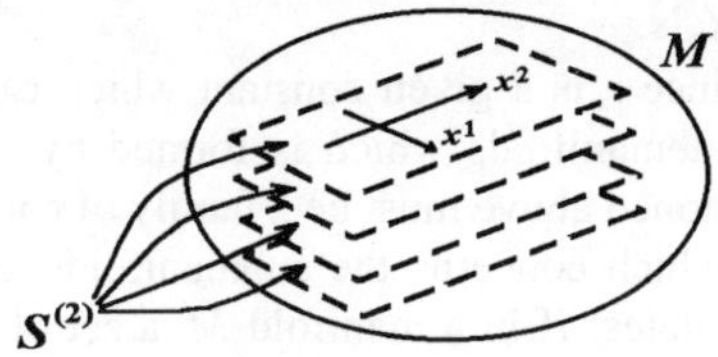

Figure 2.2.1

Now we look at the operators of angular momentum in 3-dimensional spaces:

$$L_x = -z\frac{\partial}{\partial y} + y\frac{\partial}{\partial z}, \quad L_y = -x\frac{\partial}{\partial z} + z\frac{\partial}{\partial x}, \quad L_z = -y\frac{\partial}{\partial x} + x\frac{\partial}{\partial y} \tag{2.2.1}$$

which have the following commutative relations:

$$\left[L_x, L_y\right] = y\frac{\partial}{\partial x} - x\frac{\partial}{\partial y} = -L_z$$

$$\left[L_y, L_z\right] = z\frac{\partial}{\partial y} - y\frac{\partial}{\partial z} = -L_x$$

$$\left[L_z, L_x\right] = x\frac{\partial}{\partial z} - z\frac{\partial}{\partial x} = -L_y \tag{2.2.2}$$

The operators of angular momentum L_x, L_y, and L_z are in fact the tangent vector fields X_1, X_2, and X_3 in 3-dimensional spaces.

$$X_1 = L_x, \quad X_2 = L_y, \quad X_3 = L_z \tag{2.2.3}$$

Let $\Sigma^3 = \{X_1, X_2, X_3\}$ represents the set of tangent vector fields, then Eq.2.2.2 indicates that the Lie brackets $\left[X_i, X_j\right]$ are, without exception, linear combinations of the three tangent vector fields in Σ^3.

Let us also notice that L_x, L_y, and L_z are *not* linearly independent, since

$$xL_x + yL_y + zL_z = 0 \tag{2.2.4}$$

Therefore at every point (*x*, *y*, *z*), there are not three, but only two independent X_i's, which indicates that the integral submanifolds obtained from $\Sigma^3 = \{X_1, X_2, X_3\}$ are 2-dimensional surfaces.

Take

$$r = (x^2 + y^2 + z^2)^{1/2} \tag{2.2.5}$$

we find that

$$L_x r = 0, \quad L_y r = 0, \quad L_z r = 0 \tag{2.2.6}$$

Hence moving along the direction of any tangent vector field, $r = (x^2 + y^2 + z^2)^{1/2}$ neither increases nor decreases. Therefore in a 3-manifold *M*, the submanifolds formed by the tangent vectors are not just 2-dimensional surfaces, but have to be *spheres* with $r = (x^2 + y^2 + z^2)^{1/2} = \text{constant}$. The dimension 2 is consistent with the number of independent X_i's.

At the mean time, since *r* is a given constant which can be taken as a series of numbers, the family of submanifolds which is formed by the integral curves of the tangent vector fields mentioned above must be a family of concentric spheres.

There is a theorem which concerns the submanifolds and foliations, namely the Frobenius' Theorem. It states: If in a manifold *M*, a set of *m* smooth tangent vector fields is defined as

$$\Sigma^m = \{X_1, X_2, \cdots, X_m\}, \tag{2.2.7}$$

and if the Lie bracket of any pair of vector fields in the set can be written as a linear combination of $X_1, \cdots, X_m$, then the integral curves of these vector fields can form a family of submanifolds, of which the dimension of each submanifold is less than (if $X_1, \cdots, X_m$ are not all linearly independent) or equal to *m*.

In the previous example of $\Sigma^3 = \{X_1, X_2, X_3\}$ where *m*=3, the submanifolds are concentric spheres which are of dimension 2 (<3), since X_1, X_2, X_3 are not all linearly independent.

Frobenius' Theorem has two forms. The form mentioned above is demonstrated with the aid of vector fields. There is another form which is demonstrated with the aid of cotangent vector fields (1-form). It can be shown that the two forms are equivalent. We will omit the other form here since it has little relevancy to the contents which we are going to discuss.

§ 2.3 Lie Derivative L_X

Suppose that a differentiable vector field exists on a manifold *M*. Then a system of coordinates (x) can be taken in a neighborhood of an arbitrary point *p* of *M*. A tangent vector field can be introduced as

$$X = \xi^i(x)\frac{\partial}{\partial x^i}.$$

Suppose also that $K(x)$ is a tensor field (such as a scalar field, a vector field, etc.). Then the action of the Lie derivative L_X to $K(x)$ can be defined as (with *p* represents the point x)

$$L_X K\big|_p = \lim_{t\to 0}\frac{1}{t}\big(K(x+\xi(x)t) - K(x)\big) \tag{2.3.1}$$

where x and $\xi(x)$ are both vectors, while t is a parameter. Let

$$\frac{dx^i}{dt} = \xi^i(x), \qquad (i = 1, 2, \cdots, n) \tag{2.3.2}$$

we can obtain the integral curves by integration:

$$x^i = x^i(x_p, t) \qquad\qquad (x_p^i = x^i(x_p, t=0)) \tag{2.3.3}$$

1. Suppose that K is a field of scalars $f(x)$.

$$L_X f(x)\big|_p = \lim_{t\to 0}\frac{1}{t}\big(f(x+\xi t) - f(x)\big) = \lim_{t\to 0}\frac{\xi^i t}{t}\frac{\partial f}{\partial x^i}$$

$$= \xi^i \frac{\partial f}{\partial x^i}\bigg|_{t=0} = X f\,\big|_p \tag{2.3.4}$$

2. Suppose that K is a field of vectors $Y(x) = \eta^i(x)\dfrac{\partial}{\partial x^i}$ $(x^{i\,\prime} = x^i + \xi^i t$, $x^i = x^{i\,\prime} - \xi^i t)$.

$$L_X Y(x)\big|_p = \lim_{t\to 0}\frac{1}{t}\left(\eta^i(x+\xi t)\frac{\partial x^j}{\partial x^{i\,\prime}}\frac{\partial}{\partial x^j} - \eta^i(x)\frac{\partial}{\partial x^i}\right)\bigg|_p$$

$$\left[\frac{\partial x^j}{\partial x^{i\,\prime}} = \delta_i^j - \frac{\partial \xi^j}{\partial x^{i\,\prime}}t \approx \delta_i^j - \frac{\partial \xi^j}{\partial x^i}t, \text{ the } t^2 \text{ term is omitted}\right]$$

$$= \lim_{t\to 0}\frac{1}{t}\left(\eta^i(x+\xi t)\left(\delta_i^j - \frac{\partial \xi^j}{\partial x^i}t\right)\frac{\partial}{\partial x^j} - \eta^i(x)\frac{\partial}{\partial x^i}\right)\bigg|_p$$

$$= \lim_{t\to 0}\frac{1}{t}\left(\left(\eta^i(x) + \frac{\partial \eta^i}{\partial x^k}\xi^k(x)t\right)\left(\delta_i^j - \frac{\partial \xi^j}{\partial x^i}t\right)\frac{\partial}{\partial x^j} - \eta^i(x)\frac{\partial}{\partial x^i}\right)\bigg|_p$$

$$= \lim_{t\to 0}\frac{1}{t}\left(\left(\eta^i(x) - \eta^i(x)\frac{\partial \xi^j}{\partial x^i}t + \frac{\partial \eta^i}{\partial x^k}\xi^k(x)t\right)\frac{\partial}{\partial x^j} - \eta^i(x)\frac{\partial}{\partial x^i}\right)\bigg|_p$$

$$= \xi^k(x)\frac{\partial \eta^i}{\partial x^k}\frac{\partial}{\partial x^j} - \eta^i(x)\frac{\partial \xi^j}{\partial x^i}\frac{\partial}{\partial x^j} = [X, Y]\,\big|_p \tag{2.3.5}$$

3. Suppose that K is a 1-form $\omega(x) = \omega_i(x)dx^i$.

$$L_X\,\omega_i(x)dx^i = \lim_{t\to 0}\frac{1}{t}\left(\left(\omega_i(x) + \frac{\partial \omega_i(x)}{\partial x^k}\xi^k t\right)\frac{dx^i}{dx^j}dx^j - \omega_i(x)dx^i\right)$$

$$= \lim_{t\to 0}\frac{1}{t}\left(\left(\omega_i(x)+\frac{\partial\omega_i(x)}{\partial x^k}\xi^k t\right)\left(\delta^i_j+\frac{\partial\xi^i}{\partial x^j}t\right)dx^j-\omega_i(x)dx^i\right)$$

$$=\left(\omega_i(x)\frac{\partial\xi^i}{\partial x^j}dx^j+\frac{\partial\omega_j(x)}{\partial x^k}\xi^k dx^j\right)$$

$$=\left(\omega_i(x)\frac{\partial\xi^i}{\partial x^j}+\frac{\partial\omega_j(x)}{\partial x^k}\xi^k\right)dx^j \tag{2.3.6}$$

4. Suppose that K is an arbitrary tensor field $T^{i_1\cdots i_p}_{k_1\cdots k_q}$. It can be shown in a similar way that the action of L_X for each subscript can be treated as a 1-form in Eq.2.3.4, and for each superscript the action can be treated as a vector field in Eq.2.3.3. For the part that does not fit as either indices, the action can be treated as a scalar field in Eq.2.3.2.

Besides, there are three other interesting properties:

1. L_X's action on a scalar field still gives a scalar field. L_X's action on a vector field still gives a vector field. L_X's action on a 1-form still gives a 1-form. Thus L_X is a linear operator which does not change the type of a tensor field.

2. L_X is a differential operator which can act on both tensor fields and differential forms, while the exterior differential operator d can only act on differential forms.

3. The existence of L_X requires a given vector field X on a manifold, while d does not require any additional structure on a manifold.

§ 2.4 Interior Product i_X and Three Cartan Formulas

The interior product of a vector $X=\xi^j\frac{\partial}{\partial x^j}$ and a differential form $\omega=\omega_{i_1,i_2,\cdots,i_p}dx^{i_1}\wedge dx^{i_2}\wedge\cdots\wedge dx^{i_p}$ is a kind of operation, which is denoted by $i_X\omega$ or $i(X)\omega$ or $X\lrcorner\omega$. The properties of i_X are:

1. p-forms are reduced to $(p-1)$-forms by i_X;

2. $i_X f=0$ (where f is an ordinary function) (2.4.1)

3. $i_X dx^j=\xi^j$ is equivalent to $<dx^j,X>=\xi^j$ (see Eq.1.3.7) (2.4.2)

Thus i_X is a kind of generalization of the contraction of a vector $\frac{\partial}{\partial x^i}$ and an 1-form dx^j mentioned in Chapter 1.

4. $i_X i_X=0$ (2.4.3)

Proof: Suppose we have $\omega_{i_1i_2i_3\cdots}dx^{i_1}\wedge dx^{i_2}\wedge dx^{i_3}\wedge\cdots$, where $\omega_{i_1i_2i_3\cdots}$ is completely anti-symmetric with respect to $i_1,i_2,i_3,\cdots$. Then

$$i_X\omega = \xi^j \omega_{ji_2i_3\cdots} dx^{i_2} \wedge dx^{i_3} \wedge \cdots$$
$$i_X i_X \omega = \xi^j \xi^k \omega_{jki_3\cdots} dx^{i_3} \wedge \cdots = 0$$

The proof is done since $\xi^j \xi^k$ is symmetric with respect to j and k and $\omega_{jki_3\cdots}$ is anti-symmetric with respect to j and k. □

5. i_X is an odd differential operator, with the following property:

$$\begin{aligned}
&i_X(\alpha_p \wedge \beta_q) \\
&= (i_X\alpha_p) \wedge \beta_q + (i_X\beta_q)(-1)^{pq} \wedge \alpha_p \quad \text{(exchanging } \alpha_p \text{ and } \beta_q) \\
&= (i_X\alpha_p) \wedge \beta_q + (-1)^{pq}(-1)^{p(q-1)} \alpha_p \wedge (i_X\beta_q) \quad \text{(exchanging } \alpha_p \text{ and } i_X\beta_q) \\
&= (i_X\alpha_p) \wedge \beta_q + (-1)^{p} \alpha_p \wedge (i_X\beta_q)
\end{aligned} \tag{2.4.4}$$

Now we prove the three Cartan's formulas.

a) $[L_X, d] = 0$.

Proof: 1. Apply to f:

$$df = \frac{\partial f}{\partial x^i} dx^i$$

$$L_X df = \xi^k \frac{\partial^2 f}{\partial x^k \partial x^i} dx^i + \frac{\partial f}{\partial x^i}\frac{\partial \xi^i}{\partial x^k} dx^k \quad \text{(see Eq.2.3.4)}$$

$$L_X f = \xi^i \frac{\partial f}{\partial x^i}$$

$$dL_X f = \xi^i \frac{\partial^2 f}{\partial x^i \partial x^k} dx^k + \frac{\partial \xi^i}{\partial x^k}\frac{\partial f}{\partial x^i} dx^k$$

$$\therefore \quad [L_X, d] f = 0 \tag{2.4.5}$$

2. Apply to an 1-form $\omega = \omega_j dx^j$:

$$d\omega = d(\omega_j dx^j) = \frac{\partial \omega_j}{\partial x^i} dx^i \wedge dx^j$$

$$\begin{aligned}
L_X d\omega &= \frac{\partial^2 \omega_j}{\partial x^k \partial x^i} \xi^k dx^i \wedge dx^j + \frac{\partial \omega_j}{\partial x^i}\frac{\partial \xi^i}{\partial x^k} dx^k \wedge dx^j + \frac{\partial \omega_j}{\partial x^i}\frac{\partial \xi^j}{\partial x^k} dx^i \wedge dx^k \\
&= \left(\frac{\partial^2 \omega_j}{\partial x^i \partial x^k} \xi^k dx^i \wedge dx^j + \frac{\partial \omega_j}{\partial x^k}\frac{\partial \xi^k}{\partial x^i} dx^i \wedge dx^j + \frac{\partial \omega_k}{\partial x^i}\frac{\partial \xi^k}{\partial x^j} dx^i \wedge dx^j \right)
\end{aligned}$$

$$L_X \omega = \left(\omega_k \frac{\partial \xi^k}{\partial x^j} + \frac{\partial \omega_j}{\partial x^k} \xi^k \right) dx^j \quad \text{(see Eq.2.3.4)}$$

$$dL_X \omega = \left(\frac{\partial \omega_k}{\partial x^i}\frac{\partial \xi^k}{\partial x^j} + \omega_k \frac{\partial^2 \xi^k}{\partial x^i \partial x^j} + \frac{\partial^2 \omega_j}{\partial x^k \partial x^i} \xi^k + \frac{\partial \omega_j}{\partial x^k}\frac{\partial \xi^k}{\partial x^i} \right) dx^i \wedge dx^j$$

(the second term is zero since $\partial x^i \partial x^j$ is symmetric).

$$\therefore \left[L_X, d\right]\omega = 0 \qquad (2.4.6)$$

3. Apply to $\omega^A \wedge \omega^B$ with the assumptions that $\left[L_X, d\right]\omega^A = 0$ and $\left[L_X, d\right]\omega^B = 0$:

$$L_X d(\omega^A \wedge \omega^B) = L_X(d\omega^A \wedge \omega^B + \omega^A \wedge d\omega^B)$$

$$= L_X d\omega^A \wedge \omega^B + d\omega^A \wedge L_X\omega^B + L_X\omega^A \wedge d\omega^B + \omega^A \wedge L_X d\omega^B$$

$$dL_X(\omega^A \wedge \omega^B) = d(L_X\omega^A \wedge \omega^B + \omega^A \wedge L_X\omega^B)$$

$$= dL_X\omega^A \wedge \omega^B + L_X\omega^A \wedge d\omega^B + d\omega^A \wedge L_X\omega^B + \omega^A \wedge dL_X\omega^B$$

$$\therefore \left[L_X, d\right](\omega^A \wedge \omega^B) = \left[L_X, d\right]d\omega^A \wedge \omega^B + \omega^A \wedge \left[L_X, d\right]\omega^B = 0$$

Therefore the following equation holds for any *n*-form ω^n (including the 0-form):

$$\left[L_X, d\right]\omega^n = 0 \qquad (2.4.7)$$

$$\therefore \left[L_X, d\right] = 0 \qquad (2.4.8)$$

□

b) $L_X = d \cdot i_X + i_X \cdot d$.

Proof: 1. Apply to f: (Take $X = \xi^i \dfrac{\partial}{\partial x^i}$ and use Eq.2.4.2)

$$L_X f = \xi^i \frac{\partial f}{\partial x^i}; \quad (d \cdot i_X + i_X \cdot d)f = i_X \cdot df = i_X \frac{\partial f}{\partial x^i} dx^i = \xi^i \frac{\partial f}{\partial x^i}.$$

$$\therefore \quad L_X f = (d \cdot i_X + i_X \cdot d)f \qquad (2.4.9)$$

2. Apply to an 1-form:

$$L_X \omega_i dx^i = \left(\frac{\partial \omega_j}{\partial x^k}\xi^k + \omega_i \frac{\partial \xi^i}{\partial x^j}\right)dx^j;$$

$$(d \cdot i_X + i_X \cdot d)\omega_i dx^i = d \cdot \omega_i \xi^i + i_X \cdot \frac{\partial \omega_i}{\partial x^k} dx^k \wedge dx^i$$

$$= \frac{\partial \omega_i}{\partial x^k}\xi^i dx^k + \omega_i \frac{\partial \xi^i}{\partial x^k} dx^k + \frac{\partial \omega_i}{\partial x^k}\xi^k dx^i - \frac{\partial \omega_i}{\partial x^k} dx^k \xi^i$$

$$= \left(\frac{\partial \omega_i}{\partial x^k}\xi^k + \omega_j \frac{\partial \xi^j}{\partial x^i}\right)dx^i$$

$$\therefore \quad L_X \omega_i dx^i = (d \cdot i_X + i_X \cdot d)\omega_i dx^i \qquad (2.4.10)$$

3. Apply to $\omega_{p+q} = \alpha_p \wedge \beta_q$ where α_p is a *p*-form, β_q is a *q*-form, and it is known that $L_X \alpha_p = (d \cdot i_X + i_X \cdot d)\alpha_p$, and $L_X \beta_q = (d \cdot i_X + i_X \cdot d)\beta_q$. We obtain:

$$L_X(\alpha_p \wedge \beta_q) = (L_X\alpha_p) \wedge \beta_q + (L_X\beta_q)(-1)^{pq} \wedge \alpha_p \quad \text{(see the derivation of Eq.2.4.4)}$$

$$= [(d \cdot i_X + i_X \cdot d)\alpha_p] \wedge \beta_q + [(d \cdot i_X + i_X \cdot d)\beta_q](-1)^{pq} \wedge \alpha_p$$

It is known that both L_X and $(d \cdot i_X + i_X \cdot d)$ will not change the tensor types of α_p

and β_q. Hence

$$L_X(\alpha_p \wedge \beta_q) = [(d \cdot i_X + i_X \cdot d)\alpha_p] \wedge \beta_q + \alpha_p \wedge [(d \cdot i_X + i_X \cdot d)\beta_q]$$
$$= [(d \cdot i_X + i_X \cdot d)(\alpha_p \wedge \beta_q)]$$

That is,

$$L_X \omega_{p+q} = (d \cdot i_X + i_X \cdot d)\omega_{p+q} \tag{2.4.11}$$

Therefore the following equation holds for any form ω^n (including the 0-form):

$$L_X = (d \cdot i_X + i_X \cdot d) \tag{2.4.12}$$

□

c) $[L_X, i_Y] = i_{[X,Y]}$.

Proof: 1. Apply to f:

$$L_X \cdot i_Y f - i_Y \cdot L_X f = 0 - i_Y \cdot \xi^i \frac{\partial f}{\partial x^i} = 0 = i_{[X,Y]} f$$

2. Apply to the 1-form $\omega = \omega_i dx^i$:

$$L_X \cdot i_Y \omega_i dx^i - i_Y \cdot L_X \omega_i dx^i = L_X \omega_i \eta^i - i_Y \cdot \left(\frac{\partial \omega_i}{\partial x^k} \xi^k + \omega_k \frac{\partial \xi^k}{\partial x^i} \right) dx^i$$

$$= \xi^j \eta^i \frac{\partial \omega_i}{\partial x^j} + \xi^j \frac{\partial \eta^i}{\partial x^j} \omega_i - \left(\frac{\partial \omega_i}{\partial x^k} \xi^k + \omega_k \frac{\partial \xi^k}{\partial x^i} \right) \eta^i$$

$$= \left(\xi^j \frac{\partial \eta^i}{\partial x^j} - \eta^j \frac{\partial \xi^i}{\partial x^j} \right) \omega_i = i_{[X,Y]} \omega_i dx^i \tag{2.4.13}$$

$$\because \quad [X,Y] = \left[\xi^k \frac{\partial}{\partial x^k}, \eta^l \frac{\partial}{\partial x^l} \right] = \xi^k \frac{\partial \eta^l}{\partial x^k} \frac{\partial}{\partial x^l} - \eta^l \frac{\partial \xi^k}{\partial x^l} \frac{\partial}{\partial x^k}$$

$$= \left(\xi^k \frac{\partial \eta^l}{\partial x^k} - \eta^k \frac{\partial \xi^l}{\partial x^k} \right) \frac{\partial}{\partial x^l}$$

$$\therefore \quad i_{[X,Y]} \omega_l dx^l = \left(\xi^k \frac{\partial \eta^l}{\partial x^k} - \eta^k \frac{\partial \xi^l}{\partial x^k} \right) \omega_l$$

which gives us Eq.2.4.13.

3. Apply to $\omega_{p+q} = \omega_p \wedge \omega_q$, where ω_p and ω_q are *p*-form and *q*-form respectively, and it is known that $[L_X, i_Y]\omega_p = i_{[X,Y]}\omega_p$ and $[L_X, i_Y]\omega_q = i_{[X,Y]}\omega_q$. Thus

$$[L_X, i_Y](\omega_p \wedge \omega_q) = ([L_X, i_Y]\omega_p) \wedge \omega_q + ([L_X, i_Y]\omega_q)(-1)^{pq} \wedge \omega_p$$
$$= (i_{[X,Y]}\omega_p) \wedge \omega_q + (i_{[X,Y]}\omega_q)(-1)^{pq} \wedge \omega_p$$
$$= i_{[X,Y]}(\omega_p \wedge \omega_q)$$

That is, $[L_X, i_Y]\omega_{p+q} = i_{[X,Y]}\omega_{p+q}$. Therefore the following equation holds in any case.

$$[L_X, i_Y] = i_{[X,Y]} \tag{2.4.14}$$

The three Cartan's formulas given in Eq.2.4.8, Eq.2.4.12 and Eq.2.4.14 are quite useful in the analysis of the differential forms.

§ 2.5 Group Spaces of Lie Groups

If M is a differential manifold, and Lie group G acts transitively on M, then G is called a *transitive transformation group.* M is called a *group manifold of G.* "Transitive" means that for any two points x and y in M, there exists an element g in G which transforms x to y, i.e., $gx=y$. For example:

Let us consider $SU(2)$ group. All the matrices of $SU(2)$ group have the following form:

$$g = \begin{pmatrix} \alpha & \beta \\ -\bar{\beta} & \bar{\alpha} \end{pmatrix}, \qquad \alpha\bar{\alpha} + \beta\bar{\beta} = 1 \tag{2.5.1}$$

The three generating elements of $SU(2)$ group are:

$$\sigma_1 = \begin{pmatrix} 0 & 1 \\ 1 & 0 \end{pmatrix}, \ \sigma_2 = \begin{pmatrix} 0 & -i \\ i & 0 \end{pmatrix}, \text{ and } \sigma_3 = \begin{pmatrix} 1 & 0 \\ 0 & -1 \end{pmatrix} \tag{2.5.2}$$

Thus each group element g can be written as ($n_1^2 + n_2^2 + n_3^2 = 1$):[1]

$$\begin{aligned} g &= e^{i\theta\vec{n}\cdot\vec{\sigma}/2} \\ &= \begin{pmatrix} 1 & 0 \\ 0 & 1 \end{pmatrix} + \left(\frac{i\theta\vec{n}\cdot\vec{\sigma}}{2}\right) + \frac{1}{2!}\left(\frac{i\theta\vec{n}\cdot\vec{\sigma}}{2}\right)^2 + \frac{1}{3!}\left(\frac{i\theta\vec{n}\cdot\vec{\sigma}}{2}\right)^3 \\ &\quad + \frac{1}{4!}\left(\frac{i\theta\vec{n}\cdot\vec{\sigma}}{2}\right)^4 + \frac{1}{5!}\left(\frac{i\theta\vec{n}\cdot\vec{\sigma}}{2}\right)^5 + \cdots \end{aligned} \tag{2.5.3}$$

Plug

$$\begin{aligned} (\vec{n}\cdot\vec{\sigma})^2 &= (n_1\sigma_1 + n_2\sigma_2 + n_3\sigma_3)(n_1\sigma_1 + n_2\sigma_2 + n_3\sigma_3) \\ &= (n_1^2 + n_2^2 + n_3^2)\begin{pmatrix} 1 & 0 \\ 0 & 1 \end{pmatrix} = \begin{pmatrix} 1 & 0 \\ 0 & 1 \end{pmatrix} \end{aligned} \tag{2.5.4}$$

into Eq.2.5.3, we obtain:

$$\begin{aligned} g &= \begin{pmatrix} 1 & 0 \\ 0 & 1 \end{pmatrix}\left(1 - \frac{1}{2!}\left(\frac{\theta}{2}\right)^2 + \frac{1}{4!}\left(\frac{\theta}{2}\right)^4 - \cdots\right) \\ &\quad + \left(i(\vec{n}\cdot\vec{\sigma})\left(\frac{\theta}{2}\right) - \frac{i}{3!}(\vec{n}\cdot\vec{\sigma})\left(\frac{\theta}{2}\right)^3 + \frac{i}{5!}(\vec{n}\cdot\vec{\sigma})\left(\frac{\theta}{2}\right)^5 - \cdots\right) \end{aligned}$$

---

[1] Any element of $SU(2)$ group can be written as $\exp(i\alpha_a X_a)$, where α_a $(a = 1, 2, \cdots, n)$ are real parameters, and X_a's are linearly independent Hermitian operators.

$$= \cos\left(\frac{\theta}{2}\right)\begin{pmatrix}1 & 0\\ 0 & 1\end{pmatrix} + i\,\vec{n}\cdot\vec{\sigma}\,\sin\left(\frac{\theta}{2}\right) \qquad (2.5.5)$$

Note that θ here satisfies $0 \le \theta \le 2\pi$. Eq.2.5.5 can be re-written as:

$$g = \begin{pmatrix}1 & 0\\ 0 & 1\end{pmatrix} x_0 + i\sigma_1 x_1 + i\sigma_2 x_2 + i\sigma_3 x_3$$

where

$$x_0 = \cos\left(\frac{\theta}{2}\right),\quad x_1 = n_1 \sin\left(\frac{\theta}{2}\right),\quad x_2 = n_2 \sin\left(\frac{\theta}{2}\right),\quad x_3 = n_3 \sin\left(\frac{\theta}{2}\right) \qquad (2.5.6)$$

$$\therefore\ x_0^2 + x_1^2 + x_2^2 + x_3^2 = 1 \qquad (0 \le \theta \le 2\pi) \qquad (2.5.7)$$

Thus we see that the parameters of g, (x_0, x_1, x_2, x_3) given by Eq.2.5.6, must be the coordinates of some point on S^3. Hence the hypersphere S^3 represented by Eq.2.5.7 is a group manifold of the $SU(2)$ group.

Also we shall see that if given two group elements g_1 and g_2, there must be two points $x = (x_0, x_1, x_2, x_3)$ and $x' = (x_0', x_1', x_2', x_3')$ on S^3 corresponding to g_1 and g_2 respectively. Further more, one must be able to find a g such that

$$g \cdot g_1 = g_2 \qquad (2.5.8)$$

g has 4 unknown parameters, and $g \cdot g_1 = g_2$ provides exactly 4 equations, therefore the solution g is uniquely determined.

Hence the S^3 manifold in Eq.2.5.7 is just the "group manifold" of the Lie group $SU(2)$; $SU(2)$ is thus called the "*transitive transformation group*" of the S^3 manifold.

Here we introduce several concepts concerning group manifolds (which will be used later):

Suppose that x is a point on M, and g is an element of group G.

1. If there exists a $g \in G$ such that $gx = x$ for $\forall x \in M$, then $g = e$ is an identity element, and G is said to act on M "*effectively*" (the identity element e is unique).
2. If there is one point $x_0 \in M$ (but not $\forall x \in M$), and a group of g (not a single g) satisfying $gx_0 = x_0$, then this group of g forms a subgroup G_{x_0} of G:

$$G_{x_0} = \{ g \in G \mid gx_0 = x_0 \in M \} \qquad (2.5.9)$$

   G_{x_0} is called an "*isotropic subgroup*" at x_0.
3. If there is an isotropic subgroup G_{x_0} at x_0, then $G_{x_0} x_0 = x_0$. Taking another point $gx_0 = x$, we must have

$$\left(g G_{x_0} g^{-1}\right) x = g G_{x_0} x_0 = g x_0 = x$$

   Thus there must be an isotropic subgroup G_x at the point $gx_0 = x$:

$$G_x = g G_{x_0} g^{-1} = \{ g' \in G \mid g' x = x \in M \} \qquad (2.5.9)_1$$

   Here g is uniquely determined by $gx_0 = x$. Thus $G_x = g G_{x_0} g^{-1}$ and G_{x_0} are isomorphic. By analogy, if there is an isotropic subgroup G_{x_0} at some point x_0, then the isotropic subgroups G_x exist at every point *x*, and G_{x_0}, G_{x_1}, G_{x_2}, $\cdots$ are isomorphic to each other.

4. For some point $x_0 \in M$, the set O_{x_0} of all the points which x_0 can possibly reach under the actions of every element *g* of a group *G* is called the "*orbit of the group G*", denoted by

$$O_{x_0} \equiv G_{x_0} \equiv \{gx_0 \mid g \in G\} \in M .$$

5. If two orbits intersect: $O_{x_0} \cap O_{x_0'} \neq \phi$. Then the points which can be reached by G_{x_0} must also be reached by $G_{x_0'}$, that is, $O_{x_0} = O_{x_0'}$.

   Proof: Take any point *x* from $O_{x_0} \cap O_{x_0'}$, we must have $gx_0 = x$ and $g'x_0' = x$. Hence $x_0 = g^{-1}x = g^{-1}g'x_0' = g''x_0'$. Therefore the points that can be reached by G_{x_0} must also be reached by $G_{x_0'}$. Thus $O_{x_0} = O_{x_0'}$ when $O_{x_0} \cap O_{x_0'} \neq \phi$.

6. Since the orbits do not intersect, they form a division of the manifold *M*. If each orbit is treated as a point, then the manifold consisting those points is called the "*orbit space*" *M/G*. For example:

$$O(n)\big/O(n-1) = S^{n-1} \tag{2.5.10}$$

   Remark: It is known that when *R* is given, the orbit of the *n*-dimensional orthogonal group $O(n) = \{A \mid A = n \times n$ matrix, $A^T A = E\}$ is S^{n-1} with radius *R* ($S^{n-1} = \{x \mid |x| = R\}$ is a sphere of dimension $n-1$). But $O(n)$: $S^{n-1} \to S^{n-1}$ does not map injectively to each point of S^{n-1}. There is an $O(n-1)$ isotropic group (e.g. take any point x_0 of S^2 as the south or north pole, and do an $O(2)$ rotation. Then $O(2)$: $x_0 \to x_0$. Thus each point of S^2 has an isotropic subgroup $O(2)$.). Therefore each point of S^{n-1} corresponds to each point of $O(n)\big/O(n-1)$ ($O(n-1)$ is an isotropic subgroup in $O(n)$).

§ 2.6 *Invariant Tangent and Cotangent Fields on a Lie Group Manifold M*

Take an arbitrary group element *g* of a Lie group *G*. The left translation L_a of the Lie group manifold *G* is defined as the following:

$$L_a:\ G \to G \qquad \text{(where } a \text{ and } g \text{ are both group elements)} \tag{2.6.1}$$
$$g \to L_a(g) = ag \in G$$

If the tangent vector $X(g)$ is *left invariant*, then when L_a acts on the tangent vector $X(g)$ at the point *g* of *M*, it *must* satisfy

$$L_a X(g) = X(L_a(g)) = X(ag) \tag{2.6.2}$$

Thus for a left invariant tangent vector field, we must have:

$$L_g X(e) = X(L_g(e)) = X(ge) = X(g) \qquad (\forall g \in G) \tag{2.6.3}$$

In other words, when g travels all over G, a vector $X(e)$ at the identity element e can give a vector field $X(g)$ on the group through a left homeomorphic transformation, and obviously this vector field is left invariant since we may obtain from Eq.2.6.3 that

$$L_a X(g) = X(L_a(g)) = X(ag)$$

which fits exactly the requirement of Eq.2.6.2 about left invariance.

The left invariant tangent vector fields on the group manifold M of a Lie group G are defined globally on M, and are all differentiable vector fields.

Since a Lie group G of dimension n has n independent group elements g_1, g_2, $\cdots$, g_n, there must be n left invariant tangent vector fields $X(g_i)$ $(i = 1,2,\cdots,n)$ satisfying Eq.2.6.2:

$$L_a X(g_i) = X(L_a(g_i)) = X(ag_i)$$

The n left invariant tangent vector fields $X(g_i)$ at the point e of the group manifold M span an n-dimensional left invariant tangent space $T_e(G)$.

$$L_g\colon\ T_e(G) \to T_g(G) \tag{2.6.4}$$

When g travels all over G, we may obtain a left invariant tangent bundle $T(G)$ of G (the collection of all $T_g^{'}(G)$).

We would not repeat the discussion with respect to the right invariant tangent vector fields since it is similar to the one with respect to the left invariant tangent vector fields.

Above we have discussed the invariant tangent vector fields, which is distinguished by its connecting the group element $g \in G$ and the corresponding tangent vector $X(g)$ at g on a group manifold M, and building up an 1-1 relationship which holds for each $g \in G$ (see Eq.2.6.2 and Eq.2.6.3).

The cases of invariant cotangent vector fields are similar. Suppose that ω is a cotangent vector (1-form) on a group manifold. ω is called a left invariant form if it satisfies

$$L_a \omega = \omega\,^{1} \qquad \forall a \in G \quad \text{(where } G \text{ is a matrix group)} \tag{2.6.5}$$

For example, the *Maurer-Cartan form*:

$$\omega = g^{-1} dg \tag{2.6.6}$$

This is a 1-form with the group element g changing continuously. Apply a constant group element g_0 to the left of g: $g \to g_0 g$, we obtain

$$L_{g_0} \omega = (g_0 g)^{-1} d(g_0 g) = g^{-1} g_0^{-1} g_0 dg = g^{-1} dg = \omega \tag{2.6.7}$$

which indicates that ω is indeed a left invariant 1-form.

The left invariant 1-form ω can be expanded by the constant matrices λ_a of a Lie algebra G:

1 In some books of mathematics, the L_a acting on a tangent vector is sometimes denoted as L_{a*} and the L_a acting on a cotangent vector as L_a^{*}. Here the *'s are all omitted.

$$\omega = g^{-1}dg = \Phi_a \frac{\lambda_a}{2i} \tag{2.6.8}$$

where λ_a's in Eq.2.6.8 are independent, and thus each Φ_a is a left invariant 1-form.

$d(g^{-1}g) = 0 \to dg^{-1} \cdot g + g^{-1} \cdot dg = 0$ gives

$$dg^{-1} + g^{-1}dg\, g^{-1} = 0 \tag{2.6.9}$$

From $d(d(g^{-1}g)) = d(dg^{-1} \cdot g) + d(g^{-1} \cdot dg) = 0$ we obtain

$$-dg^{-1} \wedge dg + d(g^{-1}dg) = 0 \tag{2.6.10}$$

Plug Eq.2.6.9 into Eq.2.6.10 we have $g^{-1}dg\, g^{-1} \wedge dg + d(g^{-1}dg) = 0$, which can be rearranged as

$$g^{-1}dg \wedge g^{-1}dg + d(g^{-1}dg) = 0 \tag{2.6.11}$$

Now plug Eq.2.6.8 into Eq.2.6.11 we have

$$d\Phi_a \frac{\lambda_a}{2i} + \Phi_b \frac{\lambda_b}{2i} \wedge \Phi_c \frac{\lambda_c}{2i} = 0 \tag{2.6.12}$$

Eq.2.6.12 is the first form of the Maurer-Cartan structural equations. It illustrates the characters of a left invariant cotangent vector field.

Eq.2.6.12 can also be written as

$$0 = d\Phi_a \frac{\lambda_a}{2i} + \frac{1}{2}\Phi_b \frac{\lambda_b}{2i} \wedge \Phi_c \frac{\lambda_c}{2i} + \frac{1}{2}\Phi_c \frac{\lambda_c}{2i} \wedge \Phi_b \frac{\lambda_b}{2i}$$

$$= d\Phi_a \frac{\lambda_a}{2i} + \frac{1}{2}\Phi_b \wedge \Phi_c \left[\frac{\lambda_b}{2i}, \frac{\lambda_c}{2i}\right]_-$$

$$\to d\Phi_a + \frac{1}{2} f_{abc} \Phi_b \wedge \Phi_c = 0 \tag{2.6.13}$$

Eq.2.6.13 is the second form of the Maurer-Cartan structural equations. f_{abc} is the structural constant related to the constant matrices $[\lambda_a, \lambda_b]$ of a Lie group G.

Now let us consider the right invariant case. Take a right invariant Maurer-Cartan form

$$\omega' = dg \cdot g^{-1}, \tag{2.6.14}$$

$$R_{g_0}\omega' = d(g\,g_0) \cdot (g\,g_0)^{-1} = dg\,g_0\,g_0^{-1}g^{-1} = dg \cdot g^{-1} = \omega' \tag{2.6.15}$$

indicates that ω' is a right invariant 1-form. Expand by the constant matrices λ_a:

$$\omega' = dg \cdot g^{-1} = \Phi'_a \frac{\lambda_a}{2i} \tag{2.6.16}$$

where each Φ'_a is a right invariant 1-form.

$d(g\,g^{-1}) = 0 \to dg \cdot g^{-1} + g \cdot dg^{-1} = 0$ gives

$$dg^{-1} + g^{-1}dg\, g^{-1} = 0 \quad \text{which is the same as Eq.2.6.9.} \tag{2.6.9$_1$}$$

From $d(d(g\,g^{-1})) = d(dg \cdot g^{-1}) + d(g \cdot dg^{-1}) = 0$ we obtain

$$-dg \wedge dg^{-1} + d(g \cdot dg^{-1}) = 0 \tag{2.6.17}$$

which can also be written as $-dg \wedge dg^{-1} - d(dg \cdot g^{-1}) = 0$. Use Eq.2.6.9$_1$ we have

$$dg\,g^{-1}\wedge dg\,g^{-1}-d(dg\cdot g^{-1})=0 \tag{2.6.18}$$

Now plug in Eq.2.6.16 we obtain

$$\Phi'_a\frac{\lambda_a}{2i}\wedge\Phi'_b\frac{\lambda_b}{2i}-d\left(\Phi'_c\frac{\lambda_c}{2i}\right)=0 \tag{2.6.19}$$

$$\rightarrow d\Phi'_c\frac{\lambda_c}{2i}-\frac{1}{2}\Phi'_a\wedge\Phi'_b\left[\frac{\lambda_a}{2i},\frac{\lambda_b}{2i}\right]_-$$

$$\rightarrow d\Phi'_a-\frac{1}{2}f_{abc}\Phi'_b\wedge\Phi'_c=0 \tag{2.6.20}$$

It is thus clear that the Maurer-Cartan structural equations Eq.2.6.13 and Eq.2.6.20 which were obtained from left invariance and right invariance respectively are similar, except that the terms of $\frac{1}{2}f_{abc}\Phi_b\wedge\Phi_c$ are differed by a $\pm$ sign. There's another way of deduction which is more efficient:

From $g^{-1}g=I$ we have $(dg^{-1})g+g^{-1}(dg)=0\rightarrow dg^{-1}=-g^{-1}dg\,g^{-1}$.

$\omega=g^{-1}dg$:

$$d\omega=dg^{-1}\wedge dg=-g^{-1}dg\,g^{-1}\wedge dg=-g^{-1}dg\wedge g^{-1}dg=-\omega\wedge\omega$$

$$\therefore\quad d\omega+\omega\wedge\omega=0$$

which gives us Eq.2.6.11.

$\omega'=dg\,g^{-1}$:

$$d\omega'=-dg\wedge dg^{-1}=dg\,g^{-1}\wedge dg\,g^{-1}=\omega'\wedge\omega'$$

$$\therefore\quad d\omega'-\omega'\wedge\omega'=0$$

which gives us Eq.2.6.18.

Chapter 3. Curvature Tensors, Torsion Tensors, Covariant Differentials and Adjoint Exterior Differentials

§ 3.1 Covariant Differentials and Connections

The relations of the covariant properties of the tangent spaces of two neighboring points on a differential manifold cannot be described by either the differential operator d or the Lie derivative L_X previously introduced. Now in order to define the covariant differential on a tangent bundle $T(M)$, a cotangent bundle $T^*(M)$, and a tensor bundle $T_s^r(M)$, we introduce the connections.

As we have discussed in the previous sections that at an arbitrary point p of a differential manifold M, a group of base vectors of a tangent space $T_p(M)$ is $\left\{\frac{\partial}{\partial x^i}\right\}$, and a group of base vectors of a cotangent space $T_p^*(M)$ is $\left\{dx^i\right\}$. These two groups of base vectors are called the natural frames.

In a natural frame, dx^i is a small distance from p yet without a definite measure. In other words, differs from the other geometric quantities of the point p, the position of dx^i includes not only the point p, but also the neighboring points of p. Since we will need dx^i to represent a small displacement in derivation of the curvature tensors, for the sake of convenience, we keep using the natural frame.

With the natural frame we can define the tensor space $T_{p_s}^{\ r}(M)$, where any tensor has the following form:

$$W_{i_1 i_2 \cdots i_s}^{j_1 j_2 \cdots j_r}(x)\partial_{j_1} \otimes \partial_{j_2} \otimes \cdots \partial_{j_r} \otimes dx^{i_1} \otimes dx^{i_2} \otimes \cdots dx^{i_s} \tag{3.1.1}$$

The union of the tensor spaces at every point p of M can be written as

$$T_s^r(M) = \bigcup_p T_{p_s}^{\ r}(M) \tag{3.1.2}$$

That is a tensor bundle. The following table gives a comparison:

Tangent vector bundle: $T(M) = \bigcup_p T_p(M)$, $T_p(M)$ *is a tangent vector space at p.*

Cotangent vector bundle: $T^*(M) = \bigcup_p T_p^*(M)$, $T_p^*(M)$ *is a cotangent vector space at p.*

Tensor bundle: $T_s^r(M) = \bigcup_p T_{p_s}^{\ r}(M)$, $T_{p_s}^{\ r}(M)$ *is a tensor space at p.*

Now we introduce the covariant differential ∇. ∇ is also a kind of mapping:

$$\nabla:\left.\begin{array}{ll} T_{p_s}^{\ r}(M) & \rightarrow T_{p_{s+1}}^{\ r}(M) \\ K \in T_{p_s}^{\ r}(M) & \rightarrow \nabla K \in T_{p_{s+1}}^{\ r}(M) \end{array}\right\} \tag{3.1.3}$$

where ∇K is the covariant differential of K. ∇ has the following properties:

1. Linearity:
$$\nabla(aK+bK')=a\nabla K+b\nabla K' \qquad (a \text{ and } b \text{ are constants}) \tag{3.1.4}$$
2. Obeying the rules of Leibniz, e.g.
$$\nabla(K\otimes K')=(\nabla K)\otimes K'+K\otimes(\nabla K') \tag{3.1.5}$$
3. For some ordinary function f,
$$\nabla f=df=\frac{\partial f}{\partial x^i}dx^i \tag{3.1.6}$$
4. For a tangent vector field $X=\xi^a\frac{\partial}{\partial x^a}$, we have that

(i) $\xi^a(x)$ is a function, and according to Eq.3.1.6, $\nabla\xi^a(x)=d\xi^a(x)=\frac{\partial\xi^a}{\partial x^i}dx^i$.

(ii) The connection $\Gamma_a^b(x)$ is introduced for $\frac{\partial}{\partial x^a}$: $\nabla\frac{\partial}{\partial x^a}=\Gamma_a^b(x)\frac{\partial}{\partial x^b}$.

$$\therefore\quad \nabla X=\frac{\partial\xi^a}{\partial x^i}dx^i\frac{\partial}{\partial x^a}+\xi^a\cdot\Gamma_a^b(x)\frac{\partial}{\partial x^b}$$

— the covariant differential adds a dx^i to the first term, hence dx^i has to be introduced into $\Gamma_a^b(x)$ as well such that $\Gamma_a^b(x)=\Gamma_{ia}^b(x)dx^i$ becomes a 1-form. Thus the equation can also be written as

$$\nabla X=\frac{\partial\xi^a}{\partial x^i}dx^i\frac{\partial}{\partial x^a}+\xi^b\cdot\Gamma_{ib}^a dx^i\frac{\partial}{\partial x^a}=\left(\frac{\partial\xi^a}{\partial x^i}+\Gamma_{ib}^a\xi^b\right)dx^i\frac{\partial}{\partial x^a} \tag{3.1.7}$$

where

$$\Gamma_a^b(x)=\Gamma_{ia}^b(x)dx^i \tag{3.1.8}$$

is called a connection 1-form. There are n^2 connection 1-forms in total.

(iii) For dx^a, we use $<dx^a,\frac{\partial}{\partial x^b}>=\delta_b^a$ and $\nabla\frac{\partial}{\partial x^a}=\Gamma_a^b(x)\frac{\partial}{\partial x^b}$ to obtain

$$0=\nabla<dx^a,\frac{\partial}{\partial x^b}>=<\nabla dx^a,\frac{\partial}{\partial x^b}>+<dx^a,\nabla\frac{\partial}{\partial x^b}>$$

$$\rightarrow<\nabla dx^a,\frac{\partial}{\partial x^b}>+\Gamma_b^a(x)=0. \qquad \therefore\ \nabla dx^a=-\Gamma_b^a(x)dx^b$$

Comparing the two equations obtained from (ii) and (iii):

$$\left.\begin{aligned}\nabla\frac{\partial}{\partial x^a}&=\Gamma_a^b(x)\frac{\partial}{\partial x^b}\\ \nabla dx^a&=-\Gamma_b^a(x)dx^b\end{aligned}\right\} \tag{3.1.9}$$

we can observe some patterns. Now the question is: Will these patterns remain the same after a transformation of the coordinates?

Take a coordinate transformation matrix $\Lambda_b^a(x)$. Let the transformations of dx^a

and $\frac{\partial}{\partial x^a}$ be:

$$\left.\begin{aligned} dx'^a &= \Lambda^{-1a}_{\ \ b}(x)dx^b \\ \frac{\partial}{\partial x'^a} &= \Lambda^b_a(x)\frac{\partial}{\partial x^b} \end{aligned}\right\} \tag{3.1.10}$$

$T'^a_b(x)$ after the transformation can be obtained by using $\Lambda^{-1}(x)\Lambda(x) = I$ $\rightarrow \nabla\Lambda^{-1}(x) = -\Lambda^{-1}(x)(\nabla\Lambda(x))\Lambda^{-1}(x)$ and $\nabla dx^a = -\Gamma^a_b(x)dx^b$:

$$\begin{aligned} \nabla dx'^a &= \nabla(\Lambda^{-1a}_{\ \ b}(x)dx^b) = (\nabla\Lambda^{-1a}_{\ \ b}(x))dx^b + \Lambda^{-1a}_{\ \ b}(x)\nabla dx^b \\ &= -\Lambda^{-1a}_{\ \ c}(x)(\nabla\Lambda^c_d(x))\Lambda^{-1d}_{\ \ b}(x)dx^b - \Lambda^{-1a}_{\ \ c}(x)\Gamma^c_b(x)dx^b \\ &= -\Lambda^{-1a}_{\ \ c}(x)(\nabla\Lambda^c_d(x))dx'^d - \Lambda^{-1a}_{\ \ c}(x)\Gamma^c_d(x)\Lambda^d_d dx'^d \\ &= -\left[\Lambda^{-1a}_{\ \ c}(x)(\nabla\delta^c_d + \Gamma^c_d(x))\Lambda^d_e(x)\right]dx'^e = -\Gamma'^a_e(x)dx'^e \end{aligned}$$

Now use $< dx'^a, \frac{\partial}{\partial x'^b} >= \Lambda^{-1a}_{\ \ c}(x)\Lambda^d_b(x) < dx^c, \frac{\partial}{\partial x^d} > \delta^a_b$ to get

$$0 =< \nabla dx'^a, \frac{\partial}{\partial x'^b} > + < dx'^a, \nabla\frac{\partial}{\partial x'^b} >= -\Gamma'^a_b(x) + < dx'^a, \nabla\frac{\partial}{\partial x'^b} >$$

$$\therefore \quad < dx'^a, \nabla\frac{\partial}{\partial x'^b} >= \Gamma'^a_b(x) \quad \rightarrow \quad \nabla\frac{\partial}{\partial x'^b} = \Gamma'^a_b(x)\frac{\partial}{\partial x'^a}$$

Hence Eq.3.1.9 after the transformation becomes:

$$\left.\begin{aligned} \nabla\frac{\partial}{\partial x'^a} &= \Gamma'^b_a(x)\frac{\partial}{\partial x'^b} \\ \nabla dx'^a &= -\Gamma'^a_b(x)dx'^b \end{aligned}\right\} \tag{3.1.11}$$

$$\Gamma'^a_b(x) = \Lambda^{-1a}_{\ \ c}(x)\left(\nabla\delta^c_d + \Gamma^c_d(x)\right)\Lambda^d_b(x) \tag{3.1.12}$$

Comparing Eq.3.1.11 and Eq.3.1.9, it is clear that the forms of the differential ∇ before and after the Λ transformation are very similar. That is one of the characteristics of the covariant differential, and that is the reason why ∇ is called the covariant differential.

When there is a tangent vector field $X(x) = \xi^a(x)\frac{\partial}{\partial x^a}$ on a manifold, we may define the directional covariant differential as the following:

(i) Suppose there is a tensor field K. The covariant differential of K is $\nabla K = dx^i\nabla_i K$, where

$$\nabla_i K =< \nabla K, \frac{\partial}{\partial x^i} > \tag{3.1.13}$$

is the covariant differential of K in the direction i.

(ii) Suppose there is a tangent vector field $X(x)=\xi^i(x)\frac{\partial}{\partial x^i}$. The covariant differential of K in the direction X is

$$\nabla_X K=<\nabla K, X>=<dx^i\nabla_i K, \xi^j(x)\frac{\partial}{\partial x^j}>=\xi^i(x)\nabla_i K \qquad (3.1.13)_1$$

(iii) The covariant differential of a tensor K in the direction $fX+gY=f\,\xi^i(x)\frac{\partial}{\partial x^i}+g\eta^i(x)\frac{\partial}{\partial x^i}$ is

$$\begin{aligned}\nabla_{fX+gY}K &=<\nabla K, fX+gY>=f<\nabla K,X>+g<\nabla K,Y>\\ &=f\,\xi^i(x)\nabla_i K+g\,\eta^i(x)\nabla_i K\end{aligned} \qquad (3.1.14)$$

(iv) Take $K=X(x)=\xi^i(x)\frac{\partial}{\partial x^i}$ together with Eq.3.1.8 and Eq.3.1.9 we have

$$\begin{aligned}\nabla X(x) &=\left(\nabla\xi^i(x)\right)\frac{\partial}{\partial x^i}+\xi^i(x)\nabla\frac{\partial}{\partial x^i}\\ &=\frac{\partial\xi^k(x)}{\partial x^j}dx^j\frac{\partial}{\partial x^k}+\xi^i(x)\Gamma_i^k(x)\frac{\partial}{\partial x^k}\\ &=\left(\frac{\partial\xi^k(x)}{\partial x^j}\frac{\partial}{\partial x^k}+\xi^i(x)\Gamma_{ji}^k(x)\frac{\partial}{\partial x^k}\right)dx^j\\ &=\left(\frac{\partial\xi^k(x)}{\partial x^j}+\xi^l(x)\Gamma_{jl}^k(x)\right)\frac{\partial}{\partial x^k}\cdot dx^j\end{aligned} \qquad (3.1.15)$$

$\therefore$ according to Eq.3.1.13

$$\begin{aligned}\nabla_i X(x) &=<\nabla X(x),\frac{\partial}{\partial x^i}>\\ &=<\left(\frac{\partial\xi^k(x)}{\partial x^j}+\xi^l(x)\Gamma_{jl}^k(x)\right)\frac{\partial}{\partial x^k}\cdot dx^j,\frac{\partial}{\partial x^i}>\\ &=\left(\frac{\partial\xi^k(x)}{\partial x^i}+\xi^l(x)\Gamma_{il}^k(x)\right)\frac{\partial}{\partial x^k}\equiv\xi^k_{;i}\frac{\partial}{\partial x^k}\end{aligned} \qquad (3.1.16)$$

$$\xi^k_{;i}=\left(\nabla_i X\right)^k=\frac{\partial\xi^k(x)}{\partial x^i}+\xi^l(x)\Gamma_{il}^k(x) \qquad (3.1.16)_1$$

For an arbitrary tensor (see Eq.3.1.1), its covariant differential can be determined in a similar way by each dx^i and each ∂_j it contains.

§ 3.2 *Curvature, Torsion, and Translation of Vectors on a Manifold*

According to the Frobenius theorem described in the Section 2.2, if in a manifold *M*, a set of *m* smooth tangent vector fields is defined as $\Sigma^m=\{X_1,X_2,\cdots,X_m\}$, and if the Lie bracket $[X_i, X_j]$ of any pair of vector fields in the set can be written as a linear combination of $X_1,\cdots,X_m$, then the integral curves of these vector fields weave a family of submanifolds, and a coordinate system can be constructed on a submanifold.

With the coordinate system, in order to keep the properties of vectors and tensors invariant after translations on a submanifold, a vector or a tensor must obey the laws of covariant differential while moving on a manifold. Let us observe the covariant translation first.

Suppose that there is a curve C through a point p whose coordinates are (x^i). The tangent vector of C at p is $X=\frac{d}{dt}=\frac{dx^i(x)}{dt}\frac{\partial}{\partial x^i}=\xi^i(x)\frac{\partial}{\partial x^i}$. $\xi^i(x)$ represents the direction of the tangent vector of C at p.

Take another tangent vector $Y(x)=\eta^i(x)\frac{\partial}{\partial x^i}$. According to Eq.3.1.15 and Eq.3.1.16 we obtain:

$$\left.\begin{aligned}
\nabla Y(x) &= \left(\frac{\partial\eta^k(x)}{\partial x^j}+\eta^l(x)\Gamma^k_{jl}(x)\right)\frac{\partial}{\partial x^k}\cdot dx^j \\
\nabla_i Y(x) &= <\nabla Y,\frac{\partial}{\partial x^i}> = \left(\frac{\partial\eta^k(x)}{\partial x^i}+\eta^l(x)\Gamma^k_{il}(x)\right)\frac{\partial}{\partial x^k} \\
\nabla_X Y(x) &= <\nabla Y,\xi^i(x)\frac{\partial}{\partial x^i}> = \xi^i(x)\left(\frac{\partial\eta^k(x)}{\partial x^i}+\eta^l(x)\Gamma^k_{il}(x)\right)\frac{\partial}{\partial x^k}
\end{aligned}\right\} \tag{3.2.1}$$

Figure 3.2.1 shows a 2-dimensional submanifold through the point A with a system of coordinates. The coordinates of A, C, C' and B are (x), $(x+du)$, $(x+dv)$ and $(x+du+dv)$ respectively, where du and dv are both constants.

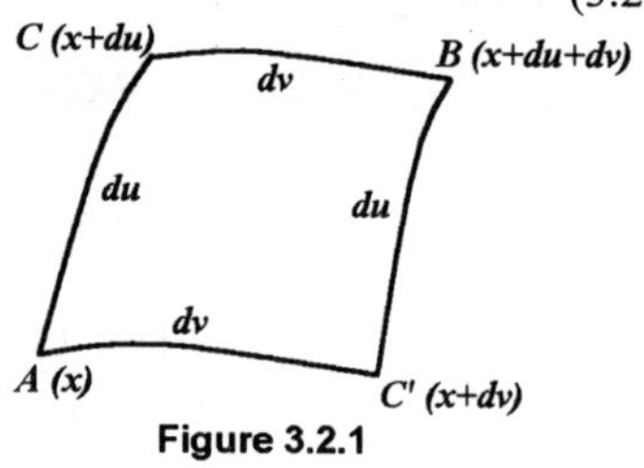

Figure 3.2.1

Take a tangent vector $a^i\frac{\partial}{\partial x^i}$ at $A(x)$ and use Eq.3.1.16, we have the covariant translations

$$A\to C:\quad a^i\frac{\partial}{\partial x^i}\to a^i\frac{\partial}{\partial x^i}+du^i\nabla_i\left(a^k\frac{\partial}{\partial x^k}\right)$$

$$C\to B:\quad a^i\frac{\partial}{\partial x^i}+du^i\nabla_i\left(a^k\frac{\partial}{\partial x^k}\right)\to a^i\frac{\partial}{\partial x^i}+du^i\nabla_i\left(a^k\frac{\partial}{\partial x^k}\right)$$
$$+dv^j\nabla_j\left(a^k\frac{\partial}{\partial x^k}\right)+dv^j du^i\nabla_j\nabla_i\left(a^k\frac{\partial}{\partial x^k}\right)$$

$$A \to C': \quad a^i \frac{\partial}{\partial x^i} \to a^i \frac{\partial}{\partial x^i} + dv^j \nabla_j \left(a^k \frac{\partial}{\partial x^k} \right)$$

$$C' \to B: \quad a^i \frac{\partial}{\partial x^i} + dv^j \nabla_j \left(a^k \frac{\partial}{\partial x^k} \right) \to a^i \frac{\partial}{\partial x^i} + dv^j \nabla_j \left(a^k \frac{\partial}{\partial x^k} \right)$$
$$+ du^i \nabla_i \left(a^k \frac{\partial}{\partial x^k} \right) + du^i dv^j \nabla_i \nabla_j \left(a^k \frac{\partial}{\partial x^k} \right)$$

$$\therefore ACB \text{ translation} - AC'B \text{ translaton} = du^i dv^j \left[(\nabla_j \nabla_i - \nabla_i \nabla_j) \left(a^k \frac{\partial}{\partial x^k} \right) \right] \tag{3.2.2}$$

From Eq.3.1.9 and Eq.3.1.11 we know that in order to keep the operations consistent, we have to follow some uniform law when applying the ∇ covariant differential. That is, for each superscript there must be an extra term of $(-\Gamma)$; for each subscript, a term of $(+\Gamma)$; and when summing up the terms whose superscript and subscript are the same, no term of Γ will be added. For example:

$$\nabla(A^a_b B^b_c) = \frac{\partial}{\partial x^i}(A^a_b B^b_c) dx^i - \Gamma^a_{id}(A^d_b B^b_c) dx^i + \Gamma^d_{ic}(A^a_b B^b_d) dx^i$$

$$\therefore \ \nabla_i (A^a_b B^b_c) = \frac{\partial}{\partial x^i}(A^a_b B^b_c) - \Gamma^a_{id}(A^d_b B^b_c) + \Gamma^d_{ic}(A^a_b B^b_d) \tag{3.2.3}$$

Now let us compute $\nabla_i \nabla_j \left(a^k \frac{\partial}{\partial x^k} \right) - \nabla_j \nabla_i \left(a^k \frac{\partial}{\partial x^k} \right)$ according to this law:

$$\nabla_i \nabla_j a^k \frac{\partial}{\partial x^k} = \nabla_i (\nabla_j a^k) \frac{\partial}{\partial x^k} + \nabla_i a^k \nabla_j \frac{\partial}{\partial x^k}$$
$$= \left(\nabla_i \nabla_j a^k \right) \frac{\partial}{\partial x^k} + (\nabla_j a^k) \left(\nabla_i \frac{\partial}{\partial x^k} \right) + (\nabla_i a^k) \left(\nabla_j \frac{\partial}{\partial x^k} \right) + a^k \left(\nabla_i \nabla_j \frac{\partial}{\partial x^k} \right)$$

$$\nabla_j \nabla_i a^k \frac{\partial}{\partial x^k} = \nabla_j (\nabla_i a^k) \frac{\partial}{\partial x^k} + \nabla_j a^k \nabla_i \frac{\partial}{\partial x^k}$$
$$= \left(\nabla_j \nabla_i a^k \right) \frac{\partial}{\partial x^k} + (\nabla_i a^k) \left(\nabla_j \frac{\partial}{\partial x^k} \right) + (\nabla_j a^k) \left(\nabla_i \frac{\partial}{\partial x^k} \right) + a^k \left(\nabla_j \nabla_i \frac{\partial}{\partial x^k} \right)$$

$$\therefore \left(\nabla_i \nabla_j - \nabla_j \nabla_i \right) a^k \frac{\partial}{\partial x^k} = \left(\left(\nabla_i \nabla_j - \nabla_j \nabla_i \right) a^k \right) \frac{\partial}{\partial x^k} + a^k \left(\left(\nabla_i \nabla_j - \nabla_j \nabla_i \right) \frac{\partial}{\partial x^k} \right) \tag{3.2.4}$$

$$\nabla_j \frac{\partial}{\partial x^k} = \langle \nabla \frac{\partial}{\partial x^k}, \frac{\partial}{\partial x^j} \rangle = \langle \Gamma^a_{lk} dx^l \frac{\partial}{\partial x^a}, \frac{\partial}{\partial x^j} \rangle = \Gamma^a_{jk} \frac{\partial}{\partial x^a}$$

$$\nabla_i \nabla_j \frac{\partial}{\partial x^k} = \langle \nabla \Gamma^a_{jk} \frac{\partial}{\partial x^a}, \frac{\partial}{\partial x^i} \rangle = \langle \frac{\partial \Gamma^a_{jk}}{\partial x^l} dx^l \frac{\partial}{\partial x^a} + \Gamma^a_{jk} \Gamma^b_{la} \frac{\partial}{\partial x^b} dx^l, \frac{\partial}{\partial x^i} \rangle$$
$$= \left(\frac{\partial \Gamma^a_{jk}}{\partial x^i} + \Gamma^b_{jk} \Gamma^a_{ib} \right) \frac{\partial}{\partial x^a}$$

$$\therefore \left(\nabla_i\nabla_j - \nabla_j\nabla_i\right)\frac{\partial}{\partial x^k} = \left(\frac{\partial\Gamma^a_{jk}}{\partial x^i} - \frac{\partial\Gamma^a_{ik}}{\partial x^j} + \Gamma^b_{jk}\Gamma^a_{ib} - \Gamma^b_{ik}\Gamma^a_{jb}\right)\frac{\partial}{\partial x^a} = R^a_{kij}\frac{\partial}{\partial x^a} \tag{3.2.5}$$

$$R^a_{kij} = \frac{\partial\Gamma^a_{jk}}{\partial x^i} - \frac{\partial\Gamma^a_{ik}}{\partial x^j} + \Gamma^b_{jk}\Gamma^a_{ib} - \Gamma^b_{ik}\Gamma^a_{jb} \tag{3.2.6}$$

where R^a_{kij} is called a *curvature tensor*.

Then compute $\left(\nabla_i\nabla_j - \nabla_j\nabla_i\right)a^k$: $\qquad \nabla_j a^k = \dfrac{\partial a^k}{\partial x^j}$,

$$\nabla_i\nabla_j a^k = \left\langle \nabla\frac{\partial a^k}{\partial x^j}, \frac{\partial}{\partial x^i}\right\rangle = \left\langle \frac{\partial^2 a^k}{\partial x^l\partial x^j}dx^l + \Gamma^m_{lj}dx^l\frac{\partial a^k}{\partial x^m}, \frac{\partial}{\partial x^i}\right\rangle$$

$$= \frac{\partial^2 a^k}{\partial x^i\partial x^j} + \Gamma^m_{ij}\frac{\partial a^k}{\partial x^m} = \frac{\partial^2 a^k}{\partial x^i\partial x^j} + \Gamma^m_{ij}\nabla_m a^k$$

$$\therefore \left(\nabla_i\nabla_j - \nabla_j\nabla_i\right)a^k = \left(\Gamma^m_{ij} - \Gamma^m_{ji}\right)\nabla_m a^k$$

$$= \left(\Gamma^m_{ij} - \Gamma^m_{ji}\right)\frac{\partial a^k}{\partial x^m} = T^m_{ji}\frac{\partial a^k}{\partial x^m} \tag{3.2.7}$$

$$T^m_{ij} = \Gamma^m_{ji} - \Gamma^m_{ij} \tag{3.2.8}$$

where T^m_{ij} is called a *torsion tensor*.

Thus Eq.3.2.4 can be written as

$$\left(\nabla_i\nabla_j - \nabla_j\nabla_i\right)a^k\frac{\partial}{\partial x^k} = \left(T^m_{ji}\frac{\partial a^k}{\partial x^m} + a^l R^k_{lij}\right)\frac{\partial}{\partial x^k} \tag{3.2.9}$$

Hence we can see that when the curvature tensor R^k_{lij} and the torsion tensor T^m_{ij} are not equal to zero, $a^i\dfrac{\partial}{\partial x^i}$ will give different results after covariant translations through *ACB* and *AC'B*. Conversely, the covariant translations through two different paths will give the same result if the curvature tensor and the torsion tensor are both zero.

Let us now consider a generalized case:

Generalize $e_a = e^i_a\dfrac{\partial}{\partial x^i}$ and $e_b = e^i_b\dfrac{\partial}{\partial x^i}$ from $\dfrac{\partial}{\partial x^i}$ and $\dfrac{\partial}{\partial x^j}$; and $[\nabla_{e_a}, \nabla_{e_b}]$ from $[\nabla_i, \nabla_j]$. The main difference here is that $\dfrac{\partial}{\partial x^i}$ and $\dfrac{\partial}{\partial x^j}$ are commutable while e_a and e_b are not.

From $e^i_a\dfrac{\partial}{\partial x^i} = e_a$ it is also clear that e^i_a is in fact the transformation matrix between the two groups of base vectors $\dfrac{\partial}{\partial x^i}$, $\dfrac{\partial}{\partial x^j}$, $\cdots$ and e_a, e_b, $\cdots$, with some relevant properties, e.g.

$$e_b^i \Gamma_{im}^l = \Gamma_{bm}^l \tag{3.2.10}$$

Therefore with the similar technique we used previously, we have

$$\nabla_{e_b} e_c = < \nabla e_c, e_b > = < dx^i \Gamma_{ic}^m e_m, e_b^j \frac{\partial}{\partial x^j} > = e_b^i \Gamma_{ic}^m e_m = \Gamma_{bc}^m e_m \tag{3.2.11}$$

$$\nabla_{e_a} \nabla_{e_b} e_c = < \nabla \Gamma_{bc}^m e_m, e_a > = < dx^i \frac{\partial \Gamma_{bc}^m}{\partial x^i} e_m + \Gamma_{bc}^m \nabla e_m, e_a^j \frac{\partial}{\partial x^j} >$$

$$= e_a^i \frac{\partial \Gamma_{bc}^m}{\partial x^i} e_m + < \Gamma_{bc}^m \Gamma_{im}^k dx^i e_k, e_a^j \frac{\partial}{\partial x^j} >$$

$$= e_a^i \frac{\partial \Gamma_{bc}^m}{\partial x^i} e_m + \Gamma_{bc}^m \Gamma_{im}^k e_k e_a^i$$

$$= e_a^i \frac{\partial \Gamma_{bc}^m}{\partial x^i} e_m + \Gamma_{bc}^m \Gamma_{am}^k e_k = \left(e_a^i \frac{\partial \Gamma_{bc}^m}{\partial x^i} + \Gamma_{bc}^n \Gamma_{an}^m \right) e_m \tag{3.2.12}$$

Immediately following Eq.3.2.12 we obtain:

$$\left(\nabla_{e_a} \nabla_{e_b} - \nabla_{e_b} \nabla_{e_a}\right) e_c = \left(e_a^i \frac{\partial \Gamma_{bc}^m}{\partial x^i} - e_b^i \frac{\partial \Gamma_{ac}^m}{\partial x^i} + \Gamma_{bc}^f \Gamma_{af}^m - \Gamma_{ac}^f \Gamma_{bf}^m \right) e_m \tag{3.2.13}$$

However, $[\nabla_{e_a}, \nabla_{e_b}]$ is not a linear differential operator since $[e_b, e_c] \neq 0$. A correcting method is to take

$$R(e_a, e_b) = [\nabla_{e_a}, \nabla_{e_b}] - \nabla_{[e_a, e_b]} \tag{3.2.14}$$

$$\nabla_{[e_a, e_b]} e_c = < \nabla e_c, f_{ab}^d e_d > = < dx^i \Gamma_{ic}^m e_m, f_{ab}^d e_d^k \frac{\partial}{\partial x^k} >$$ [1]

$$= f_{ab}^d e_d^i \Gamma_{ic}^m e_m = f_{ab}^d \Gamma_{dc}^m e_m \tag{3.2.15}$$

$$R(e_a, e_b) e_c = \left([\nabla_{e_a}, \nabla_{e_b}] - \nabla_{[e_a, e_b]} \right) e_c = R_{cab}^m e_m$$

$$= \left(e_a^i \frac{\partial \Gamma_{bc}^m}{\partial x^i} - e_b^i \frac{\partial \Gamma_{ac}^m}{\partial x^i} + \Gamma_{bc}^f \Gamma_{af}^m - \Gamma_{ac}^f \Gamma_{bf}^m - f_{ab}^d \Gamma_{dc}^m \right) e_m \tag{3.2.16}$$

When a natural frame of local coordinates is taken, $e_a^k = \delta_a^k$ and $f_{ab}^d = 0$, which makes Eq.3.2.16 the same as Eq.3.2.6. Similar to Eq.3.2.14, $T(e_a, e_b)$ has to be

$$T(e_a, e_b) = \nabla_{e_a} e_b - \nabla_{e_b} e_a - [e_a, e_b] = T_{ab}^c e_c \tag{3.2.17}$$

---

1 $[e_a, e_b] = (e_a^i \frac{\partial e_b^j}{\partial x^i} - e_b^i \frac{\partial e_a^j}{\partial x^i}) \frac{\partial}{\partial x^j} = f_{ab}^j \frac{\partial}{\partial x^j}$

§ 3.3 Structural Equations and Integrability Conditions of Curvature Tensors and Torsion Tensors

It is clear from Eq.3.2.6 that R^l_{kij} is anti-symmetric with respect to i and j. Therefore a *curvature 2-form* can be defined as:

$$\Omega^l_k \equiv \frac{1}{2} R^l_{kij} dx^i \wedge dx^j \tag{3.3.1}$$

where (dx^i) is dual to $\left(\frac{\partial}{\partial x^j}\right)$, i.e. $< dx^i, \frac{\partial}{\partial x^j} >= \delta^i_j$.

It is also clear from Eq.3.2.16 that R^m_{cab} is anti-symmetric with respect to a and b. Hence a *generalized curvature 2-form* can be defined as:

$$\Omega^a_b \equiv \frac{1}{2} R^a_{bcd} \theta^c \wedge \theta^d \qquad (3.3.1)_1$$

where (θ^a) is dual to (e_b), i.e. $< \theta^a, e_b >= \delta^a_b$.

The two dualities are consistent since $\theta^a = e^a_j dx^j$ and $e_b = e^i_b \frac{\partial}{\partial x^i}$. Thus

$$< \theta^a, e_b >= e^a_j e^i_b \delta^j_i = e^a_i e^i_b = \delta^a_b$$

Now define

$$\omega^a_b = \theta^d \Gamma^a_{db} \tag{3.3.2}$$

We have $d\omega^a_b = d(\theta^c \Gamma^a_{cb}) = (d\theta^c)\Gamma^a_{cb} - \theta^c \wedge d\Gamma^a_{cb}$ where $d\theta^c = d(e^c_j dx^j)$ $= de^c_j \wedge dx^j = \frac{\partial e^c_j}{\partial x^i} dx^i \wedge dx^j$. Since $dx^i = e^i_a \theta^a$ and $dx^j = e^j_b \theta^b$, then we have $d\theta^c = \frac{\partial e^c_j}{\partial x^i} e^i_a e^j_b \theta^a \wedge \theta^b$. Also since $e^c_j e^j_b = \delta^c_b$, then

$$\begin{aligned} d\theta^c &= -e^i_a e^c_j \frac{\partial e^j_b}{\partial x^i} \theta^a \wedge \theta^b = -e^c_j e^i_a \frac{\partial e^j_b}{\partial x^i} \theta^a \wedge \theta^b \\ &= -e^c_j \frac{1}{2}\left(e^i_a \frac{\partial e^j_b}{\partial x^i} - e^i_b \frac{\partial e^j_a}{\partial x^i} \right) \theta^a \wedge \theta^b = -\frac{1}{2} f^c_{ab} \theta^a \wedge \theta^b \end{aligned} \tag{3.3.3}$$

As well we have: $$d\Gamma^a_{cb} = \frac{\partial \Gamma^a_{cb}}{\partial x^i} dx^i = \frac{\partial \Gamma^a_{cb}}{\partial x^i} e^i_d \theta^d \tag{3.3.4}$$

$\longrightarrow \quad \theta^c \wedge d\Gamma^a_{cb} = \frac{\partial \Gamma^a_{cb}}{\partial x^i} e^i_d \theta^c \wedge \theta^d$. Plug $d\omega^a_b$ in we obtain:

$$d\omega^a_b = \left(-\frac{1}{2} f^c_{de} \theta^d \wedge \theta^e \right) \Gamma^a_{cb} - \frac{\partial \Gamma^a_{db}}{\partial x^i} e^i_e \theta^d \wedge \theta^e \tag{3.3.5}$$

Then define Ω^a_b:

$$\Omega^a_b \equiv d\omega^a_b + \omega^a_c \wedge \omega^c_b \tag{3.3.6}$$

We obtain:

$$\Omega^a_b \equiv \left(-\frac{1}{2} f^c_{de}\Gamma^a_{cb} - \frac{\partial \Gamma^a_{db}}{\partial x^i} e^i_e\right)\theta^d \wedge \theta^e + \Gamma^a_{dc}\Gamma^c_{eb}\,\theta^d \wedge \theta^e$$

$$= \frac{1}{2}\left(e^i_d \frac{\partial \Gamma^a_{eb}}{\partial x^i} - e^i_e \frac{\partial \Gamma^a_{db}}{\partial x^i} + \Gamma^a_{dc}\Gamma^c_{eb} - \Gamma^a_{ec}\Gamma^c_{db} - f^c_{de}\Gamma^a_{cb}\right)\theta^d \wedge \theta^e$$

$$= \frac{1}{2} R^a_{bde}\theta^d \wedge \theta^e$$

which is consistent with Eq.3.3.1.

Eq.3.3.1$_1$ and Eq.3.3.6 are called the *structural equations of a curvature.*

From Eq.3.2.8 we can see that T^m_{ij} is anti-symmetric with respect to i and j. Therefore a *torsion 2-form* can be defined as:

$$\tau^i = \frac{1}{2} T^i_{jk} dx^i \wedge dx^k \tag{3.3.7}$$

Calculating $T(e_a, e_b)$ in Eq.3.2.17,

$$T(e_a, e_b) = <\nabla e_b, e_a> - <\nabla e_a, e_b> - f^c_{ab}e_c$$

$$= <\theta^d\Gamma^c_{db}e_c, e_a> - <\theta^d\Gamma^c_{da}e_c, e_b> - f^c_{ab}e_c$$

$$= (\Gamma^c_{ab} - \Gamma^c_{ba} - f^c_{ab})e_c = T^c_{ab}e_c \tag{3.3.8}$$

$$\therefore\quad T^c_{ab} = (\Gamma^c_{ab} - \Gamma^c_{ba} - f^c_{ab}) \tag{3.3.9}$$

we then obtain the *generalized torsion 2-form*:

$$\tau^a = \frac{1}{2} T^a_{bc}\theta^b \wedge \theta^c = \frac{1}{2}(\Gamma^a_{bc} - \Gamma^a_{cb} - f^a_{bc})\theta^b \wedge \theta^c \tag{3.3.10}$$

From Eq.3.3.2 and Eq.3.3.8 we have

$$d\theta^a + \omega^a_b \wedge \theta^b = -\frac{1}{2} f^a_{bc}\theta^b \wedge \theta^c + \theta^b\Gamma^a_{bc} \wedge \theta^c$$

$$= \frac{1}{2}(\Gamma^a_{bc} - \Gamma^a_{cb} - \frac{1}{2} f^a_{bc})\theta^b \wedge \theta^c \tag{3.3.11}$$

and hence obtain

$$\tau^a \equiv d\theta^a + \omega^a_b \wedge \theta^b \tag{3.3.12}$$

Eq.3.3.11 and Eq.3.3.12 are called the *structural equations of a torsion.*

Differentiate the generalized curvature 2-form Eq.3.3.6:

$$d\Omega^a_b = d\omega^a_c \wedge \omega^c_b - \omega^a_c \wedge d\omega^c_b$$

$$= (\Omega^a_c - \omega^a_d \wedge \omega^d_c) \wedge \omega^c_b - \omega^a_c \wedge (\Omega^c_b - \omega^c_d \wedge \omega^d_b)$$

$$= \Omega^a_c \wedge \omega^c_b - \omega^a_c \wedge \Omega^c_b - \omega^a_d \wedge \omega^d_c \wedge \omega^c_b + \omega^a_c \wedge \omega^c_d \wedge \omega^d_b$$

$$d\Omega^a_b = \Omega^a_c \wedge \omega^c_b - \omega^a_c \wedge \Omega^c_b \tag{3.3.13}$$

Eq.3.3.13 is called the *integrability condition of a curvature.*

Differentiate the generalized torsion 2-form Eq.3.3.12 and use Eq.3.3.6:

$$\begin{aligned} d\tau^a &= d\omega^a_b \wedge \theta^b - \omega^a_c \wedge d\theta^c \\ &= (\Omega^a_b - \omega^a_d \wedge \omega^d_b) \wedge \theta^b - \omega^a_c \wedge d\theta^c \\ &= \Omega^a_b \wedge \theta^b - \omega^a_c \wedge (\omega^c_b \wedge \theta^b + d\theta^c) \end{aligned}$$

Again use Eq.3.3.12 we obtain

$$d\tau^a = \Omega^a_b \wedge \theta^b - \omega^a_c \wedge \tau^c$$

$$\therefore \quad d\tau^a + \omega^a_c \wedge \tau^c = \Omega^a_b \wedge \theta^b \tag{3.3.14}$$

Eq.3.3.14 is called the *integrability condition of a torsion.*

*§ 3.4 Hodge * and Adjoint Exterior Differentials*

Suppose there is an n-dimensional space with a completely anti-symmetric tensor field of order p:

$$f_{i_1 i_2 \cdots i_p}(x) \tag{3.4.1}$$

which is a smooth function of x. Then the p-form at the point x can be written as:

$$f_{i_1 i_2 \cdots i_p}(x)\, dx^{i_1} \wedge dx^{i_2} \wedge \cdots \wedge dx^{i_p} \tag{3.4.2}$$

Since there are $\dfrac{n(n-1)(n-2)\cdots(n-p+1)}{p!}$ ways of taking p distinct dx^i's out of n distinct dx^i's, the dimension of this kind of p-form must be:

$$\frac{1}{p!} n(n-1)(n-2)\cdots(n-p+1) = \frac{n!}{p!(n-p)!}$$

For the same reason, the dimension of an $(n-p)$-form in an n-dimensional space must be:

$$\frac{1}{(n-p)!} n(n-1)(n-2)\cdots(p+1) = \frac{n!}{(n-p)!\,p!}$$

The dimensions of a p-form and an $(n-p)$-form are the same, therefore they must have some kind of dual relationship, which can be expressed by the Hodge * operator as the following:

$$*(dx^{i_1} \wedge \cdots \wedge dx^{i_p}) = \frac{1}{(n-p)!} \varepsilon^{i_1 i_2 \cdots i_p}{}_{i_{p+1} \cdots i_n}\, dx^{i_{p+1}} \wedge dx^{i_{p+2}} \wedge \cdots \wedge dx^{i_n} \tag{3.4.3}$$

where $\varepsilon^{i_1 i_2 \cdots i_p}{}_{i_{p+1} \cdots i_n} = \varepsilon_{i_1 i_2 \cdots i_p i_{p+1} \cdots i_n}$ is a completely anti-symmetric tensor.[1]

---

1 The property of completely anti-symmetric tensor of $\varepsilon^{i_1 i_2 \cdots i_p}{}_{i_{p+1} \cdots i_n} = \varepsilon_{i_1 i_2 \cdots i_p i_{p+1} \cdots i_n}$ also applies to Eq.3.4.5$_1$, Eq.3.4.6$_1$, and Eq.3.4.7.

When p=0, we have:

$$*1=\frac{1}{n!}\varepsilon_{i_1\cdots i_n}dx^{i_1}\wedge\cdots\wedge dx^{i_n} \tag{3.4.4}$$

Taking any r-form α_r as

$$\alpha_r=\frac{1}{r!}f_{i_1\cdots i_r}dx^{i_1}\wedge\cdots\wedge dx^{i_r} \tag{3.4.5}$$

we have:

$$*\alpha_r=\frac{1}{r!}f_{i_1\cdots i_r}*(dx^{i_1}\wedge\cdots\wedge dx^{i_r})$$

$$=\frac{1}{r!}f_{i_1\cdots i_r}\frac{1}{(n-r)!}\varepsilon^{i_1 i_2\cdots i_r}{}_{i_{r+1}\cdots i_n}dx^{i_{r+1}}\wedge dx^{i_{r+2}}\wedge\cdots\wedge dx^{i_n} \qquad (3.4.5)_1$$

Suppose that there is another r-form β_r as

$$\beta_r=\frac{1}{r!}h_{i_1\cdots i_r}dx^{i_1}\wedge\cdots\wedge dx^{i_r} \tag{3.4.6}$$

Then again we have:

$$*\beta_r=\frac{1}{r!}h_{j_1\cdots j_r}*(dx^{j_1}\wedge\cdots\wedge dx^{j_r})$$

$$=\frac{1}{r!}h_{j_1\cdots j_r}\frac{1}{(n-r)!}\varepsilon^{j_1 j_2\cdots j_r}{}_{j_{r+1}\cdots j_n}dx^{j_{r+1}}\wedge dx^{j_{r+2}}\wedge\cdots\wedge dx^{j_n} \qquad (3.4.6)_1$$

Therefore,

$$\alpha_r\wedge*\beta_r=\frac{1}{r!}f_{i_1\cdots i_r}\frac{1}{r!}h_{j_1\cdots j_r}\frac{1}{(n-r)!}\varepsilon^{j_1 j_2\cdots j_r}{}_{j_{r+1}\cdots j_n}dx^{i_1}\wedge\cdots\wedge dx^{i_r}$$

$$\wedge dx^{i_{r+1}}\wedge dx^{i_{r+2}}\wedge\cdots\wedge dx^{i_n},$$

$$\beta_r\wedge*\alpha_r=\frac{1}{r!}h_{j_1\cdots j_r}\frac{1}{r!}f_{i_1\cdots i_r}\frac{1}{(n-r)!}\varepsilon^{j_1 j_2\cdots j_r}{}_{j_{r+1}\cdots j_n}dx^{i_1}\wedge\cdots\wedge dx^{i_r}$$

$$\wedge dx^{i_{r+1}}\wedge dx^{i_{r+2}}\wedge\cdots\wedge dx^{i_n}.$$

$$\therefore\quad \alpha_r\wedge*\beta_r=\beta_r\wedge*\alpha_r \tag{3.4.7}$$

Define the interior product of α_p and β_p as

$$(\alpha_p,\beta_p)=\int_M\alpha_p\wedge*\beta_p \tag{3.4.8}$$

From Eq.3.4.7 we know that

$$(\alpha_p,\beta_p)=(\beta_p,\alpha_p) \tag{3.4.9}$$

Now we introduce the "*adjoint exterior differential operator*" δ.

Define: $\delta=(-1)^{np+n+1}*d*\ \Leftrightarrow\ \delta\alpha_p=(-1)^{np+n+1}*d*\alpha_p$ (3.4.10)

$\delta=-*d*$ when n is even, and $\delta=(-1)^p*d*$ when n is odd.

For the sake of convenience of the following computations, we first calculate $**\alpha_r$ and $\delta\alpha_r$ here:

1. $$**\alpha_r = *\frac{1}{r!}\frac{1}{(n-r)!} f_{i_1\cdots i_r}\varepsilon^{i_1 i_2\cdots i_r}{}_{i_{r+1}\cdots i_n}\, dx^{i_{r+1}}\wedge dx^{i_{r+2}}\wedge\cdots\wedge dx^{i_n}$$

(according to Eq.3.4.5$_1$)

$$= \frac{1}{r!} f_{i_1\cdots i_r}\frac{1}{(n-r)!}\varepsilon^{i_1 i_2\cdots i_r}{}_{i_{r+1}\cdots i_n} * dx^{i_{r+1}}\wedge dx^{i_{r+2}}\wedge\cdots\wedge dx^{i_n}$$

$$= \frac{1}{r!} f_{i_1\cdots i_r}\frac{1}{(n-r)!}\varepsilon^{i_1 i_2\cdots i_r}{}_{i_{r+1}\cdots i_n}\frac{1}{r!}\varepsilon^{i_{r+1}\cdots i_n}{}_{k_1\cdots k_r}\, dx^{k_1}\wedge\cdots\wedge dx^{k_r}$$

$$= \frac{1}{r!}\frac{1}{r!}\frac{1}{(n-r)!} f_{i_1\cdots i_r}\varepsilon^{i_1 i_2\cdots i_r}{}_{i_{r+1}\cdots i_n}(-1)^{r(n-r)}\varepsilon_{k_1\cdots k_r}{}^{i_{r+1}\cdots i_n}\, dx^{k_1}\wedge\cdots\wedge dx^{k_r}$$

$$= \frac{1}{r!}\frac{1}{r!}\frac{1}{(n-r)!} f_{i_1\cdots i_r}(-1)^{r(n-r)}(n-r)!\,\delta^{i_1 i_2\cdots i_r}_{k_1\cdots k_r}\, dx^{k_1}\wedge\cdots\wedge dx^{k_r}$$

$$= \frac{1}{r!}\frac{1}{r!}\frac{1}{(n-r)!} f_{i_1\cdots i_r}(-1)^{r(n-r)}(n-r)!\,r!\, dx^{i_1}\wedge\cdots\wedge dx^{i_r}$$

$$= \frac{1}{r!}(-1)^{r(n-r)} f_{i_1\cdots i_r}\, dx^{i_1}\wedge\cdots\wedge dx^{i_r}$$

$$= (-1)^{r(n-r)}\alpha_r = \begin{cases} +\alpha_r & \text{if } n \text{ is odd} \\ (-1)^r\alpha_r & \text{if } n \text{ is even}\end{cases} \qquad (3.4.11)$$

2. $\delta\alpha_r = (-1)^{nr+n+1} * d * \alpha_r$.

According to Eq.3.4.5$_1$ and $d f_{i_1\cdots i_r} = f_{i_1\cdots i_r;k}dx^k$, we obtain:

$$d*\alpha_r = \frac{1}{r!} f_{i_1\cdots i_r;k}\frac{1}{(n-r)!}\varepsilon^{i_1 i_2\cdots i_r}{}_{i_{r+1}\cdots i_n}\, dx^k\wedge dx^{i_{r+1}}\wedge\cdots\wedge dx^{i_n}$$

$$*d*\alpha_r = \frac{1}{r!(n-r)!} f_{i_1\cdots i_r;k}\varepsilon^{i_1 i_2\cdots i_r}{}_{i_{r+1}\cdots i_n} * dx^k\wedge dx^{i_{r+1}}\wedge\cdots\wedge dx^{i_n}$$

$$= \frac{1}{r!(n-r)!} f_{i_1\cdots i_r;k}\varepsilon^{i_1 i_2\cdots i_r}{}_{i_{r+1}\cdots i_n}\frac{1}{(n-(n-r+1))!}\varepsilon_{k\underbrace{i_{r+1}\cdots i_n}_{n-r}\underbrace{k_1 k_2\cdots k_{r-1}}_{r-1}}\, dx^{k_1}\wedge\cdots\wedge dx^{k_{r-1}}$$

$$= \frac{1}{r!(n-r)!(r-1)!} f_{i_1\cdots i_r;k}\varepsilon^{i_1 i_2\cdots i_r}{}_{i_{r+1}\cdots i_n}(-1)^{(r-1)(n-r)}\varepsilon_{k k_1 k_2\cdots k_{r-1} i_{r+1}\cdots i_n}\, dx^{k_1}\wedge\cdots\wedge dx^{k_{r-1}}$$

$$= \frac{1}{r!(n-r)!(r-1)!} f_{i_1\cdots i_r;k}(-1)^{(r-1)(n-r)}\delta^{i_1 i_2\cdots i_r i_{r+1}\cdots i_n}_{k k_1\cdots k_{r-1} i_{r+1}\cdots i_n}\, dx^{k_1}\wedge\cdots\wedge dx^{k_{r-1}}$$

$$= \frac{1}{r!(n-r)!(r-1)!} f_{i_1\cdots i_r;k}(-1)^{(r-1)(n-r)}(n-r)!\,\delta^{i_1 i_2\cdots i_r}_{k k_1\cdots k_{r-1}}\, dx^{k_1}\wedge\cdots\wedge dx^{k_{r-1}}$$

$$= \frac{1}{r!(r-1)!} f_{i_1\cdots i_r;k}(-1)^{(r-1)(n-r)}\delta^{i_1 i_2\cdots i_r}_{k k_1\cdots k_{r-1}}\, dx^{k_1}\wedge\cdots\wedge dx^{k_{r-1}}$$

$$\delta\alpha_r=(-1)^{nr+n+1}*d*\alpha_r$$

$$=\frac{1}{r!(r-1)!}(-1)^{nr+n+1+(r-1)(n-r)}f_{i_1\cdots i_r;k}\,\delta^{i_1i_2\cdots i_r}_{k\,k_1\cdots k_{r-1}}dx^{k_1}\wedge\cdots\wedge dx^{k_{r-1}}$$

$$=-\frac{1}{r!(r-1)!}f_{i_1\cdots i_r;k}\,\delta^{i_1i_2\cdots i_r}_{k\,k_1\cdots k_{r-1}}dx^{k_1}\wedge dx^{k_2}\wedge\cdots\wedge dx^{k_{r-1}} \tag{3.4.12}$$

$(nr+n+1+(r-1)(n-r)=2nr+1+r(1-r)=\text{an odd number})$

Here we list selectively some essential properties of δ:

[1] $(\alpha_r,d\beta_{r-1})=(\delta\alpha_r,\beta_{r-1})$. (3.4.13)

Proof: It is known from (3.4.12) that $\delta\alpha_r$ is a $(r-1)$-form. Use (3.4.9):

$$(\delta\alpha_r,\beta_{r-1})=(\beta_{r-1},\delta\alpha_r)=\int_M\beta_{r-1}\wedge{*\delta\alpha_r}$$

Then according to Eq.3.4.10: $\delta\alpha_r=(-1)^{nr+n+1}*d*\alpha_r$,

$\therefore\ (\delta\alpha_r,\beta_{r-1})=\int_M\beta_{r-1}\wedge(-1)^{nr+n+1}**d*\alpha_r$ where $d*\alpha_r$ is an $(n-r+1)$-form.

We now obtain according to Eq.3.4.11 that

$$(\delta\alpha_r,\beta_{r-1})=\int_M\beta_{r-1}\wedge(-1)^{nr+n+1}\cdot(-1)^{(n-r+1)(r-1)}d*\alpha_r$$

$$=(-1)^{-r^2+2r}\int_M\beta_{r-1}\wedge d*\alpha_r \tag{3.4.14}$$

Denote $*\alpha_r$ as $a_{n-r}=a_{j_1j_2\cdots j_{n-r}}dx^{j_1}\wedge dx^{j_2}\wedge\cdots\wedge dx^{j_{n-r}}$, we also have

$$\beta_{r-1}=\beta_{i_1\cdots i_{r-1}}dx^{i_1}\wedge\cdots\wedge dx^{i_{r-1}}$$

$$d*\alpha_r=da_{n-r}=a_{j_1j_2\cdots j_{n-r};k}dx^k\wedge dx^{j_1}\wedge\cdots\wedge dx^{j_{n-r}}$$

Plug the above equations into Eq.3.4.13 and take M as a manifold *without boundary*, then after integration by parts we have:

$$(\delta\alpha_r,\beta_{r-1})=(-1)^{-r^2+2r}\int_M\beta_{i_1\cdots i_{r-1}}dx^{i_1}\wedge\cdots\wedge dx^{i_{r-1}}\wedge a_{j_1j_2\cdots j_{n-r};k}dx^k\wedge dx^{j_1}\wedge\cdots\wedge dx^{j_{n-r}}$$

$$=(-1)^{-r^2+2r}\int_M-\beta_{i_1\cdots i_{r-1};k}(-1)^{r-1}dx^k\wedge dx^{i_1}\wedge\cdots\wedge dx^{i_{r-1}}\wedge a_{j_1j_2\cdots j_{n-r}}dx^{j_1}\wedge\cdots\wedge dx^{j_{n-r}}$$

$$=(-1)^{-r^2+2r}\int_M\beta_{i_1\cdots i_{r-1};k}dx^k\wedge dx^{i_1}\wedge\cdots\wedge dx^{i_{r-1}}\wedge a_{j_1j_2\cdots j_{n-r}}dx^{j_1}\wedge\cdots\wedge dx^{j_{n-r}}$$

$$=(-1)^{-r(r-1)}\int_M d\beta_{r-1}\wedge a_{n-r}=\int_M d\beta_{r-1}\wedge{*\alpha_r}=(\alpha_r,d\beta_{r-1})$$

□

[2] $\delta^2=0$ (3.4.15)

Proof: $\delta\alpha_p=(-1)^{np+n+1}*d*\alpha_p$ (according to Eq.3.4.10) where $\delta\alpha_p$ and $*d*\alpha_p$ are both $(p-1)$-forms.

$$\delta\cdot\delta\alpha_p=(-1)^{n(p-1)+n+1}*d*\delta\alpha_p$$

$$=(-1)^{n(p-1)+n+1}*d*(-1)^{np+n+1}*d*\alpha_p$$

$$= (-1)^n * d * * d * \alpha_p$$

where $d * \alpha_p$ is an $(n-p+1)$-form. Then according to Eq.3.4.11 we obtain:

$$* * d * \alpha_p = (-1)^{(n-p+1)(p-1)} d * \alpha_p$$

Thus $\delta \cdot \delta \alpha_p = (-1)^n (-1)^{(n-p+1)(p-1)} * dd * \alpha_p = 0$ (since dd=0) holds for any p. Therefore $\delta \cdot \delta = 0$. □

[3] $\Delta = (d+\delta)^2 = d\delta + \delta d$ (3.4.16)

This is obvious since dd=0 and $\delta\delta$=0.

Here we list selectively some essential properties of Δ:

(i) $*\Delta = \Delta *$ (3.4.17)

Proof: $*\Delta\alpha_p = *d\delta\alpha_p + *\delta d\alpha_p$ where

$$*d\delta\alpha_p = *d(-1)^{np+n+1} * d * \alpha_p = (-1)^{np+n+1} * d * d * \alpha_p, \text{ and}$$

$$*\delta d\alpha_p = (-1)^{n(p+1)+n+1} * * d * d\alpha_p = (-1)^{n(p+1)+n+1} (-1)^{(n-p)(n-(n-p))} d * d\alpha_p$$

$$= (-1)^{1-p^2} d * d\alpha_p$$

$\Delta * \alpha_p = d\delta * \alpha_p + \delta d * \alpha_p$ where

$$\delta d * \alpha_p = (-1)^{n(n-p+1)+n+1} * d * d * \alpha_p = (-1)^{np+n+1} * d * d * \alpha_p, \text{ and}$$

$$d\delta * \alpha_p = d * d * (-1)^{n(n-p)+n+1} * \alpha_p = d * d(-1)^{n(n-p)+n+1} * * \alpha_p$$

$$= d * (-1)^{n(n-p)+n+1} d(-1)^{p(n-p)} \alpha_p = (-1)^{1-p^2} d * d\alpha_p$$

$\therefore$ $(*d\delta + *\delta d)\alpha_p = (d\delta * + \delta d *)\alpha_p$ holds for any p.

$\therefore$ $*\Delta\alpha_p = \Delta * \alpha_p$ holds for any α_p. Hence $*\Delta = \Delta *$. □

(ii) $(\alpha_r, \Delta\beta_r) = (\Delta\alpha_r, \beta_r)$ (3.4.18)

Proof: According to Eq.3.4.16, $(\alpha_r, \Delta\beta_r) = (\alpha_r, d\delta\beta_r) + (\alpha_r, \delta d\beta_r)$. Then with Eq.3.4.13 and Eq.3.4.9, the equation becomes

$$(\alpha_r, \Delta\beta_r) = (\delta\alpha_r, \delta\beta_r) + (d\alpha_r, d\beta_r)$$

$$= (d\delta\alpha_r, \beta_r) + (\delta d\alpha_r, \beta_r) = (\Delta\alpha_r, \beta_r)$$

□

(iii) $(\alpha_r, \Delta\alpha_r) \geq 0$ (3.4.19)

Proof: According to Eq.3.4.18 we have

$$(\alpha_r, \Delta\alpha_r) = (d\alpha_r, d\alpha_r) + (\delta\alpha_r, \delta\alpha_r)$$

$$= \int d\alpha_r \wedge *d\alpha_r + \int \delta\alpha_r \wedge *\delta\alpha_r \qquad (3.4.19)_1$$

Then according to Eq.3.4.6 and Eq.3.4.6$_1$ we obtain

$$\gamma_r = \frac{1}{r!} g_{i_1 \cdots i_r} dx^{i_1} \wedge \cdots \wedge dx^{i_r}$$

$$*\gamma_r = \frac{1}{r!} g_{i_1 \cdots i_r} \frac{1}{(n-r)!} \varepsilon^{i_1 i_2 \cdots i_r}{}_{i_{r+1} \cdots i_n} dx^{i_{r+1}} \wedge \cdots \wedge dx^{i_n}$$

$$\gamma_r \wedge {}^*\gamma_r = \left(\frac{1}{r!}\right)^2 \frac{1}{(n-r)!}\left(g_{i_1\cdots i_r}\right)^2 \varepsilon_{i_1 i_2\cdots i_r i_{r-1}\cdots i_n} dx^{i_1}\wedge\cdots\wedge dx^{i_r}\wedge dx^{i_{r-1}}\wedge\cdots\wedge dx^{i_n}$$

Use Eq.1.4.13 we have

$$\gamma_r \wedge {}^*\gamma_r = \left(\frac{1}{r!}\right)^2 \frac{1}{(n-r)!}\left(g_{i_1\cdots i_r}\right)^2 n!\cdot(n\text{-dimensional volume element}) \geq 0 \quad (3.4.20)$$

$\therefore$ $\int d\alpha_r \wedge {}^*d\alpha_r$ and $\int \delta\alpha_r \wedge {}^*\delta\alpha_r$ in Eq.3.4.19$_1$ are both nonnegative. Hence $(\alpha_r, \Delta\alpha_r) \geq 0$. □

It can also be observed from Eq.3.4.19$_1$ that if $\Delta\alpha_r = 0$, then $d\alpha_r = \delta\alpha_r = 0$ which gives another set of definitions:

$\Delta\omega_p = 0$, ω_p is a harmonic form ($d\omega_p = \delta\omega_p = 0$).

$d\omega_p = 0$, ω_p is a closed form.

$\delta\omega_p = 0$, ω_p is a coclosed form.

Other definitions are:

$\alpha = d\beta$, α is an exact form.

$\alpha = \delta\beta$, α is a co-exact form.

Hodge Decomposition Theorem: Any *p*-form ω_p on a compact manifold *M* without boundary can be decomposed into the sum of an exact form, a co-exact form, and a harmonic form (the proof of uniqueness is omitted here):

$$\omega_p = d\alpha_{p-1} + \delta\beta_{p+1} + \gamma_p \quad (3.4.21)$$

where γ_p is the harmonic form which maintains the main properties of ω_p. Furthermore, $d\alpha_{p-1}$, $\delta\beta_{p+1}$ and γ_p are orthogonal to each other. The proof of the othogonality is the following.

Proof: According to Eq.3.4.13 and the fact that γ_p is a harmonic form ($d\gamma_p = \delta\gamma_p = 0$),

$$(d\alpha_{p-1}, \delta\beta_{p+1}) = (dd\alpha_{p-1}, \beta_{p+1}) = (\alpha_{p-1}, \delta\delta\beta_{p+1}) = 0$$

Since $d\gamma_p = \delta\gamma_p = 0$, then

$$(\alpha_{p-1}, \delta\gamma_p) = (d\alpha_{p-1}, \gamma_p) = 0 \quad \text{and} \quad (\beta_{p+1}, d\gamma_p) = (\delta\beta_{p+1}, \gamma_p) = 0$$

□

Chapter 4. Riemannian Geometry

§ 4.1 Riemannian Metric

In Riemannian geometry, the square of the distance between two neighboring points on an n-manifold is

$$ds^2 = \sum_{i,j=1}^{n} g_{ij}(x)\,dx^i \cdot dx^j \qquad (g_{ij}(x) = g_{ji}(x)) \tag{4.1.1}$$

where $g_{ij}(x)$ is called the Riemannian metric which is a tensor field differentiable everywhere with $\det|g_{ij}| \neq 0$.

A manifold with Riemannian metric is called a Riemannian manifold. Any point p on an n-Riemannian manifold M has also a tangent space $T_p(M)$ of which any two tangent vectors $X_p = \xi^i \partial_i$ and $Y_p = \eta^j \partial_j$ have their scalar product as

$$X_p \cdot Y_p = g_{ij}(p)\,\xi^i(p)\,\eta^j(p) \tag{4.1.2}$$

Thus the length of a tangent vector can be obtained:

$$|X_p| = \sqrt{X_p \cdot X_p} = \sqrt{g_{ij}(p)\,\xi^i(p)\,\xi^j(p)} \tag{4.1.3}$$

The angle θ between the two tangent vectors X_p and Y_p at p is:

$$\cos\theta = \frac{X_p \cdot Y_p}{\sqrt{|X_p| \cdot |Y_p|}} \tag{4.1.4}$$

The Hodge * operation on a Riemannian manifold contains $g_{\mu\nu}$, e.g. Eq.3.4.3 of previous chapter with Riemannian metric should be written as

$$*dx^{i_1} \wedge \cdots \wedge dx^{i_r} = \frac{1}{(n-r)!}\,\varepsilon^{i_1 \cdots i_r}{}_{i_{r+1} \cdots i_n}\, dx^{i_{r+1}} \wedge \cdots \wedge dx^{i_n} \tag{4.1.5}$$

where $\varepsilon^{i_1 \cdots i_r}{}_{i_{r+1} \cdots i_n} = g^{i_1 k_1} g^{i_2 k_2} \cdots g^{i_r k_r} \cdot \varepsilon_{k_1 k_2 \cdots k_r i_{r+1} \cdots i_n}$. (4.1.6)

Hence appear the super- and sub-scripts, which make the contraction of the identical indices more convenient, and also make it easier to elevate a subscript to a superscript.

In order to determine the relationship between $\varepsilon_{i_1 \cdots i_n}$ and $\delta^{1 \cdots n}_{i_1 \cdots i_n}$, we first use Eq.1.4.3$_1$:

$$\delta^{1 \cdots n}_{i_1 \cdots i_n} = \begin{vmatrix} \delta^1_{i_1} & \delta^1_{i_2} & \cdots & \delta^1_{i_n} \\ \delta^2_{i_1} & \delta^2_{i_2} & \cdots & \delta^2_{i_n} \\ \cdots & \cdots & \cdots & \cdots \\ \delta^n_{i_1} & \delta^n_{i_2} & \cdots & \delta^n_{i_n} \end{vmatrix} \tag{4.1.7}$$

and let

$$\varepsilon_{i_1 \cdots i_n} = \gamma\, \delta^{1 \cdots n}_{i_1 \cdots i_n} \tag{4.1.8}$$

Since $\varepsilon_{i_1 \cdots i_n}$ is a tensor, under the coordinate transformation we should have:

$$\varepsilon_{i_1 \cdots i_n} \frac{\partial x^{i_1}}{\partial x'^{k_1}} \cdots \frac{\partial x^{i_n}}{\partial x'^{k_n}} = \varepsilon'_{k_1 \cdots k_n} \tag{4.1.9}$$

According to Eq.4.1.8, we obtain from Eq.4.1.9 that

$$\gamma\, \delta^{1 \cdots n}_{i_1 \cdots i_n} \frac{\partial x^{i_1}}{\partial x'^{k_1}} \cdots \frac{\partial x^{i_n}}{\partial x'^{k_n}} = \gamma'\, \delta^{1 \cdots n}_{k_1 \cdots k_n} \tag{4.1.10}$$

where $$\delta^{1 \cdots n}_{i_1 \cdots i_n} \frac{\partial x^{i_1}}{\partial x'^{k_1}} \cdots \frac{\partial x^{i_n}}{\partial x'^{k_n}} = \begin{vmatrix} \frac{\partial x^1}{\partial x'^{k_1}} & \frac{\partial x^1}{\partial x'^{k_2}} & \cdots & \frac{\partial x^1}{\partial x'^{k_n}} \\ \frac{\partial x^2}{\partial x'^{k_1}} & \frac{\partial x^2}{\partial x'^{k_2}} & \cdots & \frac{\partial x^2}{\partial x'^{k_n}} \\ \cdots & \cdots & \cdots & \cdots \\ \frac{\partial x^n}{\partial x'^{k_1}} & \frac{\partial x^n}{\partial x'^{k_2}} & \cdots & \frac{\partial x^n}{\partial x'^{k_n}} \end{vmatrix}$$

$$= \begin{vmatrix} \frac{\partial x^1}{\partial x'^1} & \frac{\partial x^1}{\partial x'^2} & \cdots & \frac{\partial x^1}{\partial x'^n} \\ \frac{\partial x^2}{\partial x'^1} & \frac{\partial x^2}{\partial x'^2} & \cdots & \frac{\partial x^2}{\partial x'^n} \\ \cdots & \cdots & \cdots & \cdots \\ \frac{\partial x^n}{\partial x'^1} & \frac{\partial x^n}{\partial x'^2} & \cdots & \frac{\partial x^n}{\partial x'^n} \end{vmatrix} \delta^{12 \cdots n}_{k_1 k_2 \cdots k_n} = J^{-1}\, \delta^{12 \cdots n}_{k_1 k_2 \cdots k_n}, \tag{4.1.11}$$

and $$J = \begin{vmatrix} \frac{\partial x'^1}{\partial x^1} & \frac{\partial x'^2}{\partial x^1} & \cdots & \frac{\partial x'^n}{\partial x^1} \\ \frac{\partial x'^1}{\partial x^2} & \frac{\partial x'^2}{\partial x^2} & \cdots & \frac{\partial x'^n}{\partial x^2} \\ \cdots & \cdots & \cdots & \cdots \\ \frac{\partial x'^1}{\partial x^n} & \frac{\partial x'^2}{\partial x^n} & \cdots & \frac{\partial x'^n}{\partial x^n} \end{vmatrix} = \begin{vmatrix} \frac{\partial x'^1}{\partial x^1} & \frac{\partial x'^1}{\partial x^2} & \cdots & \frac{\partial x'^1}{\partial x^n} \\ \frac{\partial x'^2}{\partial x^1} & \frac{\partial x'^2}{\partial x^2} & \cdots & \frac{\partial x'^2}{\partial x^n} \\ \cdots & \cdots & \cdots & \cdots \\ \frac{\partial x'^n}{\partial x^1} & \frac{\partial x'^n}{\partial x^2} & \cdots & \frac{\partial x'^n}{\partial x^n} \end{vmatrix} \tag{4.1.12}$$

($J^{-1}J = J J^{-1} = 1$).

Then according to Eq.4.1.9, Eq.4.1.10 and Eq.4.1.11 we have:

$$\gamma\, \delta^{1 \cdots n}_{i_1 \cdots i_n} \frac{\partial x^{i_1}}{\partial x'^{k_1}} \cdots \frac{\partial x^{i_n}}{\partial x'^{k_n}} = \gamma\, J^{-1}\, \delta^{12 \cdots n}_{k_1 k_2 \cdots k_n} = \gamma'\, \delta^{12 \cdots n}_{k_1 k_2 \cdots k_n}$$

$$\therefore\ \gamma\, J^{-1} = \gamma' \longrightarrow \frac{\gamma}{\gamma'} = J \tag{4.1.13}$$

Define: g = absolute value of $\det(g_{ij})$, $\qquad g'$ = absolute value of $\det(g'_{ij})$.

$$J = \text{absolute value of } \det\left(\frac{\partial x'^i}{\partial x^k}\right), \quad J^{-1} = \text{absolute value of } \det\left(\frac{\partial x^i}{\partial x'^k}\right).$$

Since $\det\left(\frac{\partial x^i}{\partial x'^k}\right)\det\left(g_{ij}\right)\det\left(\frac{\partial x^j}{\partial x'^l}\right) = \det\left(g'_{kl}\right)$, we obtain $J^{-1} g J^{-1} = g'$.

$$\therefore \quad \frac{g}{g'} = J^2 \tag{4.1.14}$$

Comparing to Eq.4.1.13, it is clear that

$$\frac{\gamma}{\gamma'} = J = \sqrt{\frac{g}{g'}} \tag{4.1.15}$$

Therefore, taking $\gamma = \sqrt{g}$ and $\gamma' = \sqrt{g'}$, it is guaranteed that

$$\varepsilon_{i_1 \cdots i_n} = \sqrt{g}\, \delta^{1 \cdots n}_{i_1 \cdots i_n} \tag{4.1.16}$$

is a tensor.

Furthermore, $\varepsilon^{i_1 \cdots i_r}{}_{i_{r+1} i_{r+2} \cdots i_n}$ can be defined by g_{ij}:

$$\begin{aligned} \varepsilon^{i_1 \cdots i_r}{}_{i_{r+1} i_{r+2} \cdots i_n} &= g^{i_1 k_1} \cdots g^{i_r k_r} \cdot \varepsilon_{k_1 \cdots k_r i_{r+1} \cdots i_n} \\ &= g^{i_1 k_1} \cdots g^{i_r k_r}\, \gamma\, \delta^{1 \cdots r\, r+1 \cdots n}_{k_1 \cdots k_r i_{r+1} \cdots i_n} \\ &= g^{i_1 k_1} \cdots g^{i_r k_r} \sqrt{g}\, \delta^{1 \cdots r\, r+1 \cdots n}_{k_1 \cdots k_r i_{r+1} \cdots i_n} \end{aligned} \tag{4.1.17}$$

Then we have:

$$* dx^{i_1} \wedge \cdots \wedge dx^{i_r} = \frac{1}{(n-r)!} g^{i_1 k_1} \cdots g^{i_r k_r} \sqrt{g}\, \delta^{1 \cdots r\, r+1 \cdots n}_{k_1 \cdots k_r i_{r+1} \cdots i_n} dx^{i_{r+1}} \wedge \cdots \wedge dx^{i_n} \tag{4.1.18}$$

When $r=0$:

$$*1 = \frac{1}{n!} \sqrt{g}\, \delta^{1 \cdots n}_{i_1 \cdots i_n} dx^{i_1} \wedge \cdots \wedge dx^{i_n} = \sqrt{g}\, dx^1 \wedge \cdots \wedge dx^n$$

Since $dx'^1 \wedge \cdots \wedge dx'^n = J\, dx^1 \wedge \cdots \wedge dx^n$ and $J = \sqrt{\frac{g}{g'}}$, thus

$$\sqrt{g'}\, dx'^1 \wedge \cdots \wedge dx'^n = \sqrt{g}\, dx^1 \wedge \cdots \wedge dx^n .$$

Therefore

$$*1 = \sqrt{g}\, dx^1 \wedge \cdots \wedge dx^n = \sqrt{g'}\, dx'^1 \wedge \cdots \wedge dx'^n \tag{4.1.19}$$

According to Eq.4.1.16 and also since that $\varepsilon_{i_1 \cdots i_n}$ has the properties of a tensor, we may introduce η here:

$$\varepsilon^{j_1 \cdots j_n} = \eta\, \delta^{j_1 \cdots j_n}_{1 \cdots n} \tag{4.1.20}$$

Similar to Eq.4.1.11, it can be obtained that

$$\varepsilon_{i_1 \cdots i_n} = \sqrt{g}\, \delta^{1 \cdots n}_{i_1 \cdots i_n} = g_{i_1 j_1} \cdots g_{i_n j_n} \varepsilon^{j_1 \cdots j_n}$$

$$= g_{i_1 j_1} \cdots g_{i_n j_n} \eta \delta^{j_1 \cdots j_n}_{1 \cdots n} = \eta \begin{vmatrix} g_{i_1 1} & g_{i_1 2} & \cdots & g_{i_1 n} \\ g_{i_2 1} & g_{i_2 2} & \cdots & g_{i_2 n} \\ \cdots & \cdots & \cdots & \cdots \\ g_{i_n 1} & g_{i_n 2} & \cdots & g_{i_n n} \end{vmatrix}$$

$$= \eta \begin{vmatrix} g_{11} & g_{12} & \cdots & g_{1n} \\ g_{21} & g_{22} & \cdots & g_{2n} \\ \cdots & \cdots & \cdots & \cdots \\ g_{n1} & g_{n2} & \cdots & g_{nn} \end{vmatrix} \delta^{1 \cdots n}_{i_1 \cdots i_n} = \eta \det\left(g_{ij}\right) \delta^{1 \cdots n}_{i_1 \cdots i_n} = \eta\, g \,\mathrm{sign(g)} \delta^{1 \cdots n}_{i_1 \cdots i_n}$$

where g is the absolute value of $\det\left(g_{ij}\right)$ and sign(g) is the ± sign of $\det\left(g_{ij}\right)$.

If Eq.4.1.16 is required to be true, then

$$\eta = \mathrm{sign(g)} \frac{1}{\sqrt{g}} \tag{4.1.21}$$

$$\therefore \quad \varepsilon^{i_1 \cdots i_n} = \mathrm{sign(g)} \frac{1}{\sqrt{g}} \delta^{i_1 \cdots i_n}_{1 \cdots n} \tag{4.1.22}$$

Hence the relationship between $\varepsilon_{i_1 \cdots i_n}$, $\varepsilon^{i_1 \cdots i_n}$ and $\delta^{1 \cdots n}_{i_1 \cdots i_n}$, $\delta^{i_1 \cdots i_n}_{1 \cdots n}$ is determined by Eq.4.1.16 and Eq.4.1.22.

Compare to what has been discussed in Section 3.4, sign(g) will surely appear after the Riemannian metric is introduced, e.g. it can be obtained from Eq.4.1.16 and Eq.4.1.22 that

$$\varepsilon^{i_1 \cdots i_n} \varepsilon_{j_1 \cdots j_n} = \mathrm{sign(g)} \frac{1}{\sqrt{g}} \delta^{i_1 \cdots i_n}_{1 \cdots n} \sqrt{g} \delta^{1 \cdots n}_{j_1 \cdots j_n} = \mathrm{sign(g)} \delta^{i_1 \cdots i_n}_{j_1 \cdots j_n} \tag{4.1.23}$$

After the Riemannian metric is introduced, Eq.3.4.11 becomes:

$$** \alpha_r = \mathrm{sign(g)}(-1)^{r(n-r)} \alpha_r = \mathrm{sign(g)} \cdot \begin{cases} \alpha_r & \text{if } n \text{ is odd} \\ (-1)^r \alpha_r & \text{if } n \text{ is even} \end{cases} \tag{4.1.24}$$

The definition of the adjoint exterior differential operator δ also changes from Eq.3.4.10 to

$$\delta \alpha_r = \mathrm{sign(g)}(-1)^{nr+n+1} * d * \alpha_r \tag{4.1.25}$$

with some of the particular properties of δ being invariant, e.g.

$$(\alpha_r, d\beta_{r-1}) = (\delta \alpha_r, \beta_{r-1}) \tag{4.1.26$_1$}$$

$$\delta^2 = 0 \tag{4.1.26$_2$}$$

$$\Delta = (d + \delta)^2 = d\delta + \delta d \tag{4.1.26$_3$}$$

Some properties of Δ, e.g.

$$*\Delta = \Delta * \tag{4.1.27}$$

$$(\alpha_r, \Delta\beta_r) = (\Delta\alpha_r, \beta_r)$$
$$(\alpha_r, \Delta\alpha_r) = 0$$

are all invariant. The procedures of the proof are the same as that of Section 3.4.

§ 4.2 Levi-Civita Parallel Transport, Riemannian Connection and Curvature Tensor

It has been known that a Riemannian metric tensor field $g_{ij}(x) = g_{ji}(x)$ which is differentiable everywhere can be defined on a smooth *n*-manifold. The square of the infinitesimal arc length *ds* at the point *x* can be expressed by $g_{ij}(x)$ as:

$$ds^2 = g_{ij}(x)dx^i dx^j \tag{4.2.1}$$

The most essential point of Levi-Civita parallel transport is to let

$$\nabla ds^2 = 0 = \nabla(g_{ij}(x)dx^i dx^j)$$
$$= dx^k\left(\frac{\partial}{\partial x^k} g_{ij} - \Gamma^l_{ki} g_{lj} - \Gamma^l_{kj} g_{il}\right) dx^i dx^j = g_{ij,k} dx^k dx^i dx^j$$

which leads to

$$g_{ij,k} = \frac{\partial}{\partial x^k} g_{ij} - \Gamma^l_{ki} g_{lj} - \Gamma^l_{kj} g_{il} = 0 \tag{4.2.2}$$

That is, if the parallel transport of ds^2 is invariant (i.e. $\nabla ds^2 = 0$), then Eq.4.2.2 must be true. The connection Γ^i_{jk} satisfying Eq.4.2.2 is called the *allowable connection*. From a geometrical point of view, the allowable connection means that the lengths and the angles of the tangent vectors are invariant under parallel transports. For example: Suppose that there are two tangent vectors $A^i(x)\frac{\partial}{\partial x^i}$ and $B^j(x)\frac{\partial}{\partial x^j}$. It is required under parallel transport that

$$\nabla\left(A^i(x)\frac{\partial}{\partial x^i}\right) = 0 = \left(\nabla A^i(x)\right)\frac{\partial}{\partial x^i} + A^i(x)\nabla\frac{\partial}{\partial x^i}$$
$$= \left(\frac{\partial A^b(x)}{\partial x^a} + A^i(x)\Gamma^b_{ai}\right)dx^a \frac{\partial}{\partial x^b} = 0$$
$$\nabla\left(B^j(x)\frac{\partial}{\partial x^j}\right) = 0 = \left(\nabla B^j(x)\right)\frac{\partial}{\partial x^j} + B^j(x)\nabla\frac{\partial}{\partial x^j}$$
$$= \left(\frac{\partial B^c(x)}{\partial x^d} + B^j(x)\Gamma^c_{dj}\right)dx^d \frac{\partial}{\partial x^c} = 0$$

which is equivalent to require

$$\frac{\partial A^b(x)}{\partial x^a} = -A^i(x)\Gamma^b_{ai} \quad \text{and} \quad \frac{\partial B^c(x)}{\partial x^d} = -B^j(x)\Gamma^c_{dj} \tag{4.2.3}$$

Thus for ($g_{ij}A^i B^j$) we have (according to Eq.4.2.3 and Eq.4.2.2)

$$\nabla\left(g_{ij}A^iB^j\right)=dx^k\left(\frac{\partial g_{ij}}{\partial x^k}A^iB^j+g_{ij}\frac{\partial A^i}{\partial x^k}B^j+g_{ij}A^i\frac{\partial B^j}{\partial x^k}\right)$$

$$=dx^kA^iB^j\left(\frac{\partial g_{ij}}{\partial x^k}-g_{lj}\Gamma^l_{ki}-g_{il}\Gamma^l_{kj}\right)=0 \tag{4.2.4}$$

under parallel transport. Therefore $\nabla\left(g_{ij}A^iB^j\right)=\nabla\left(g_{ij}A^iA^j\right)=\nabla\left(g_{ij}B^iB^j\right)=0$.

Hence we can see that if the allowable connection is taken, then the lengths and the angles between the tangent vectors are all invariant under parallel transport. That corresponds to the invariance of ds^2 under parallel transport ($\nabla ds^2 = \nabla(g_{ij}(x)dx^idx^j)=0$) when an allowable connection is taken.

With allowable connection as the prerequisite, Eq.4.2.2 also implies:

$$\frac{\partial}{\partial x^j}g_{ik}+\frac{\partial}{\partial x^i}g_{jk}-\frac{\partial}{\partial x^k}g_{ij}=(\Gamma^l_{ji}+\Gamma^l_{ij})g_{lk}+(\Gamma^l_{jk}-\Gamma^l_{kj})g_{li}+(\Gamma^l_{ik}-\Gamma^l_{ki})g_{lj}$$

The torsion tensor also has to be zero on a Riemannian manifold (see Eq.3.2.8):

$$\Gamma^k_{ij}-\Gamma^k_{ji}=T^k_{ij}=0 \tag{4.2.5}$$

Thus we have

$$\Gamma^m_{ij}=\frac{1}{2}g^{km}\left(\frac{\partial}{\partial x^j}g_{ik}+\frac{\partial}{\partial x^i}g_{jk}-\frac{\partial}{\partial x^k}g_{ij}\right) \tag{4.2.6}$$

where Γ^m_{ij} is called the Riemannian connection. Riemann's curvature tensor R^a_{kij} is obtained by plugging Γ^m_{ij} of Eq.4.2.6 into Eq.3.2.6. According to Eq.3.2.6, R^a_{kij} is anti-symmetric with respect to i and j. With this fact we can write the curvature 2-form of Riemannian geometry. This curvature 2-form has the same form of Eq.3.3.1, which indicates the variation of a vector on a Riemannian manifold after a parallel transport along an infinitesimal circuit. The variation comes from the bending of the Riemannian manifold (of which the torsion is zero).

Several main formulas (or the symmetries) concerning the Riemann's curvatures are derived as the following:

From Eq.3.2.6:

$$R^a_{bcd}=\frac{\partial\Gamma^a_{bd}}{\partial x^c}-\frac{\partial\Gamma^a_{bc}}{\partial x^d}+\Gamma^e_{bd}\Gamma^a_{ce}-\Gamma^e_{bc}\Gamma^a_{de} \tag*{(4.2.6)$_1$}$$

From Eq.4.2.6, Γ^a_{bc} and Γ_{abc} can be expressed by g_{ab} and g^{ab} on a Riemannian manifold:

$$\left.\begin{aligned}\Gamma_{abc}&=\frac{1}{2}\left(\frac{\partial g_{ab}}{\partial x^c}+\frac{\partial g_{ac}}{\partial x^b}-\frac{\partial g_{bc}}{\partial x^a}\right)\\ \Gamma^d{}_{bc}&=g^{da}\Gamma_{abc}=\frac{1}{2}g^{da}\left(\frac{\partial g_{ab}}{\partial x^c}+\frac{\partial g_{ac}}{\partial x^b}-\frac{\partial g_{bc}}{\partial x^a}\right)=\Gamma^d{}_{cb}\end{aligned}\right\} \tag{4.2.7}$$

$$\frac{\partial g_{ab}}{\partial x^c} = \Gamma_{abc} + \Gamma_{bac} \tag{4.2.8}$$

Use Eq.4.2.6$_1$, Eq.4.2.7 and Eq.4.2.8 we obtain:

$$\left.\begin{aligned} R_{abcd} &= g_{ae} R^e{}_{bcd}\,, \\ R_{abcd} &= \frac{1}{2}\left(\frac{\partial^2 g_{ad}}{\partial x^b \partial x^c} + \frac{\partial^2 g_{bc}}{\partial x^a \partial x^d} - \frac{\partial^2 g_{bd}}{\partial x^a \partial x^c} - \frac{\partial^2 g_{ac}}{\partial x^b \partial x^d}\right) \\ &\quad + g_{ef}\left(\Gamma^e_{ad}\Gamma^f_{bc} - \Gamma^e_{ac}\Gamma^f_{bd}\right) \end{aligned}\right\} \tag{4.2.9}$$

The following three formulas are immediately derived from Eq.4.2.9:

(i) $R_{abcd} = -R_{bacd}$. (4.2.10)

(ii) $R_{abcd} = -R_{abdc}$. (4.2.11)

(iii) $R_{abcd} = R_{cdba}$. (4.2.12)

It may be worth to mention that the formula (i) can be derived from the prerequisite equations of the allowable connections Eq.4.2.3. According to Eq.4.2.3 and $g_{ij} = g_{ji}$, the prerequisite equations of the allowable connections can also be written as:

$$dg_{ij} = \Gamma^l_i g_{lj} + \Gamma^l_j g_{li} \qquad \text{(where } \Gamma^i_j = \Gamma^i_{kj} dx^k \text{ is a 1-form)} \tag{4.2.13}$$

G, Γ and Γ^T can be expressed by matrices as the following:

$$G = \begin{pmatrix} g_{11} & g_{12} & \cdots & g_{1m} \\ g_{21} & g_{22} & \cdots & g_{2m} \\ \cdots & \cdots & \cdots & \cdots \\ g_{m1} & g_{m2} & \cdots & g_{mm} \end{pmatrix} \quad \Gamma = \begin{pmatrix} \Gamma^1_1 & \Gamma^2_1 & \cdots & \Gamma^m_1 \\ \Gamma^1_2 & \Gamma^2_2 & \cdots & \Gamma^m_2 \\ \cdots & \cdots & \cdots & \cdots \\ \Gamma^1_m & \Gamma^2_m & \cdots & \Gamma^m_m \end{pmatrix} \quad \Gamma^T = \begin{pmatrix} \Gamma^1_1 & \Gamma^1_2 & \cdots & \Gamma^1_m \\ \Gamma^2_1 & \Gamma^2_2 & \cdots & \Gamma^2_m \\ \cdots & \cdots & \cdots & \cdots \\ \Gamma^m_1 & \Gamma^m_2 & \cdots & \Gamma^m_m \end{pmatrix}$$

$$\begin{aligned} ddg_{ij} &= 0 = d\Gamma^l_i \cdot g_{lj} - \Gamma^l_i \wedge dg_{lj} + d\Gamma^l_j \cdot g_{li} - \Gamma^l_j \wedge dg_{li} \\ &= d\Gamma^l_i \cdot g_{lj} - \Gamma^l_i \wedge \left(\Gamma^k_l g_{kj} + \Gamma^k_j g_{kl}\right) \\ &\quad + d\Gamma^l_j \cdot g_{li} - \Gamma^l_j \wedge \left(\Gamma^k_l g_{ki} + \Gamma^k_i g_{kl}\right) \\ &= \left(d\Gamma^k_i + \Gamma^k_l \wedge \Gamma^l_i\right) g_{kj} + \left(d\Gamma^k_j + \Gamma^k_l \wedge \Gamma^l_j\right) g_{ki} \\ &\quad - \Gamma^l_i \wedge \Gamma^k_j g_{kl} - \Gamma^k_j \wedge \Gamma^l_i g_{kl} \\ &= \left(d\Gamma^k_i + \Gamma^k_l \wedge \Gamma^l_i\right) g_{kj} + \left(d\Gamma^k_j + \Gamma^k_l \wedge \Gamma^l_j\right) g_{ki} = 0 \end{aligned} \tag{4.2.14}$$

(since $\Gamma^l_i \wedge \Gamma^k_j = -\Gamma^k_j \wedge \Gamma^l_i$, therefore the second line is equal to zero)

According to Eq.3.3.6, and since $\Gamma^i_j = \omega^i{}_j$, we have

$$\Omega^k_i g_{kj} + \Omega^k_j g_{ki} = \Omega_{ji} + \Omega_{ij} = 0 \tag{4.2.15}$$

Also according to Eq.3.3.1,

$$\Omega_{ab} = g_{ae}\Omega^e_b = g_{ae}\frac{1}{2}R^e_{bcd}dx^c \wedge dx^d = \frac{1}{2}R_{abcd}dx^c \wedge dx^d$$

$$\Omega_{ba} = g_{be}\Omega^e_a = g_{be}\frac{1}{2}R^e_{acd}dx^c \wedge dx^d = \frac{1}{2}R_{bacd}dx^c \wedge dx^d$$

$\therefore \quad (R_{abcd} + R_{bacd}) = 0$ which gives that formula (i).

There are still two more formulas:

(iv) $R_{l[kij]} = R_{lkij} + R_{lijk} + R_{ljki} = 0$ (4.2.16)

Proof: Since the torsion of a Riemannian manifold is zero, i.e. $T^i_{jk} = 0$, Eq.3.3.7 gives $\tau^i = 0$. It follows immediately the integrability condition of torsion in Eq.3.3.14 that:

$$\Omega^a_b \wedge \theta^b = 0 \quad \rightarrow \quad \Omega^l_k \wedge dx^k = 0$$

Then use Eq.3.3.1 we have:

$$\Omega^l_k \wedge dx^k = \frac{1}{2}R^l_{kij}dx^i \wedge dx^j \wedge dx^k = 0$$

$$R^l_{kij}dx^i \wedge dx^j \wedge dx^k = 0$$

Permute *k*, *i*, and *j*: $\quad R^l_{ijk}dx^j \wedge dx^k \wedge dx^i = 0$

$$R^l_{jki}dx^k \wedge dx^i \wedge dx^j = 0$$

Now add up: $\quad \left(R^l_{kij} + R^l_{ijk} + R^l_{jki}\right)dx^i \wedge dx^j \wedge dx^k = 0$

Since each group of $dx^i \wedge dx^j \wedge dx^k$ is an independent basis which includes all possible permutations of dx^i, dx^j and dx^k, therefore we have

$$R^l_{[kij]} = R^l_{kij} + R^l_{ijk} + R^l_{jki} = 0$$

$\rightarrow \quad R_{l[kij]} = g_{lh}R^h_{[kij]} = 0$ which gives us Eq.4.2.16. □

(v) $R_{lk[ij;m]} = 0$. (4.2.17)

Proof: First, by the rule of covariant differentiation we discussed in Chapter 3 (see Eq.3.2.3), the covariant derivative of R^l_{kij} is obtained as the following:

$$\nabla_m R^l_{kij} = \frac{\partial R^l_{kij}}{\partial x^m} + \Gamma^l_{mh}R^h_{kij} - \Gamma^h_{mk}R^l_{hij} - \Gamma^h_{mi}R^l_{khj} - \Gamma^h_{mj}R^l_{kih} \tag{4.2.18}$$

Use Eq.3.3.13 to replace Ω^a_b by $\frac{1}{2}R^a_{bcd}dx^c \wedge dx^d$ and ω^a_b by $dx^m\Gamma^a_{mb}$, we have

$$\begin{aligned} d\Omega^a_b &= \frac{1}{2}dR^a_{bcd}dx^c \wedge dx^d = \frac{1}{2}\frac{\partial R^a_{bcd}}{\partial x^m}dx^m \wedge dx^c \wedge dx^d \\ &= \Omega^a_e \wedge \omega^e_b - \omega^a_e \wedge \Omega^e_b \\ &= \frac{1}{2}\left(R^a_{ecd}\Gamma^e_{mb} - R^e_{bcd}\Gamma^a_{me}\right)dx^m \wedge dx^c \wedge dx^d \end{aligned} \tag{4.2.19}$$

$$\therefore \quad \frac{\partial R^l_{kij}}{\partial x^m}dx^m \wedge dx^i \wedge dx^j = \left(\Gamma^h_{mk}R^l_{hij} - \Gamma^l_{mh}R^h_{kij}\right)dx^m \wedge dx^i \wedge dx^j \tag{4.2.20}$$

Then define $\nabla_m R^l_{kij}$ in Eq.4.2.18 as

$$R^l_{kij;m} = \nabla_m R^l_{kij} \tag{4.2.21}$$

According to Eq.4.2.18 and Eq.4.2.20, we have

$$R^l_{kij;m} dx^m \wedge dx^i \wedge dx^j$$
$$= \left(\Gamma^h_{mk} R^l_{hij} - \Gamma^l_{mh} R^h_{kij} + \Gamma^l_{mh} R^h_{kij} - \Gamma^h_{mk} R^l_{hij} - \Gamma^h_{mi} R^l_{khj} - \Gamma^h_{mj} R^l_{kih}\right) dx^m \wedge dx^i \wedge dx^j$$
$$= \left(-\Gamma^h_{mi} R^l_{khj} - \Gamma^h_{mj} R^l_{kih}\right) dx^m \wedge dx^i \wedge dx^j = 0 \tag{4.2.22}$$

The reason of Eq.4.2.22 being zero is that Γ^h_{mi} is symmetric with respect to *m* and *i*, and Γ^h_{mj} is symmetric with respect to *m* and *j*.

Then permute *i*, *j* and *m* and sum them up, we obtain:

$$\left(R^l_{kij;m} + R^l_{kjm;i} + R^l_{kmi;j}\right) dx^m \wedge dx^i \wedge dx^j = 0$$

Again, since each group of $dx^m \wedge dx^i \wedge dx^j$ is an independent basis, we then have

$$R^l_{k[ij;m]} = R^l_{kij;m} + R^l_{kjm;i} + R^l_{kmi;j} = 0$$

$$\rightarrow \quad R_{lk[ij;m]} = g_{lh} R^h_{k[ij;m]} = 0 \quad \text{which gives us Eq.4.2.17.} \qquad \square$$

The contraction of Riemann's curvature tensors can also form some other kinds of tensors, mainly being the Ricci curvature tensors:

$$R_{kj} = R^l_{klj} \tag{4.2.23}$$

and the Ricci curvature scalars:

$$R = g^{ij} R_{ij} \tag{4.2.24}$$

According to the formula (i), i.e. Eq.4.2.10, we obtain:

$$R_{lkij} = R_{ijlk} \quad \rightarrow \quad g^{li} R_{lkij} = g^{li} R_{ijlk} \quad \rightarrow \quad R^i_{kij} = R^l_{jlk}$$

$$\therefore \quad R_{kj} = R_{jk} \tag{4.2.25}$$

Suppose that there is a tangent vector $X = \xi^i \dfrac{\partial}{\partial x^i}$ at the point *x*, and $R_{ij}(x)$ is the Ricci curvature tensor at *x*, then

$$\frac{R_{ij}\xi^i \xi^j}{g^{ij}\xi^i \xi^j} \tag{4.2.26}$$

is called the Ricci curvature in the direction of *X* at the point *x*.

§ 4.3 Two Interesting Examples:

A. Geodesic Curves

Suppose that there is a curve *C*(*t*), of which each point has its coordinates $x^i(t)$,

i=1, 2, ⋯, n. The tangent vector of $C(t)$ is $X = \xi^i \dfrac{\partial}{\partial x^i}$ where $\xi^i = \dfrac{dx^i}{dt}$.

Let X moves from t to $t+dt$ along $C(t)$ by a parallel transport. Then by the definition of Levi-Civita parallel transport, the change in the tangent vector X must be:

$$\nabla X = \left[dx^k \left(\frac{\partial}{\partial x^k} \xi^i \right) \frac{\partial}{\partial x^i} + \xi^i dx^j \Gamma^k_{ji} \frac{\partial}{\partial x^k} \right]$$

$$= dx^l \left(\frac{\partial}{\partial x^l} \xi^i + \xi^k \Gamma^i_{lk} \right) \frac{\partial}{\partial x^i} = dt \cdot \xi^l \left(\frac{\partial}{\partial x^l} \frac{dx^i}{dt} + \xi^k \Gamma^i_{lk} \right) \frac{\partial}{\partial x^i}$$

$$= dt \cdot \left(\frac{d}{dt} \frac{dx^i}{dt} + \Gamma^i_{lk} \xi^l \xi^k \right) \frac{\partial}{\partial x^i} = dt \cdot \left(\frac{d^2 x^i}{dt^2} + \Gamma^i_{lk} \frac{dx^l}{dt} \frac{dx^k}{dt} \right) \frac{\partial}{\partial x^i} \quad (4.3.1)$$

Thus if the tangent vector X of any point of the curve $C(t)$ has that ∇X=0 after a parallel transport along $C(t)$, then $x^i(t)$ (i=1, 2, ⋯, n) of each point of $C(t)$ must satisfy the following equation:

$$\frac{d^2 x^i}{dt^2} + \Gamma^i_{lk} \frac{dx^l}{dt} \frac{dx^k}{dt} = 0 \quad (4.3.2)$$

$C(t)$ that satisfies the above equation is the geodesic curve, and Eq.4.3.2 is called the geodesic equation.

Since $\Gamma^i_{lk} = 0$ in an Euclidean space, the solution of Eq.4.3.2 connecting two points P and Q must be a straight line. In a non-Euclidean space, the solution $C(t)$ of Eq.4.3.2 is called a geodesic curve which is the locally shortest curve between two points P and Q.

From Eq.4.3.1 it is also clear that writing X as $\xi^j \dfrac{\partial}{\partial x^j}$, we can obtain that

$$\nabla_X X = < dx^l \left(\frac{\partial}{\partial x^l} \xi^i + \xi^k \Gamma^i_{lk} \right) \frac{\partial}{\partial x^i}, \xi^j \frac{\partial}{\partial x^j} >$$

$$= \xi^j \left(\frac{\partial}{\partial x^j} \xi^i + \xi^k \Gamma^i_{jk} \right) \frac{\partial}{\partial x^i} = \left(\frac{dx^j}{dt} \frac{\partial}{\partial x^j} \xi^i + \Gamma^i_{jk} \xi^j \xi^k \right) \frac{\partial}{\partial x^i}$$

$$= \left(\frac{d^2 x^i}{dt^2} + \Gamma^i_{jk} \xi^j \xi^k \right) \frac{\partial}{\partial x^i} = 0 \quad (4.3.3)$$

Thus a geodesic curve can as well be defined as the following: A curve $C(t)$ is a geodesic curve if the tangent vector X of any point of $C(t)$ is invariant under parallel transports along $C(t)$.

Notice that only the symmetric part of Γ^i_{lk} in the term $\Gamma^i_{lk} \dfrac{dx^l}{dt} \dfrac{dx^k}{dt}$ of Eq.4.3.2 makes contributions while the anti-symmetric part does not.

B. *Killing Vectors*

Let x be a given point on the manifold M. Its coordinates become x' after a coordinate transformation. Then

$$g'_{ij}(x') = \frac{\partial x^k}{\partial x'^i}\frac{\partial x^l}{\partial x'^j} g_{kl}(x) \tag{4.3.4}$$

If the function form of the metric is invariant under this transformation, i.e. $g'_{ij}(x')$ as a function of x' and $g_{ij}(x)$ as a function of x have the same function forms, then g'_{ij} and g_{ij} shall have the following relations:

$$g'_{ij}(x) = g_{ij}(x), \qquad g'_{ij}(x') = g_{ij}(x') \tag{4.3.5}$$

Then according to Eq.4.3.4, we have

$$g_{ij}(x) = \frac{\partial x'^k}{\partial x^i}\frac{\partial x'^l}{\partial x^j} g'_{kl}(x') = \frac{\partial x'^k}{\partial x^i}\frac{\partial x'^l}{\partial x^j} g_{kl}(x'), \tag{4.3.6}$$

and from Eq.4.3.6 we have

$$g_{ij}(x)dx^i dx^j = \frac{\partial x'^k}{\partial x^i}\frac{\partial x'^l}{\partial x^j} g_{kl}(x')dx^i dx^j = g_{kl}(x')dx'^k dx'^l \tag{4.3.7}$$

Eq.4.3.7 shows that ds^2 and ds'^2 before and after the transformation are the same. Therefore a transformation satisfying Eq.4.3.6 is called an isometric transformation.

The length and angle of a tangent vector, and the distance between two points are all preserved under an isometric transformation.

Now let us observe the infinitesimal coordinate transformations

$$x'^k = x^k + \varepsilon\xi^k(x), \qquad |\varepsilon| << 1. \tag{4.3.8}$$

which is equivalent to a small movement of the point x in the direction of the vector field $X = \xi^i(x)\frac{\partial}{\partial x^i}$. Plug Eq.4.3.8 into Eq.4.3.6 we obtain:

$$\begin{aligned} g_{ij}(x) &= \left(\frac{\partial(x^k + \varepsilon\xi^k(x))}{\partial x^i}\right)\left(\frac{\partial(x^l + \varepsilon\xi^l(x))}{\partial x^j}\right)\left(g_{kl}(x) + \frac{\partial g_{kl}(x)}{\partial x^m}\varepsilon\xi^m(x)\right) \\ &= \left(\delta^k_i + \varepsilon\frac{\partial\xi^k(x)}{\partial x^i}\right)\left(\delta^l_j + \varepsilon\frac{\partial\xi^l(x)}{\partial x^j}\right)\left(g_{kl}(x) + \frac{\partial g_{kl}(x)}{\partial x^m}\varepsilon\xi^m(x)\right) \\ &= g_{ij}(x) + \varepsilon\left(\frac{\partial\xi^k(x)}{\partial x^i}g_{kj}(x) + \frac{\partial\xi^l(x)}{\partial x^j}g_{il}(x) + \frac{\partial g_{ij}(x)}{\partial x^m}\xi^m(x)\right) \end{aligned}$$

$$\longrightarrow \qquad \frac{\partial\xi^k(x)}{\partial x^i}g_{kj}(x) + \frac{\partial\xi^l(x)}{\partial x^j}g_{il}(x) + \frac{\partial g_{ij}(x)}{\partial x^m}\xi^m(x) = 0 \tag{4.3.9}$$

Replace $\xi^k(x)$ and $\xi^l(x)$ by $\xi_i(x)$ and $\xi_j(x)$ in Eq.4.3.9, then

$$
\begin{aligned}
0 &= \left[\frac{\partial \xi^k(x) g_{kj}(x)}{\partial x^i} - \xi^k(x)\frac{\partial g_{kj}(x)}{\partial x^i}\right] + \left[\frac{\partial \xi^l(x) g_{il}(x)}{\partial x^j} - \xi^l(x)\frac{\partial g_{il}(x)}{\partial x^j}\right] \\
&\quad + g^{ml}(x)\xi_l(x)\frac{\partial g_{ij}(x)}{\partial x^m} \\
&= \frac{\partial \xi_j(x)}{\partial x^i} + \frac{\partial \xi_i(x)}{\partial x^j} + \xi_l(x) g^{ml}(x)\left(\frac{\partial g_{ij}(x)}{\partial x^m} - \frac{\partial g_{mj}(x)}{\partial x^i} - \frac{\partial g_{im}(x)}{\partial x^j}\right) \\
&= \frac{\partial \xi_j(x)}{\partial x^i} + \frac{\partial \xi_i(x)}{\partial x^j} - 2\xi_l(x)\Gamma^l_{ij}(x) \qquad \text{(according to Eq.4.2.6)} \\
&= \left(\frac{\partial \xi_j(x)}{\partial x^i} - \xi_l(x)\Gamma^l_{ij}(x)\right) + \left(\frac{\partial \xi_i(x)}{\partial x^j} - \xi_l(x)\Gamma^l_{ij}(x)\right) \\
&= \nabla_i \xi_j(x) + \nabla_j \xi_i(x) = \xi_{j,i}(x) + \xi_{i,j}(x) = 0
\end{aligned} \tag{4.3.10}
$$

Thus we can see that if Eq.4.3.10 holds, then the coordinate transformation of Eq.4.3.8 will lead to an isometric transformation. $X = \xi^i(x)\dfrac{\partial}{\partial x^i}$ is called the killing vector field of the metric field $g_{ij}(x)$.

It can be proven that the Lie bracket $[X, Y]=Z$ is also a killing vector field if X and Y are both killing vector fields satisfying Eq.4.3.10. Therefore the set of killing vectors forms a Lie group which is called the isometric transformation group. The number of parameters of the largest isometric transformation group is $\frac{1}{2}n(n+1)$ on an n-dimensional Riemannian manifold (the proof is omitted).

§ 4.4 *Vierbeins on an n-Dimensional Riemannian Manifold*

Vierbein and spinor connection have to be introduced when there is a spinor (take the half-integer spin) field on an n-dimensional Riemannian manifold. In this section we briefly demonstrate the concepts and the features of vierbein and spinor connection.

At each point x of a Riemannian manifold, a coordinate system of the group of four orthogonal $\xi^a(x)$ (a=1, 2, 3, 4) and the metric of the system $\eta_{ab}(x)$ (a, b=1, 2, 3, 4) can be constructed, with the following required relations with x^μ and $g_{\mu\nu}(x)$ of the n-dimensional Riemannian manifold:

$$
\begin{aligned}
ds^2 &= g_{\mu\nu} dx^\mu dx^\nu = \eta_{ab}\, d\xi^a\, d\xi^b \\
&= g_{\mu\nu}\frac{\partial x^\mu}{\partial \xi^a}\frac{\partial x^\nu}{\partial \xi^b} d\xi^a\, d\xi^b = \eta_{ab}\frac{\partial \xi^a}{\partial x^\mu}\frac{\partial \xi^b}{\partial x^\nu} dx^\mu dx^\nu
\end{aligned} \tag{4.4.1}
$$

Hence derives

$$g_{\mu\nu} = \eta_{ab} \frac{\partial\xi^a}{\partial x^\mu} \frac{\partial\xi^b}{\partial x^\nu}; \qquad \eta_{ab} = g_{\mu\nu} \frac{\partial x^\mu}{\partial\xi^a} \frac{\partial x^\nu}{\partial\xi^b} \tag{4.4.2}$$

Definition 4.4.1: $$e^\mu_a = \frac{\partial x^\mu}{\partial\xi^a(x)}. \tag{4.4.3}$$

Therefore,

$$\left. \begin{aligned} e_a &= e^\mu_a \frac{\partial}{\partial x^\mu} = \frac{\partial x^\mu}{\partial\xi^a(x)} \frac{\partial}{\partial x^\mu} = \frac{\partial}{\partial\xi^a(x)} \\ \eta_{ab} &= e^\mu_a e^\nu_b g_{\mu\nu} \end{aligned} \right\} \tag{4.4.4}$$

Definition 4.4.2: $$\theta^a_\mu = \frac{\partial\xi^a(x)}{\partial x^\mu}. \tag{4.4.5}$$

Thus,

$$\left. \begin{aligned} \theta^a &= \theta^a_\mu dx^\mu = \frac{\partial\xi^a(x)}{\partial x^\mu} dx^\mu = d\xi^a(x) \\ \eta^{ab} &= \theta^a_\mu \theta^b_\nu g^{\mu\nu} \end{aligned} \right\} \tag{4.4.6}$$

Then from Eq.4.4.5 and Eq.4.4.3 we have

$$\theta^a_\mu e^\mu_b = \frac{\partial\xi^a(x)}{\partial x^\mu} \cdot \frac{\partial x^\mu}{\partial\xi^b(x)} = \delta^a_b \tag{4.4.7}$$

Thus we have obtained four orthogonal 1-forms $\theta^a = \theta^a_\mu dx^\mu = d\xi^a(x)$ (a=1, 2, 3, 4), which are the basic vectors of Lorentz system of coordinates at the point x, and form a vierbein of x.

There are two kinds of indices after the vierbein is introduced: μ, ν =1, 2, $\cdots$, n and a, b =1, 2, 3, 4. That implicates the coordinate transformation groups of two local regions: one is the local coordinate transformation group $x \to x'$ on an n-manifold M, e.g.

$$e^\mu_a(x') = \frac{\partial x'^\mu}{\partial x^\nu} e^\nu_a(x), \qquad \theta^a_\mu(x') = \frac{\partial x^\nu}{\partial x'^\mu} \theta^a_\nu(x); \tag{4.4.8}$$

the other is the Lorentz rotation group of the orthogonal vierbein at each point x of M, e.g.

$$e'^\mu_a(x) = L^b_a(x) e^\mu_b(x), \qquad \theta'^a_\mu(x) = L^{-1}{}^b{}_a(x) \theta^b_\mu(x) \tag{4.4.9}$$

The values of η_{ab} and η^{ab} are taken as

$$\eta_{ab} = \eta^{ab} = \pm\delta^a_b \tag{4.4.10}$$

where in a Minkowski 4-dimensional space the + sign is taken when a=1, 2, or 3, and the — sign is taken when a=0 (or 4).

The purpose of introducing the vierbein is to introduce the spin wave functions and the γ matrix acting on the spin wave functions. They are both essential to the research of the theory of quantum supergravity. But it is too bad that it has been proven in the middle 80's that same as the theory of quantum gravity, the theory of quantum supergravity is not renormalizable. Therefore the introduction of the vierbein and its properties ends here.

§ 4.5 *Conformal Group on a Riemannian Manifold (n=3, 4, 5, ⋯)*

Observe an n-dimensional Riemannian manifold M, on which a coordinate transformation is taken and is required to satisfy the following:

$$g_{\mu\nu}(x) \to g'_{\mu\nu}(x') = \frac{\partial x^{\alpha}}{\partial x'^{\mu}}\frac{\partial x^{\beta}}{\partial x'^{\nu}} g_{\alpha\beta}(x) = \Omega(x) g_{\mu\nu}(x) \tag{4.5.1}$$

where x and x' represent the coordinates of the same point before and after the transformation respectively.

Such is called the conformal transformation, since it keeps the angle between two vectors u and v unchanged. That is, the $\cos\theta$ in the following equation

$$\frac{u \cdot v}{(u^2 v^2)^{1/2}} = \cos\theta \tag{4.5.2}$$

is invariant under the transformation. According to Eq.4.5.1, the following relations are true:

$$\begin{aligned} u\cdot v = g_{\mu\nu}(x)u^{\mu}v^{\nu} &\longrightarrow \Omega(x)g_{\mu\nu}(x)u^{\mu}v^{\nu}, \\ u^2 = g_{\mu\nu}(x)u^{\mu}u^{\nu} &\longrightarrow \Omega(x)g_{\mu\nu}(x)u^{\mu}u^{\nu}, \\ v^2 = g_{\mu\nu}(x)v^{\mu}v^{\nu} &\longrightarrow \Omega(x)g_{\mu\nu}(x)v^{\mu}v^{\nu}. \end{aligned}$$

Translation and rotation are both conformal transformations, and hence form a subgroup of the conformal group.

In order to illustrate the restrictions which the coordinate transformations have to obey for the sake of conformal invariance, we transform $g_{\mu\nu}(x)$ in a neighborhood of x into the Cartesian metric η_{ab} through a local coordinate transformation. To be clear, the super- and subscripts of the Cartesian metric are denoted by $a, b, c, \cdots$. Thus we have

$$\begin{aligned} g_{ab}(x) &= \eta_{ab} = \delta_{ab} \\ ds^2 = g_{ab}dx^a dx^b &= \eta_{ab}dx^a dx^b = \delta_{ab}dx^a dx^b \end{aligned} \tag{4.5.3}$$

Then write down the infinitesimal transformation of the coordinates:

$$x^a \to x'^a = x^a + \varepsilon^a, \quad x^a = x'^a - \varepsilon^a \tag{4.5.4}$$

Differentiate once and we obtain:

$$\begin{aligned} g'_{ab}(x') &= \frac{\partial x^f}{\partial x'^a}\frac{\partial x^g}{\partial x'^b} g_{fg}(x) = (\delta_a^f - \partial_a\varepsilon^f)(\delta_b^g - \partial_b\varepsilon^g)\delta_{fg} \\ &= \delta_{ab} - (\partial_a\varepsilon_b + \partial_b\varepsilon_a) \end{aligned} \tag{4.5.5}$$

If Eq.4.5.1 is required to be satisfied, then the restriction is

$$(\partial_a \varepsilon_b + \partial_b \varepsilon_a) = \Lambda \delta_{ab} \tag{4.5.6}$$

Hence,

$$\delta^{ab}(\partial_a \varepsilon_b + \partial_b \varepsilon_a) = \Lambda \delta^{ab} \delta_{ab} = \Lambda n$$

$$\longrightarrow \quad \Lambda = \frac{1}{n}\delta^{ab}(\partial_a \varepsilon_b + \partial_b \varepsilon_a) = \frac{1}{n}(\partial_a \varepsilon^a + \partial_b \varepsilon^b)$$

$$= \frac{2}{n}\partial_a \varepsilon^a = \frac{2}{n}(\partial \cdot \varepsilon) \tag{4.5.7}$$

Use Eq.4.5.6:

$$\partial_a \varepsilon_b + \partial_b \varepsilon_a = \frac{2}{n}(\partial \cdot \varepsilon)\delta_{ab} \tag{4.5.8}$$

Use Eq.4.5.5:

$$g'_{ab}(x') = \delta_{ab}\left(1 - \frac{2}{n}(\partial \cdot \varepsilon)\right)$$

Compare with Eq.4.5.1, we have

$$\Omega(x) = 1 - \frac{2}{n}(\partial \cdot \varepsilon) \tag{4.5.9}$$

Now begin with Eq.4.5.8:

$$\partial_a \partial_b: \quad \partial_a \partial_b \left(\partial_c \varepsilon_d + \partial_d \varepsilon_c\right) = \frac{2}{n}\partial_a \partial_b (\partial \cdot \varepsilon)\delta_{cd}$$

Multiply by η^{bd}: $\partial_a \partial_b \eta^{bd}\left(\partial_c \varepsilon_d + \partial_d \varepsilon_c\right) = \frac{2}{n}\partial_a \partial_b \eta^{bd}(\partial \cdot \varepsilon)\delta_{cd}$

$$\longrightarrow \quad \partial_a \partial_c \partial_b \varepsilon^b + \partial_a \partial_b \eta^{bd} \partial_d \varepsilon_c = \partial_a \partial_c (\partial \cdot \varepsilon) + \partial_a \Box \varepsilon_c = \frac{2}{n}\partial_a \partial_c (\partial \cdot \varepsilon)$$

$$\longrightarrow \quad (n-2)\partial_a \partial_c (\partial \cdot \varepsilon) + n \Box \partial_a \varepsilon_c = 0 \tag{4.5.10}$$

Switch a and c in Eq.4.5.10 then add to Eq.4.5.10, we obtain:

$$2(n-2)\partial_a \partial_c (\partial \cdot \varepsilon) + n \Box \left(\partial_c \varepsilon_a + \partial_a \varepsilon_c\right) = 0 \tag{4.5.11}$$

Use Eq.4.5.8:

$$\Box \left(\partial_c \varepsilon_a + \partial_a \varepsilon_c\right) = \frac{2}{n}\Box (\partial \cdot \varepsilon)\delta_{ac}$$

Plug the above relation into Eq.4.5.11, we have

$$(n-2)\partial_a \partial_c (\partial \cdot \varepsilon) + \Box (\partial \cdot \varepsilon)\delta_{ac} = 0 \tag{4.5.12}$$

It can be seen from Eq.4.5.10 and Eq.4.5.12 that n=2 is a special case, which we will discuss in the next section.

Now let us look at the case where n>2 (n=3, 4, 5, ⋯). When n>2, $(n - 2) \neq 0$. Apply $\partial_g \partial_h$ to Eq.4.5.8 we have

$$\partial_g \partial_h \partial_a \varepsilon_b + \partial_g \partial_h \partial_b \varepsilon_a = \frac{2}{n}\partial_g \partial_h \partial_c \varepsilon_c \delta_{ab} \tag*{(4.5.8)$_1$}$$

$a \neq b, g \neq h$: $\partial_g \partial_h \partial_a \varepsilon_b + \partial_g \partial_h \partial_b \varepsilon_a = 0$

The two terms being independent implies that $\partial_g \partial_h \partial_a \varepsilon_b = 0$, $\partial_g \partial_h \partial_b \varepsilon_a = 0$, and also $\partial_b \partial_h \partial_b \varepsilon_a = 0$, etc..

$a \neq b, g = h$: $\partial_g \partial_g \partial_a \varepsilon_b + \partial_g \partial_g \partial_b \varepsilon_a = 0$

Again the two terms are independent, thus $\partial_g \partial_g \partial_a \varepsilon_b = 0$.

a=b, sum up with respect to c, but *not* to the repetitive a's:

$$\partial_g \partial_g \partial_a \varepsilon_a + \partial_g \partial_g \partial_a \varepsilon_a = \frac{2}{n} \partial_g \partial_g \partial_c \varepsilon_c .$$

On the right hand side of the equation, $\partial_g \partial_g \partial_1 \varepsilon_1$, $\partial_g \partial_g \partial_2 \varepsilon_2$, $\cdots$, $\partial_g \partial_g \partial_n \varepsilon_n$ are independent. Hence $\partial_g \partial_g \partial_a \varepsilon_a = 0$, $\partial_g \partial_g \partial_b \varepsilon_b = 0$, $\cdots$, $\partial_g \partial_g \partial_g \varepsilon_g = 0$.

g=h=a=b, without summing up with respect to the repetitive a's we obtain $\partial_g \partial_g \partial_g \varepsilon_g = 0$.

In short, differentiating any ε_g (or ε_a $\cdots$) three times yields zero, which can also be proven by using Eq.4.5.12. Therefore we may conclude that the power series expansion of ε_a consists of only three terms, the constant term, the linear term, and the quadratic term. That is, ε_a can be written in general as

$$\varepsilon_a = \alpha_a + \alpha_{ab} x^b + \frac{1}{2} \alpha_{abc} x^b x^c \tag{4.5.13}$$

Plug Eq.4.5.13 into Eq.4.5.8 and use Eq.4.5.7, we can find the expressions of those coefficients α_a, α_{ab}, and α_{abc} which satisfy the requirements of conformal invariance:

(i) $\varepsilon_a = \alpha_a$ where the n α_a's are constants, representing the ordinary translation. The relevant generators are (P_a).

(ii) $\varepsilon_a = \alpha_{ab}$ where $\alpha_{ab} = -\alpha_{ba}$ represents the rotation. There are $\frac{n(n-1)}{2}$ α_{ab}'s, and the relevant generators are (M_{ab}).

(iii) $\varepsilon_a = bx^c$ where b represents the global dilation. The unique relevant generator is (D).

(iv) $\varepsilon_a = (2\delta_{ab} k_a - \delta_{ad} k_b) x^b x^d$ where the n k_a's represent special conformal transformations. The relevant generators are (K_a).

Thus when n>2 (n=3, 4, 5, $\cdots$), then number of the relevant generators of the conformal transformations is

$$\underset{(P_a)}{n} + \underset{(M_{ab})}{\frac{n(n-1)}{2}} + \underset{(D)}{1} + \underset{(K_a)}{n} = \frac{(n+1)(n+2)}{2}$$

which is infinite.

§ 4.6 Conformal Group on a Riemannian Manifold (n=2)

Take n=2, $g_{ab} = \delta_{ab}$, then Eq.4.5.8 yields to the following system of equations:

$$\left.\begin{array}{ll}\partial_1\varepsilon_2 + \partial_2\varepsilon_1 = 0 & \longrightarrow \partial_1\varepsilon_2 = -\partial_2\varepsilon_1 \\ \partial_1\varepsilon_1 + \partial_1\varepsilon_1 = \partial_1\varepsilon_1 + \partial_2\varepsilon_2 & \longrightarrow \partial_1\varepsilon_1 = \partial_2\varepsilon_2\end{array}\right\} \tag{4.6.1}$$

Eq.4.6.1 indicates that ε_1 and ε_2 exactly satisfy the Cauchy-Riemann condition when n=2. Therefore $\varepsilon(z) = \varepsilon^1 + i\varepsilon^2$ and $\bar{\varepsilon}(\bar{z}) = \varepsilon^1 - i\varepsilon^2$ can certainly be given in the complex coordinate system $z = x^1 + ix^2$ and $\bar{z} = x^1 - ix^2$. It is then indicated that the correspondence between the conformal transformation $\varepsilon^1 \to \varepsilon^{1\prime}$ and $\varepsilon^2 \to \varepsilon^{2\prime}$ on a 2-dimensional Riemannian manifold and the analytic coordinate transformation of z

$$z \to z' = f(z), \qquad \bar{z} \to \bar{z}' = f(\bar{z}) \tag{4.6.2}$$

on a complex manifold is 1-1. Before and after the transformation, ds^2 becomes

$$ds^2 = dz\,d\bar{z} \quad \to \quad ds'^2 = \left|\frac{\partial f}{\partial z}\right|^2 dz\,d\bar{z} \tag{4.6.3}$$

Compare to Eq.4.5.1, Eq.4.6.3 makes it clear that Eq.4.6.2 is indeed a conformal transformation, where

$$\Omega = \left|\frac{\partial f}{\partial z}\right|^2 \tag{4.6.4}$$

Now let us take a look at the generators and their commutative relations of the infinitesimal transformation of Eq.4.6.2. Suppose that $\varepsilon(z) = \sum_n \varepsilon_n z^{n+1}$. Then

$$\begin{aligned} f(z) \to f(z+\varepsilon(z)) &= f(z) + \varepsilon(z)\frac{\partial}{\partial z}f(z) \\ &= f(z) + \sum_n \varepsilon_n z^{n+1}\frac{\partial}{\partial z}f(z) \end{aligned} \tag{4.6.5}$$

Therefore there are infinitely many generators:

$$L_n = z^{n+1}\frac{\partial}{\partial z} \qquad (n \in Z) \tag{4.6.6}$$

ε_n's are the group parameters of the infinitesimal transformation.

The commutative relation is:

$$[L_m, L_n] = z^{m+1}\frac{\partial}{\partial z}z^{n+1}\frac{\partial}{\partial z} - z^{n+1}\frac{\partial}{\partial z}z^{m+1}\frac{\partial}{\partial z} = (n-m)L_{m+n} \tag{4.6.7}$$

Similarly,

$$\bar{L}_n = \bar{z}^{n+1}\frac{\partial}{\partial \bar{z}} \qquad (n \in Z) \tag{4.6.8}$$

$$[\overline{L}_m, \overline{L}_n] = \overline{z}^{m+1} \frac{\partial}{\partial \overline{z}} \overline{z}^{n+1} \frac{\partial}{\partial \overline{z}} - \overline{z}^{n+1} \frac{\partial}{\partial \overline{z}} \overline{z}^{m+1} \frac{\partial}{\partial \overline{z}} = (n-m)\overline{L}_{m+n} \tag{4.6.9}$$

$$[\overline{L}_m, L_n] = \overline{z}^{m+1} \frac{\partial}{\partial \overline{z}} z^{n+1} \frac{\partial}{\partial z} - z^{n+1} \frac{\partial}{\partial z} \overline{z}^{m+1} \frac{\partial}{\partial \overline{z}} = 0 \tag{4.6.10}$$

The set of the above L_m and $\overline{L}_m$, together with their commutative relations Eq.4.6.7, Eq.4.6.9 and Eq.4.6.10 form the well-known Virasoro algebra. This algebra was derived from the invariance of the conformal group on a 2-dimensional Riemannian manifold, and thus is quite essential to those theories of physics, including the theory of 2-dimensional σ model, the theory of 2-dimensional statistical physics, etc., which possess conformal invariance in the 2-dimensional temporospace.

By the way, the n=1 case is trivial since there is only one ε_1, and $g_{11} = 0$. Eq.4.5.5 is simplified as $g'_{11} = 1 - 2\partial_1\varepsilon_1 = 1 - \Lambda$, which gives $\Lambda = 2\partial_1\varepsilon_1$, and $\Omega = 1 - \Lambda = 1 - 2\partial_1\varepsilon_1$.

Eq.4.5.10 and Eq.4.5.12 now become

$$-\partial_1\partial_1\partial_1\varepsilon + \partial_1\partial_1\partial_1\varepsilon = 0 \quad \text{(It naturally equals zero.)}$$

Thus when n=1, Eq.4.5.1 will be satisfied no matter what value ε_1 takes. In general, take

$$\varepsilon_1(x_1) = \sum_n \varepsilon_n x_1^{n+1}$$

$$f(x_1) \to f(x_1 + \varepsilon_1(x_1)) = f(x_1) + \sum_n \varepsilon_n x_1^{n+1} \frac{\partial}{\partial x_1} f(x_1)$$

Hence the generators are $L_n = x_1^{n+1} \dfrac{\partial}{\partial x_1}$ $(n \in Z)$. The set of L_n also forms a Virasoro algebra, which differs from Eq.4.6.6 only by the fact that x is real while z is complex. The number of the generators is also infinity!

Chapter 5. Complex Manifold

§ 5.1 Complex Manifold and its Features

Complex manifold is a kind of Hausdorff space (see Definition 1.1.6) of which each point has a neighboring cover. Each of the neighborhoods is topologically homeomorphic to a cell of an n-dimensional complex space. The local coordinate transformation in the intersection of two neighborhoods is a *complex analytic transformation.*

For example, suppose that $z^1, \cdots, z^n$ are the complex coordinates on one of the neighborhoods, and $w^1, \cdots, w^n$ are the ones on the other neighborhood. They are related according to the following function in the intersection of the two neighborhoods:

$$w^i = u^i(z^1, \cdots, z^n) \qquad (i=1, 2, \cdots, n) \tag{5.1.1}$$

where each u^i is an *analytic function* of $z^1, \cdots, z^n$, and the determinant

$$\frac{\partial(w^1, \cdots, w^n)}{\partial(z^1, \cdots, z^n)} \neq 0 \tag{5.1.2}$$

The complex manifold defined above has two very special properties:

First, the dimension of a complex manifold space must be even. That is because an n-dimensional complex manifold must have exactly $2n$ variables to compose n complex coordinates.

Secondly, a complex manifold must be real oriented.

Proof: $dz^1 \wedge d\bar{z}^1 = (dx^1 + i\,dy^1) \wedge (dx^1 - i\,dy^1) = -2i\,dx^1 \wedge dy^1$

$$\longrightarrow dx^1 \wedge \cdots \wedge dx^n \wedge dy^1 \wedge \cdots \wedge dy^n = \left(\frac{i}{2}\right)^n dz^1 \wedge \cdots \wedge dz^n \wedge d\bar{z}^1 \wedge \cdots \wedge d\bar{z}^n .$$

where $dx^1 \wedge \cdots \wedge dx^n \wedge dy^1 \wedge \cdots \wedge dy^n$ and $\left(\frac{i}{2}\right)^n dz^1 \wedge \cdots \wedge dz^n \wedge d\bar{z}^1 \wedge \cdots \wedge d\bar{z}^n$ on both sides of the equation are real numbers.

Take another group of local coordinates w^k. The relation between w^k and z^k is as the following:

$$dw^1 \wedge \cdots \wedge dw^n = J\,dz^1 \wedge \cdots \wedge dz^n \tag{5.1.3}$$

where J is the Jacobian (see Eq.5.1.2) with

$$J = \frac{\partial(w^1, \cdots, w^n)}{\partial(z^1, \cdots, z^n)} \neq 0$$

and its complex conjugation

$$\bar{J} = \frac{\partial(\bar{w}^1, \cdots, \bar{w}^n)}{\partial(\bar{z}^1, \cdots, \bar{z}^n)} \neq 0 .$$

Therefore

$$dw^1\wedge\cdots\wedge dw^n\wedge d\overline{w}^1\wedge\cdots\wedge d\overline{w}^n = J\overline{J}\,dz^1\wedge\cdots\wedge dz^n\wedge d\overline{z}^1\wedge\cdots\wedge d\overline{z}^n \tag{5.1.4}$$

Note that $J\overline{J}$ here is a positive real number.

Since z and $\overline{z}$ are independent, and so are w and $\overline{w}$; and further more, w only has the function relation with z, and so has $\overline{w}$ with $\overline{z}$, therefore

$$\frac{\partial(w^1,\cdots,w^n,\overline{w}^1,\cdots,\overline{w}^n)}{\partial(z^1,\cdots,z^n,\overline{z}^1,\cdots,\overline{z}^n)} = \begin{vmatrix} \dfrac{\partial(w^1,\cdots,w^n)}{\partial(z^1,\cdots,z^n)} & 0 \\ 0 & \dfrac{\partial(\overline{w}^1,\cdots,\overline{w}^n)}{\partial(\overline{z}^1,\cdots,\overline{z}^n)} \end{vmatrix} = J\overline{J} > 0 \tag{5.1.5}$$

Thus we see that the Jacobian of any pair of local coordinates, $(w^k,\overline{w}^k)$ and $(z^k,\overline{z}^k)$, of two neighborhoods with non-empty intersections on a complex manifold must be positive. That is, a complex manifold must be real oriented. □

Notice that the above two features are the necessary but not sufficient conditions for complex manifolds. For example, S^4 satisfies these two conditions, but Hopf and Ehresman have proven that it is impossible to construct a complex structure on S^4. Hence S^4 is not a complex manifold.

§ 5.2 *Complex Structures and Almost Complex Manifolds on Vector Spaces*

According to the definition in Section 5.1, the coordinate transformation on the intersection of any two neighborhoods of a complex manifold must be a *complex analytic transformation.* Hence we can derive the complex structure.

Take an *n*-dimensional complex manifold, of which the complex coordinates are:

$$z^k = x^k + i\,y^k,\qquad \overline{z}^k = x^k - i\,y^k \qquad (k=1,2,\cdots,n) \tag{5.2.1}$$

$$dz^k = dx^k + i\,dy^k,\qquad d\overline{z}^k = dx^k - i\,dy^k \tag{5.2.2}$$

$$x^k = \frac{1}{2}(z^k+\overline{z}^k),\qquad y^k = -\frac{i}{2}(z^k-\overline{z}^k) \tag{5.2.3}$$

$$\frac{\partial}{\partial z^k} = \frac{1}{2}\left(\frac{\partial}{\partial x^k} - i\frac{\partial}{\partial y^k}\right),\qquad \frac{\partial}{\partial \overline{z}^k} = \frac{1}{2}\left(\frac{\partial}{\partial x^k} + i\frac{\partial}{\partial y^k}\right) \tag{5.2.4}$$

Suppose that *f* is a complex analytic function. Then

$$df = \left(\frac{\partial f}{\partial z^k}\right)dz^k + \left(\frac{\partial f}{\partial \overline{z}^k}\right)d\overline{z}^k \tag{5.2.5}$$

Plug Eq.5.2.2 and Eq.5.2.4 into Eq.5.2.5, we obtain

$$df = \frac{\partial f}{\partial x^k}dx^k + \frac{\partial f}{\partial y^k}dy^k \tag{5.2.6}$$

Definition 5.2.1: If f is a holomorphic function of z^k (k=1, 2, $\cdots$, n), then it is required that

$$\frac{\partial f}{\partial \bar{z}^k}=0 \longrightarrow \bar{\partial} f=\frac{\partial f}{\partial \bar{z}^k} d\bar{z}^k \tag{5.2.7}$$

Holomorphic Condition (The Cauchy-Riemann Condition):

Take $f^j=u^j+iv^j$, then the holomorphic condition is (according to Eq.5.2.4)

$$\begin{aligned}
0&=\frac{\partial(u^j+iv^j)}{\partial \bar{z}^k}=\frac{1}{2}\left(\frac{\partial}{\partial x^k}+i\frac{\partial}{\partial y^k}\right)(u^j+iv^j)\\
&=\frac{1}{2}\left(\frac{\partial u^j}{\partial x^k}-\frac{\partial v^j}{\partial y^k}\right)+\frac{i}{2}\left(\frac{\partial v^j}{\partial x^k}-\frac{\partial u^j}{\partial y^k}\right)
\end{aligned}$$

$$\longrightarrow \quad \frac{\partial u^j}{\partial x^k}=\frac{\partial v^j}{\partial y^k},\quad \frac{\partial u^j}{\partial y^k}=-\frac{\partial v^j}{\partial x^k} \tag{5.2.8}$$

Make a modification of the superscripts — take $y^k=x^{n+k}$, $v^j=u^{n+j}$ $(k, j=1, 2, \cdots, n)$. Then Eq.5.2.8 can be written as a $2n \times 2n$ matrix:

$$\left(\frac{\partial u^j}{\partial x^k}\right)_{2n\times 2n}=\begin{pmatrix}
\frac{\partial u^1}{\partial x^1} & \cdots & \frac{\partial u^n}{\partial x^1} & \frac{\partial u^{n+1}}{\partial x^1} & \cdots & \frac{\partial u^{2n}}{\partial x^1}\\
\vdots & \ddots & \vdots & \vdots & \ddots & \vdots\\
\frac{\partial u^1}{\partial x^n} & \cdots & \frac{\partial u^n}{\partial x^n} & \frac{\partial u^{n+1}}{\partial x^n} & \cdots & \frac{\partial u^{2n}}{\partial x^n}\\
\frac{\partial u^1}{\partial x^{n+1}} & \cdots & \frac{\partial u^n}{\partial x^{n+1}} & \frac{\partial u^{n+1}}{\partial x^{n+1}} & \cdots & \frac{\partial u^{2n}}{\partial x^{n+1}}\\
\vdots & \ddots & \vdots & \vdots & \ddots & \vdots\\
\frac{\partial u^1}{\partial x^{2n}} & \cdots & \frac{\partial u^n}{\partial x^{2n}} & \frac{\partial u^{n+1}}{\partial x^{2n}} & \cdots & \frac{\partial u^{2n}}{\partial x^{2n}}
\end{pmatrix}$$

$$=\begin{pmatrix}
\frac{\partial u^1}{\partial x^1} & \cdots & \frac{\partial u^n}{\partial x^1} & \frac{\partial v^1}{\partial x^1} & \cdots & \frac{\partial v^n}{\partial x^1}\\
\vdots & \ddots & \vdots & \vdots & \ddots & \vdots\\
\frac{\partial u^1}{\partial x^n} & \cdots & \frac{\partial u^n}{\partial x^n} & \frac{\partial v^1}{\partial x^n} & \cdots & \frac{\partial v^n}{\partial x^n}\\
\frac{\partial u^1}{\partial y^1} & \cdots & \frac{\partial u^n}{\partial y^1} & \frac{\partial v^1}{\partial y^1} & \cdots & \frac{\partial v^n}{\partial y^1}\\
\vdots & \ddots & \vdots & \vdots & \ddots & \vdots\\
\frac{\partial u^1}{\partial y^n} & \cdots & \frac{\partial u^n}{\partial y^n} & \frac{\partial v^1}{\partial y^n} & \cdots & \frac{\partial v^n}{\partial y^n}
\end{pmatrix}=\begin{pmatrix} U_1 & U_2\\ U_3 & U_4\end{pmatrix} \tag{5.2.9}$$

where

$$U_1=\begin{pmatrix}\frac{\partial u^1}{\partial x^1} & \cdots & \frac{\partial u^n}{\partial x^1}\\ \vdots & \ddots & \vdots\\ \frac{\partial u^1}{\partial x^n} & \cdots & \frac{\partial u^n}{\partial x^n}\end{pmatrix}\qquad U_2=\begin{pmatrix}\frac{\partial v^1}{\partial x^1} & \cdots & \frac{\partial v^n}{\partial x^1}\\ \vdots & \ddots & \vdots\\ \frac{\partial v^1}{\partial x^n} & \cdots & \frac{\partial v^n}{\partial x^n}\end{pmatrix}$$

$$U_3=\begin{pmatrix}\frac{\partial u^1}{\partial y^1} & \cdots & \frac{\partial u^n}{\partial y^1}\\ \vdots & \ddots & \vdots\\ \frac{\partial u^1}{\partial y^n} & \cdots & \frac{\partial u^n}{\partial y^n}\end{pmatrix}\qquad U_4=\begin{pmatrix}\frac{\partial v^1}{\partial y^1} & \cdots & \frac{\partial v^n}{\partial y^1}\\ \vdots & \ddots & \vdots\\ \frac{\partial v^1}{\partial y^n} & \cdots & \frac{\partial v^n}{\partial y^n}\end{pmatrix}.$$

According to Eq.5.2.8 we have

$$U_1=U_4 \qquad U_2=-U_3 \tag{5.2.10}$$

With the $2n\times 2n$ matrix

$$J=\begin{pmatrix}0 & -I_n\\ I_n & 0\end{pmatrix} \quad \text{where } I_n \text{ is an } n\times n \text{ identity matrix,} \tag{5.2.11}$$

the holomorphic condition — Cauchy-Riemann condition can be uniformly expressed as

$$J\begin{pmatrix}U_1 & U_2\\ U_3 & U_4\end{pmatrix}=\begin{pmatrix}-U_3 & -U_4\\ U_1 & U_2\end{pmatrix}=\begin{pmatrix}U_2 & U_1\\ U_4 & -U_3\end{pmatrix}=\begin{pmatrix}U_1 & U_2\\ U_3 & U_4\end{pmatrix}J \tag{5.2.12}$$

where J satisfies

$$J^2=-I_{2n} \tag{5.2.13}$$

Let J act upon a vector space, then

$$J\begin{pmatrix}\frac{\partial}{\partial x^1}\\ \vdots\\ \frac{\partial}{\partial x^n}\\ \frac{\partial}{\partial x^{n+1}}\\ \vdots\\ \frac{\partial}{\partial x^{2n}}\end{pmatrix}=\begin{pmatrix}-\frac{\partial}{\partial x^{n+1}}\\ \vdots\\ -\frac{\partial}{\partial x^{2n}}\\ \frac{\partial}{\partial x^1}\\ \vdots\\ \frac{\partial}{\partial x^n}\end{pmatrix}\Rightarrow\begin{cases}J\frac{\partial}{\partial x^k}=-\frac{\partial}{\partial y^k}\\ J\frac{\partial}{\partial y^k}=\frac{\partial}{\partial x^k}\end{cases}.$$

Use Eq.5.2.4, we have

$$J\frac{\partial}{\partial z^k}=\frac{1}{2}J\left(\frac{\partial}{\partial x^k}-i\frac{\partial}{\partial y^k}\right)=-i\frac{\partial}{\partial z^k},\quad J\frac{\partial}{\partial \bar{z}^k}=\frac{1}{2}J\left(\frac{\partial}{\partial x^k}+i\frac{\partial}{\partial y^k}\right)=i\frac{\partial}{\partial \bar{z}^k} \quad (5.2.14)^1$$

J is the complex structure on a vector space, with two eigenvalues $\pm i$.

Definition 5.2.2: J at each point $x \in M$ is denoted by J_x, which is called the complex structure of the tangent space $T_x(M)$. The tensor field J on the manifold M is called the almost complex structure of M (J is the set of J_x ($x \in M$)).

Definition 5.2.3: A smooth manifold on which an almost complex structure can be defined is called an *almost complex manifold.*

According to Definition 5.2.3, a complex analytic manifold is an almost complex manifold, but not vice versa since it may not be possible to set up global complex coordinates on M.

If there exists an integrable almost complex structure on M, then M must be a *complex manifold* (the proof is omitted). This can be taken as the sufficient condition mentioned at the end of Section 5.1. But notice that:

1. An almost complex manifold may not be a complex manifold. For example, S^6 has almost complex structures, but the structures are not integrable. Hence S^6 is not a complex manifold (the proof is omitted).

2. An even-dimensional real oriented manifold may not be an almost complex manifold. For example, S^{2n}'s are all even-dimensional real oriented manifolds, but only S^2 is a complex manifold, which is homeomorphic to the $CP(1)$ manifold. The rest of S^n ($n \neq 1, 3$) have no almost complex structures (the proof is omitted).

Now we summarize the invariance properties under several mappings:

1. The invariance of the manifold under topological homeomorphic mappings — topological invariance.

2. The invariance of the manifold under diffeomorphic mappings — diffeomorphic invariance. (The differential structure may be varied by a topological homeomorphic mapping.)

3. The invariance of the complex manifold under holomorphic mappings — preservation of the complex structure property of a manifold (the property of the complex structure may be varied under some ordinary diffeomorphic mappings).

It is clear according to the above comparison that the holomorphic mapping is the most smooth homeomorphism which preserves the complex structure property.

---

1 The vectors with $\frac{\partial}{\partial z^k}$ as their base vectors are of (1,0) type; the vectors with $\frac{\partial}{\partial \bar{z}^k}$ as their base vectors are of (0,1) type.

§ 5.3 Almost Hermitian Manifold, Hermitian Manifold and its Connections

Assume that M is a manifold with the almost complex structure J. Take a complex local coordinate system $\{z^\alpha, \bar{z}^\alpha;\ \alpha = 1, 2, \cdots, n\}$ at a neighborhood of the point P on M. Then the base vectors $\bar{\partial}_\alpha = \frac{\partial}{\partial \bar{z}^\alpha}$ and $\partial_\alpha = \frac{\partial}{\partial z^\alpha}$ of the tangent vector field can be defined at P. According to Eq.5.2.14:

$$J\partial_\alpha = -i\partial_\alpha, \qquad J\bar{\partial}_\alpha = -i\bar{\partial}_\alpha \tag{5.3.1}$$

Now define the Hermitian metric $h_{i\bar{j}}$ which has the Hermiticity on a complex manifold ($\bar{\ }$ represents the complex conjugation):

$$h_{i\bar{j}} = \overline{h_{j\bar{i}}}\ ^{1} \tag{5.3.2}$$

The inner product of the vector $X = \xi^i \frac{\partial}{\partial z^i}$ and $Y = \bar{\eta}^j \frac{\partial}{\partial \bar{z}^j}$ is:

$$H(X,Y) = h_{i\bar{j}}\, \xi^i\, \bar{\eta}^j$$
$$H(Y,X) = h_{j\bar{i}}\, \bar{\xi}^i\, \eta^j \tag{5.3.3}$$

Hence gives $H(X,Y)$=Hermiticity:

$$H(X,Y) = \overline{H(Y,X)} \tag{5.3.4}$$

Proposition 1: $h_{ij} = h_{\bar{i}\bar{j}} = 0$. Therefore the inner product of $\xi^i \frac{\partial}{\partial z^i}$ and $\eta^j \frac{\partial}{\partial z^j}$ does not exist, neither does the inner product of $\bar{\xi}^i \frac{\partial}{\partial \bar{z}^i}$ and $\bar{\eta}^j \frac{\partial}{\partial \bar{z}^j}$.

Proposition 2: According to Eq.5.2.14, J acting on the inner product $H(X,Y)$ will not change the inner product:

$$H(JX,JY) = h_{i\bar{j}}\left(\xi^i J \frac{\partial}{\partial z^i}, \bar{\eta}^j J \frac{\partial}{\partial \bar{z}^j}\right) = h_{i\bar{j}}\left(\xi^i (-i) \frac{\partial}{\partial z^i}, \bar{\eta}^j (i) \frac{\partial}{\partial \bar{z}^j}\right)$$
$$= h_{i\bar{j}}\left(\xi^i \frac{\partial}{\partial z^i}, \bar{\eta}^j \frac{\partial}{\partial \bar{z}^j}\right) = H(X,Y) \tag{5.3.5}$$

Proposition 3: Eq.5.3.4 gives that $H(X,X) = \overline{H(X,X)}$ is real.

Definition 5.3.1: An almost complex manifold with the Hermitian metric is called *an almost Hermitian manifold.*

Definition 5.3.2: A complex manifold with *identically positive Hermitian metric* is called a *Hermitian manifold.*

Remark: According to Eq.5.3.2 and the properties of determinant:

---

1 The inverse matrix element of $h_{i\bar{j}}$ is $h^{\bar{i}j}$: $h_{i\bar{j}}\, h^{\bar{j}k} = \delta_i^k$, $h^{\bar{i}j}\, h_{j\bar{k}} = \delta_{\bar{k}}^{\bar{i}}$.

$$\det(h_{i\bar{j}}) = \begin{vmatrix} h_{1\bar{1}} & h_{1\bar{2}} & h_{1\bar{3}} & \cdots \\ h_{2\bar{1}} & h_{2\bar{2}} & h_{2\bar{3}} & \cdots \\ h_{3\bar{1}} & h_{3\bar{2}} & h_{3\bar{3}} & \cdots \\ \cdots & \cdots & \cdots & \cdots \end{vmatrix} = \begin{vmatrix} \bar{h}_{1\bar{1}} & \bar{h}_{2\bar{1}} & \bar{h}_{3\bar{1}} & \cdots \\ \bar{h}_{1\bar{2}} & \bar{h}_{2\bar{2}} & \bar{h}_{3\bar{2}} & \cdots \\ \bar{h}_{1\bar{3}} & \bar{h}_{2\bar{3}} & \bar{h}_{3\bar{3}} & \cdots \\ \cdots & \cdots & \cdots & \cdots \end{vmatrix}$$

$$= \begin{vmatrix} \bar{h}_{1\bar{1}} & \bar{h}_{1\bar{2}} & \bar{h}_{1\bar{3}} & \cdots \\ \bar{h}_{2\bar{1}} & \bar{h}_{2\bar{2}} & \bar{h}_{2\bar{3}} & \cdots \\ \bar{h}_{3\bar{1}} & \bar{h}_{3\bar{2}} & \bar{h}_{3\bar{3}} & \cdots \\ \cdots & \cdots & \cdots & \cdots \end{vmatrix} = \overline{\det(h_{i\bar{j}})} \tag{5.3.6}$$

$\therefore$ $\det(h_{i\bar{j}})$ must be a real number.

If $\det(h_{i\bar{j}})$ is not only real, but also positive, then $h_{i\bar{j}}$ is called an identically positive Hermitian metric.

In a Hermitian manifold, similar to Eq.4.1.7 and Eq.4.1.8, take

$$\delta^{1\cdots n}_{i_1\cdots i_n} = \begin{vmatrix} \delta^1_{i_1} & \cdots & \delta^1_{i_n} \\ \vdots & \ddots & \vdots \\ \delta^n_{i_1} & \cdots & \delta^n_{i_n} \end{vmatrix}, \qquad \delta^{\bar{1}\cdots\bar{n}}_{\bar{j}_1\cdots\bar{j}_n} = \begin{vmatrix} \delta^{\bar{1}}_{\bar{j}_1} & \cdots & \delta^{\bar{1}}_{\bar{j}_n} \\ \vdots & \ddots & \vdots \\ \delta^{\bar{n}}_{\bar{j}_1} & \cdots & \delta^{\bar{n}}_{\bar{j}_n} \end{vmatrix}, \tag{5.3.7}$$

and $\varepsilon_{i_1\cdots i_n} = \eta\delta^{1\cdots n}_{i_1\cdots i_n}$, $\varepsilon_{\bar{j}_1\cdots\bar{j}_n} = \bar{\eta}\delta^{\bar{1}\cdots\bar{n}}_{\bar{j}_1\cdots\bar{j}_n}$, $\eta\bar{\eta} = h$. Then with the Hodge * operation, some results concern the Hodge * in Riemannian geometry can be generalized to the Hermitian manifolds. We omit the details here since the discussion would be quite long yet the calculations are similar.

Connection can also be introduced into a complex manifold *M*. For example, assume that the local coordinate systems of some point *P* on *M* are dz^i and $d\bar{z}^i$. Then similar to Eq.3.1.9, we may define the covariant derivatives for the vectors $\dfrac{\partial}{\partial z^i}$ and $\dfrac{\partial}{\partial \bar{z}^i}$:

$$\nabla\frac{\partial}{\partial z^k} = \Gamma^i_{jk}dz^j\frac{\partial}{\partial z^i} + \Gamma^{\bar{i}}_{jk}dz^j\frac{\partial}{\partial \bar{z}^i} + \Gamma^i_{\bar{j}k}dz^{\bar{j}}\frac{\partial}{\partial z^i} + \Gamma^{\bar{i}}_{\bar{j}k}dz^{\bar{j}}\frac{\partial}{\partial \bar{z}^i} \tag{5.3.8}$$

Apply *J* to the left, and according to Eq.5.2.14, the property of the complex structure, we obtain:

$$i\nabla\frac{\partial}{\partial z^k} = i\Gamma^i_{jk}dz^j\frac{\partial}{\partial z^i} - i\Gamma^{\bar{i}}_{jk}dz^j\frac{\partial}{\partial \bar{z}^i} + i\Gamma^i_{\bar{j}k}dz^{\bar{j}}\frac{\partial}{\partial z^i} - i\Gamma^{\bar{i}}_{\bar{j}k}dz^{\bar{j}}\frac{\partial}{\partial \bar{z}^i} \tag{5.3.9}$$

Thus, to avoid conflict after applying *J* of the complex structure, i.e. without lost of the complex structure property, we must have (comparing Eq.5.3.8 to Eq.5.3.9):

$$\Gamma^{\bar{i}}_{jk} = \Gamma^{i}_{\bar{j}\bar{k}} = 0 \qquad \Gamma^{\bar{i}}_{\bar{j}k} = \Gamma^{i}_{j\bar{k}} = 0 \tag{5.3.10}$$

For a Hermitian manifold, it is also required that there must be no conflict between the connection and the metric. In other words, to compute the inner product of two vector fields $X = \xi^\alpha \frac{\partial}{\partial z^\alpha}$ and $Y = \overline{\eta}^{\bar{\beta}} \frac{\partial}{\partial \bar{z}^{\bar{\beta}}}$, we must follow the rule of Hermitian metric, that is, the inner product must be (according to Eq.5.3.3)

$$H(X,Y) = h_{\alpha\bar{\beta}}\, \xi^\alpha\, \overline{\eta}^{\bar{\beta}} \tag{5.3.11}$$

Now refer to Eq.4.2.3, if $X = \xi^i \frac{\partial}{\partial z^i}$ and $Y = \overline{\eta}^{\bar{i}} \frac{\partial}{\partial \bar{z}^{\bar{i}}}$ are invariant under the parallel transport along any curve of a Hermitian manifold, then it must be true that

$$\nabla X = \nabla\left(\xi^i \frac{\partial}{\partial z^i}\right) = \left(\nabla \xi^i\right)\frac{\partial}{\partial z^i} + \xi^i\left(\nabla \frac{\partial}{\partial z^i}\right)$$

$$= \left(d\xi^j + \xi^i \Gamma_i^j\right)\frac{\partial}{\partial z^j} = 0$$

$$\nabla Y = \nabla\left(\overline{\eta}^{\bar{j}} \frac{\partial}{\partial \bar{z}^{\bar{j}}}\right) = \left(\nabla \overline{\eta}^{\bar{j}}\right)\frac{\partial}{\partial \bar{z}^{\bar{j}}} + \overline{\eta}^{\bar{j}}\left(\nabla \frac{\partial}{\partial \bar{z}^{\bar{j}}}\right)$$

$$= \left(d\overline{\eta}^{\bar{j}} + \overline{\eta}^{\bar{i}} \Gamma_{\bar{i}}^{\bar{j}}\right)\frac{\partial}{\partial \bar{z}^{\bar{j}}} = 0 \tag{5.3.12}$$

Therefore we can see that

$$d\xi^j = -\xi^i \Gamma_i^j, \qquad d\overline{\eta}^{\bar{j}} = -\overline{\eta}^{\bar{i}} \Gamma_{\bar{i}}^{\bar{j}} \tag{5.3.13}$$

under the parallel transport.

If $H(X,Y) = h_{\alpha\bar{\beta}}\, \xi^\alpha\, \overline{\eta}^{\bar{\beta}}$ is invariant under the parallel transport along any curve of a Hermitian manifold, then (use Eq.5.3.13):

$$\nabla\left(h_{i\bar{j}}\, \xi^i\, \overline{\eta}^{\bar{j}}\right) = \left(dh_{i\bar{j}}\right)\xi^i\, \overline{\eta}^{\bar{j}} + h_{i\bar{j}}\left(d\xi^i\right)\overline{\eta}^{\bar{j}} + h_{i\bar{j}}\, \xi^i\left(d\overline{\eta}^{\bar{j}}\right)$$

$$= \left(dh_{i\bar{j}}\right)\xi^i\, \overline{\eta}^{\bar{j}} - h_{i\bar{j}}\left(\xi^k \Gamma_k^i\right)\overline{\eta}^{\bar{j}} - h_{i\bar{j}}\, \xi^i \overline{\eta}^{\bar{k}} \Gamma_{\bar{k}}^{\bar{j}}$$

$$= dh_{i\bar{j}}\, \xi^i\, \overline{\eta}^{\bar{j}} - h_{i\bar{j}} \xi^k \Gamma_k^i \overline{\eta}^{\bar{j}} - h_{i\bar{j}}\, \xi^i \overline{\eta}^{\bar{k}} \Gamma_{\bar{k}}^{\bar{j}}$$

$$= \left[dh_{i\bar{j}} - h_{k\bar{j}} \Gamma_i^k - h_{i\bar{l}} \Gamma_{\bar{j}}^{\bar{l}}\right]\xi^i\, \overline{\eta}^{\bar{j}} = 0 \tag{5.3.14}$$

where ξ^i and $\overline{\eta}^{\bar{j}}$ are arbitrary, and hence

$$dh_{i\bar{j}} - \Gamma_i^k h_{k\bar{j}} - \Gamma_{\bar{j}}^{\bar{l}} h_{i\bar{l}} = 0 \tag{5.3.15}$$

That is the condition the allowable connection $h_{i\bar{j}}$ of a Hermitian manifold must satisfy. Multiplying dx^k to the right of the equation Eq.4.2.2, we obtain an equation similar to Eq.5.3.15:

$$dg_{ij} - \Gamma_i^l g_{lj} - \Gamma_j^l g_{li} = 0$$

Therefore we can see that the allowable connection of Hermitian manifold is a generalization of the allowable connection of Riemannian manifold in Hermitian manifolds.

In the discussion of Riemannian manifolds in Chapter 4, it has been proven that the formula $\Omega_i^k g_{kj} + \Omega_j^k g_{ki} = 0$ can be derived from the allowable connection (see Eq.4.2.15). A similar formula exists in Hermitian manifolds. The demonstration follows.

Taking an exterior differential of Eq.5.3.15, we obtain

$$-d\Gamma_i^k h_{k\bar{j}} + \Gamma_i^k \wedge dh_{k\bar{j}} - d\Gamma_{\bar{j}}^{\bar{l}} h_{i\bar{l}} + \Gamma_{\bar{j}}^{\bar{l}} \wedge dh_{i\bar{l}} = 0.$$

Then according to Eq.5.3.15, plugging in $dh_{k\bar{j}} = h_{l\bar{j}}\Gamma_k^l + h_{k\bar{l}}\Gamma_{\bar{j}}^{\bar{l}}$ gives

$$-d\Gamma_i^k h_{k\bar{j}} + \Gamma_i^k \wedge \left(h_{l\bar{j}}\Gamma_k^l + h_{k\bar{l}}\Gamma_{\bar{j}}^{\bar{l}}\right) - d\Gamma_{\bar{j}}^{\bar{l}} h_{i\bar{l}} + \Gamma_{\bar{j}}^{\bar{l}} \wedge \left(h_{s\bar{l}}\Gamma_i^s + h_{i\bar{i}}\Gamma_{\bar{l}}^{\bar{i}}\right) = 0.$$

where $\Gamma_i^k \wedge h_{k\bar{l}}\Gamma_{\bar{j}}^{\bar{l}}$ and $\Gamma_{\bar{j}}^{\bar{l}} \wedge h_{s\bar{l}}\Gamma_i^s$ cancel out, and the equation then becomes

$$-d\Gamma_i^k h_{k\bar{j}} - h_{l\bar{j}}\Gamma_k^l \wedge \Gamma_i^k - d\Gamma_{\bar{j}}^{\bar{l}} h_{i\bar{l}} - h_{i\bar{i}}\Gamma_{\bar{l}}^{\bar{i}} \wedge \Gamma_{\bar{j}}^{\bar{l}} = 0.$$

After a rearrangement of the equation, we have

$$h_{l\bar{j}}\left(d\Gamma_i^l + \Gamma_k^l \wedge \Gamma_i^k\right) + h_{i\bar{l}}\left(d\Gamma_{\bar{j}}^{\bar{l}} + \Gamma_{\bar{k}}^{\bar{l}} \wedge \Gamma_{\bar{j}}^{\bar{k}}\right) = 0 \tag{5.3.16}$$

Compare to Eq.3.3.6 we may find that Γ_i^l corresponds to ω_b^a, $\left(d\Gamma_i^l + \Gamma_k^l \wedge \Gamma_i^k\right)$ corresponds to Ω_b^a, and thus Eq.5.3.16 is equivalent to $\Omega_i^k g_{kj} + \Omega_j^k g_{ki} = \Omega_{ij} + \Omega_{ji} = 0$ of Eq.4.2.15.

Recall Eq.3.2.6 of Chapter 3, the curvature R_{kij}^a is

$$R_{kij}^a = \frac{\partial\Gamma_{jk}^a}{\partial x^i} - \frac{\partial\Gamma_{ik}^a}{\partial x^j} + \Gamma_{jk}^b\Gamma_{ib}^a - \Gamma_{ik}^b\Gamma_{jb}^a$$

In case of Hermitian manifold we have

$$R_{\bar{b}cd}^a = \frac{\partial}{\partial z^c}\Gamma_{d\bar{b}}^a - \frac{\partial}{\partial z^d}\Gamma_{c\bar{b}}^a + \Gamma_{cn}^a\Gamma_{d\bar{b}}^n - \Gamma_{dn}^a\Gamma_{c\bar{b}}^n$$

$$R_{bcd}^{\bar{a}} = \frac{\partial}{\partial z^c}\Gamma_{db}^{\bar{a}} - \frac{\partial}{\partial z^d}\Gamma_{cb}^{\bar{a}} + \Gamma_{cn}^{\bar{a}}\Gamma_{db}^n - \Gamma_{dn}^{\bar{a}}\Gamma_{cb}^n \tag{5.3.17}$$

Since Eq.5.3.10 must be true to avoid destruction of the complex structure property after applying *J*, $R_{\bar{b}cd}^a$ and $R_{bcd}^{\bar{a}}$ in Eq.5.3.17 must both be zero:

$$R_{\bar{b}cd}^a = R_{bcd}^{\bar{a}} = 0 \tag{5.3.17$_1$}$$

That is the result of preserving the almost complex structure (avoiding conflict when apply *J*) as well.

§ 5.4 Kähler Manifold

As long as the Hermitian metric is defined on a complex manifold,

$$ds^2 = 2h_{\alpha\bar{\beta}}\, dz^\alpha d\bar{z}^\beta, \tag{5.4.1}$$

the differential form can be defined as well:

$$K = i\, h_{\alpha\bar{\beta}}\, dz^\alpha \wedge d\bar{z}^\beta \tag{5.4.2}$$

which is called the *Kähler 's form.* It is real since with Eq.5.3.2 we have

$$\overline{K} = -i\, h_{\beta\bar{\alpha}}\, d\bar{z}^{\bar{\alpha}} \wedge dz^\beta = i\, h_{\beta\bar{\alpha}}\, dz^\beta \wedge d\bar{z}^{\bar{\alpha}} = K$$

If the Kähler 's form on a complex manifold is *closed*, that is,

$$dK = 0, \tag{5.4.3}$$

then this complex manifold M is defined as the *Kähler 's manifold.* Since

$$dK = i\frac{\partial h_{\alpha\bar{\beta}}}{\partial z^\gamma} dz^\gamma \wedge dz^\alpha \wedge d\bar{z}^\beta + i\frac{\partial h_{\alpha\bar{\beta}}}{\partial \bar{z}^{\bar{\gamma}}} d\bar{z}^{\bar{\gamma}} \wedge dz^\alpha \wedge d\bar{z}^{\bar{\beta}}$$

$$= \frac{i}{2}\left(\frac{\partial h_{\alpha\bar{\beta}}}{\partial z^\gamma} - \frac{\partial h_{\gamma\bar{\beta}}}{\partial z^\alpha}\right) dz^\gamma \wedge dz^\alpha \wedge d\bar{z}^\beta + \frac{i}{2}\left(\frac{\partial h_{\alpha\bar{\beta}}}{\partial \bar{z}^{\bar{\gamma}}} - \frac{\partial h_{\alpha\bar{\gamma}}}{\partial \bar{z}^{\bar{\beta}}}\right) d\bar{z}^{\bar{\gamma}} \wedge dz^\alpha \wedge d\bar{z}^{\bar{\beta}},$$

thus a Hermitian manifold M is a Kähler 's manifold (satisfying $dK = 0$) if and only if

$$\frac{\partial h_{\alpha\bar{\beta}}}{\partial z^\gamma} = \frac{\partial h_{\gamma\bar{\beta}}}{\partial z^\alpha}, \qquad \frac{\partial h_{\alpha\bar{\beta}}}{\partial \bar{z}^{\bar{\gamma}}} = \frac{\partial h_{\alpha\bar{\gamma}}}{\partial \bar{z}^{\bar{\beta}}} \tag{5.4.4}$$

Now let us observe the requirements of the Kähler 's metric to the connections, starting from the covariant differential.

$$\nabla_\lambda h^{\bar{\gamma}\beta} = \frac{\partial}{\partial z^\lambda} h^{\bar{\gamma}\beta} + \Gamma^\beta_{\lambda\delta} h^{\bar{\gamma}\delta} + \Gamma^{\bar{\gamma}}_{\lambda\bar{\delta}} h^{\bar{\delta}\beta}$$

$$\nabla_\lambda h_{\alpha\bar{\gamma}} = \frac{\partial}{\partial z^\lambda} h_{\alpha\bar{\gamma}} - \Gamma^\delta_{\lambda\alpha} h_{\delta\bar{\gamma}} - \Gamma^{\bar{\delta}}_{\lambda\bar{\gamma}} h_{\alpha\bar{\delta}} \tag{5.4.5}$$

According to the requirement of the complex structure J (see Eq.5.3.10), Eq.5.4.5 can be simplified as

$$\nabla_\lambda h^{\bar{\gamma}\beta} = \frac{\partial}{\partial z^\lambda} h^{\bar{\gamma}\beta} + \Gamma^\beta_{\lambda\delta} h^{\bar{\gamma}\delta}$$

$$\nabla_\lambda h_{\alpha\bar{\gamma}} = \frac{\partial}{\partial z^\lambda} h_{\alpha\bar{\gamma}} - \Gamma^\delta_{\lambda\alpha} h_{\delta\bar{\gamma}} \tag{5.4.5$_1$}$$

It is known from Eq.5.3.15 that the Hermitian allowable connection satisfies Eq.5.3.15:

$$dh_{i\bar{j}} - h_{l\bar{j}}\Gamma^l_i - h_{i\bar{l}}\Gamma^{\bar{l}}_{\bar{j}} = 0$$

$$\rightarrow \quad \frac{\partial h_{i\bar{j}}}{\partial z^k} dz^k + \frac{\partial h_{i\bar{j}}}{\partial \bar{z}^{\bar{k}}} d\bar{z}^{\bar{k}} - h_{l\bar{j}}\Gamma^l_{ki} dz^k - h_{i\bar{l}}\Gamma^{\bar{l}}_{\bar{k}\bar{j}} d\bar{z}^{\bar{k}} = 0$$

$$\rightarrow \quad \left(\frac{\partial h_{i\bar{j}}}{\partial z^k} - h_{l\bar{j}}\Gamma^l_{ki}\right) dz^k + \left(\frac{\partial h_{i\bar{j}}}{\partial \bar{z}^{\bar{k}}} - h_{i\bar{l}}\Gamma^{\bar{l}}_{\bar{k}\bar{j}}\right) d\bar{z}^{\bar{k}} = 0$$

Thus when a Hermitian allowable connection is taken, the following relations exist (since dz^k and $d\bar{z}^{\bar{k}}$ are arbitrary):

$$\frac{\partial h_{i\bar{j}}}{\partial z^k} = h_{l\bar{j}}\Gamma^l_{ki}, \qquad \frac{\partial h_{i\bar{j}}}{\partial \bar{z}^{\bar{k}}} - h_{i\bar{l}}\Gamma^{\bar{l}}_{\bar{k}\bar{j}} \tag{5.4.6}$$

Plug Eq.5.4.6 into Eq.5.4.4, we have

$$\frac{\partial h_{\alpha\bar{\beta}}}{\partial z^\gamma} = h_{l\bar{\beta}}\Gamma^l_{\gamma\alpha} = \frac{\partial h_{\gamma\bar{\beta}}}{\partial z^\alpha} = h_{l\bar{\beta}}\Gamma^l_{\alpha\gamma} \quad \rightarrow \quad \Gamma^l_{\gamma\alpha} = \Gamma^l_{\alpha\gamma}$$

$$\frac{\partial h_{\alpha\bar{\beta}}}{\partial \bar{z}^{\bar{\gamma}}} = h_{\alpha\bar{l}}\Gamma^{\bar{l}}_{\bar{\gamma}\bar{\beta}} = \frac{\partial h_{\alpha\bar{\gamma}}}{\partial \bar{z}^{\bar{\beta}}} = h_{\alpha\bar{l}}\Gamma^{\bar{l}}_{\bar{\beta}\bar{\gamma}} \quad \rightarrow \quad \Gamma^{\bar{l}}_{\bar{\gamma}\bar{\beta}} = \Gamma^{\bar{l}}_{\bar{\beta}\bar{\gamma}} \tag{5.4.7}$$

Thus the necessary and sufficient conditions for $h_{\alpha\bar{\beta}}$ being a Kähler 's metric can be written as

$$\Gamma^l_{\gamma\alpha} = \Gamma^l_{\alpha\gamma}, \qquad \Gamma^{\bar{l}}_{\bar{\gamma}\bar{\beta}} = \Gamma^{\bar{l}}_{\bar{\beta}\bar{\gamma}} \tag{5.4.8}$$

(i.e. the necessary and sufficient condition of the Kähler 's metric is having no torsion. See Eq.3.2.8 and Eq.4.2.5.)

Since that the metric of the Kähler 's manifold has no torsion, we are able to derive some of its features to simplify the formulas. For example, take

$$\varphi = \varphi_{\alpha\bar{\beta}}\, dz^\alpha \wedge d\bar{z}^{\bar{\beta}}, \tag{5.4.9}$$

the covariant differential $\nabla_a\varphi$ of φ is (see Eq.3.1.16, the definition of ∇_i)

$$\left.\begin{aligned} \nabla_a\varphi_{\alpha\bar{\beta}}\, dz^\alpha \wedge d\bar{z}^{\bar{\beta}} &= \left(\frac{\partial\varphi_{\alpha\bar{\beta}}}{\partial z^a} - \Gamma^\sigma_{a\alpha}\varphi_{\sigma\bar{\beta}}\right) dz^a \wedge dz^\alpha \wedge d\bar{z}^{\bar{\beta}} \\ \nabla_{\bar{b}}\varphi_{\alpha\bar{\beta}}\, dz^\alpha \wedge d\bar{z}^{\bar{\beta}} &= \left(\frac{\partial\varphi_{\alpha\bar{\beta}}}{\partial \bar{z}^{\bar{b}}} - \Gamma^{\bar{\sigma}}_{\bar{b}\bar{\beta}}\varphi_{\alpha\bar{\sigma}}\right) d\bar{z}^{\bar{b}} \wedge dz^\alpha \wedge d\bar{z}^{\bar{\beta}} \end{aligned}\right\} \tag{5.4.10}$$

Because of the Kähler 's metric,

$$\Gamma^\sigma_{a\alpha} = \Gamma^\sigma_{\alpha a}, \qquad dz^a \wedge dz^\alpha = -dz^\alpha \wedge dz^a \qquad \text{(anti-symmetric)}$$

$$\Gamma^{\bar{\sigma}}_{\bar{b}\bar{\beta}} = \Gamma^{\bar{\sigma}}_{\bar{\beta}\bar{b}}, \qquad d\bar{z}^{\bar{b}} \wedge d\bar{z}^{\bar{\beta}} = -d\bar{z}^{\bar{\beta}} \wedge d\bar{z}^{\bar{b}} \qquad \text{(anti-symmetric)}$$

Thus all the terms containing $\Gamma^{\bar{\sigma}}_{\bar{b}\bar{\beta}}$ and $\Gamma^\sigma_{a\alpha}$ in Eq.5.4.10 have no contribution! Eq.5.4.10 is then simplified as

$$\left.\begin{aligned} \nabla_a\varphi_{\alpha\bar{\beta}}\, dz^\alpha \wedge d\bar{z}^{\bar{\beta}} &= \left(\frac{\partial\varphi_{\alpha\bar{\beta}}}{\partial z^a}\right) dz^a \wedge dz^\alpha \wedge d\bar{z}^{\bar{\beta}} \\ \nabla_{\bar{b}}\varphi_{\alpha\bar{\beta}}\, dz^\alpha \wedge d\bar{z}^{\bar{\beta}} &= \left(\frac{\partial\varphi_{\alpha\bar{\beta}}}{\partial \bar{z}^{\bar{b}}}\right) d\bar{z}^{\bar{b}} \wedge dz^\alpha \wedge d\bar{z}^{\bar{\beta}} \end{aligned}\right\} \tag{5.4.11}$$

The above demonstration shows that the Γ-terms will not appear in the covariant differential of a Kähler 's manifold:

$$\nabla_\alpha \varphi_{\alpha_1\cdots\alpha_p\bar\beta_1\cdots\bar\beta_q} dz^{\alpha_1}\wedge dz^{\alpha_2}\wedge\cdots\wedge dz^{\alpha_p}\wedge d\bar z^{\bar\beta_1}\wedge\cdots\wedge d\bar z^{\bar\beta_q}$$

$$= \frac{\partial\varphi_{\alpha_1\cdots\alpha_p,\bar\beta_1\cdots\bar\beta_q}}{\partial z^\alpha} dz^\alpha\wedge dz^{\alpha_1}\wedge dz^{\alpha_2}\wedge\cdots\wedge dz^{\alpha_p}\wedge d\bar z^{\bar\beta_1}\wedge\cdots\wedge d\bar z^{\bar\beta_q}$$

$$= \partial\varphi \tag{5.4.12$_1$}$$

$$\nabla_{\bar\beta} \varphi_{\alpha_1\cdots\alpha_p\bar\beta_1\cdots\bar\beta_q} dz^{\alpha_1}\wedge dz^{\alpha_2}\wedge\cdots\wedge dz^{\alpha_p}\wedge d\bar z^{\bar\beta_1}\wedge\cdots\wedge d\bar z^{\bar\beta_q}$$

$$= \frac{\partial\varphi_{\alpha_1\cdots\alpha_p,\bar\beta_1\cdots\bar\beta_q}}{\partial \bar z^{\bar\beta}} dz^\alpha\wedge dz^{\alpha_1}\wedge dz^{\alpha_2}\wedge\cdots\wedge dz^{\alpha_p}\wedge(-1)^p d\bar z^{\bar\beta}\wedge d\bar z^{\bar\beta_1}\wedge\cdots\wedge d\bar z^{\bar\beta_q}$$

$$= \bar\partial\varphi \tag{5.4.12$_2$}$$

In Eq.5.4.12$_1$ and Eq.5.4.12$_2$, we introduced two differential operators

$$\partial = \sum_{\alpha=1}^n dz^\alpha \frac{\partial}{\partial z^\alpha}, \quad \text{and} \quad \bar\partial = \sum_{\alpha=1}^n d\bar z^{\bar\alpha}\frac{\partial}{\partial \bar z^{\bar\alpha}} \tag{5.4.13}$$

Combine them we have

$$d = (\partial + \bar\partial) \tag{5.4.14}$$

Similar to $d\cdot d = 0$, we must also have (refer to Eq.1.4.17)

$$\partial\cdot\partial = 0, \quad \text{and} \quad \bar\partial\cdot\bar\partial = 0. \tag{5.4.15}$$

Since $dz^i\wedge d\bar z^{\bar j} = -d\bar z^{\bar j}\wedge dz^i$, therefore

$$\partial\cdot\bar\partial = -\bar\partial\cdot\partial \tag{5.4.16}$$

Then define the codifferential operator (see the definition of Hodge * in Eq.3.4.10):

$$\delta = -*d* = -*(\partial+\bar\partial)* = \vartheta + \bar\vartheta \tag{5.4.17}$$

where

$$\vartheta = -*\bar\partial* \quad \text{and} \quad \bar\vartheta = -*\partial*. \tag{5.4.18}$$

Similar to Eq.3.4.3, the complex Hodge * operation is introduced.

$$*\left(dz^{\alpha_1}\wedge\cdots\wedge dz^{\alpha_p}\wedge d\bar z^{\bar\beta_1}\wedge\cdots\wedge d\bar z^{\bar\beta_q}\right)$$

$$= \frac{1}{(n-p)!(n-q)!} i^n (-1)^{\frac{1}{2}n(n-1)+np} \varepsilon^{\alpha_1\cdots\alpha_p}{}_{\bar\gamma_1\cdots\bar\gamma_{n-p}}$$

$$\cdot\varepsilon^{\bar\beta_1\cdots\bar\beta_q}{}_{\delta_1\cdots\delta_{n-q}} dz^{\delta_1}\wedge\cdots\wedge dz^{\delta_{n-q}}\wedge d\bar z^{\bar\gamma_1}\wedge\cdots\wedge d\bar z^{\bar\gamma_{n-p}} \tag{5.4.19}$$

where n is the complex dimension of the complex manifold.

Example 1. $p=q=0$

$$*1 = \frac{1}{n!n!} i^n (-1)^{\frac{1}{2}n(n-1)} \varepsilon_{\bar\gamma_1\cdots\bar\gamma_n}\cdot\varepsilon_{\delta_1\cdots\delta_n} dz^{\delta_1}\wedge\cdots\wedge dz^{\delta_n}\wedge d\bar z^{\bar\gamma_1}\wedge\cdots\wedge d\bar z^{\bar\gamma_n} \tag{5.4.20}$$

Take the complex conjugation:

$$\overline{*1} = \frac{1}{n!n!}(-1)^n i^n (-1)^{\frac{1}{2}n(n-1)} \varepsilon_{\gamma_1\cdots\gamma_n}\cdot\varepsilon_{\bar\delta_1\cdots\bar\delta_n} d\bar z^{\bar\delta_1}\wedge\cdots\wedge d\bar z^{\bar\delta_n}\wedge dz^{\gamma_1}\wedge\cdots\wedge dz^{\gamma_n}$$

$$= \frac{1}{n!n!}(-1)^{n+\frac{1}{2}n(n-1)+n^2} i^n \varepsilon_{\bar{\delta}_1\cdots\bar{\delta}_n} \cdot \varepsilon_{\gamma_1\cdots\gamma_n} dz^{\gamma_1}\wedge\cdots\wedge dz^{\gamma_n}\wedge d\bar{z}^{\bar{\delta}_1}\wedge\cdots\wedge d\bar{z}^{\bar{\delta}_n}$$

$$= \frac{1}{n!n!} i^n (-1)^{\frac{1}{2}n(n-1)} \varepsilon_{\bar{\delta}_1\cdots\bar{\delta}_n} \cdot \varepsilon_{\gamma_1\cdots\gamma_n} dz^{\gamma_1}\wedge\cdots\wedge dz^{\gamma_n}\wedge d\bar{z}^{\bar{\delta}_1}\wedge\cdots\wedge d\bar{z}^{\bar{\delta}_n}$$

$$= *1 \tag{5.4.21}$$

$\therefore$ * is a real operator.

Example 2. Apply * to the left of Eq.5.4.19, we obtain (see also Eq.5.4.36):

$$**\left(dz^{\alpha_1}\wedge dz^{\alpha_2}\wedge\cdots\wedge dz^{\alpha_p}\wedge d\bar{z}^{\bar{\beta}_1}\wedge\cdots\wedge d\bar{z}^{\bar{\beta}_q}\right)$$

$$= \text{constant}\left(dz^{\alpha_1}\wedge dz^{\alpha_2}\wedge\cdots\wedge dz^{\alpha_p}\wedge d\bar{z}^{\bar{\beta}_1}\wedge\cdots\wedge d\bar{z}^{\bar{\beta}_q}\right)$$

Then use Eq.5.4.15, we have

$$\vartheta\vartheta = *\bar{\partial}**\bar{\partial}* = \text{constant} * \bar{\partial}\bar{\partial}* = 0$$

$$\bar{\vartheta}\bar{\vartheta} = *\partial**\partial* = \text{constant} * \partial\partial* = 0 \tag{5.4.22}$$

Use Eq.5.4.17:

$$\delta\delta = *d**d* = \text{constant} * dd* = 0 \tag{5.4.23}$$

Hence we have found the three operators of Laplacian type:

$$\Delta = (d+\delta)^2 = d\delta + \delta d$$

$$\Box = (\partial + \vartheta)^2 = \partial\vartheta + \vartheta\partial$$

$$\bar{\Box} = (\bar{\partial} + \bar{\vartheta})^2 = \bar{\partial}\bar{\vartheta} + \bar{\vartheta}\bar{\partial} \tag{5.4.24}$$

They have the following relationship:

$$\Delta = 2\,\Box = 2\,\bar{\Box} \tag{5.4.25}$$

To prove Eq.5.4.25, we introduce the Λ operator. For an arbitrary differential form,

$$F = \frac{1}{p!q!} f_{\alpha_1\cdots\alpha_p\bar{\beta}_1\cdots\bar{\beta}_q} dz^{\alpha_1}\wedge\cdots\wedge dz^{\alpha_p}\wedge d\bar{z}^{\bar{\beta}_1}\wedge\cdots\wedge d\bar{z}^{\bar{\beta}_q} \tag{5.4.26}$$

Apply Λ to the left of the equation, we have (that is the definition of Λ)

$$\Lambda F = \frac{1}{(p-1)!(q-1)!} ih^{\bar{\beta}\alpha} f_{\alpha\bar{\beta}\alpha_2\cdots\alpha_p\bar{\beta}_2\cdots\bar{\beta}_q} dz^{\alpha_2}\wedge\cdots dz^{\alpha_p}\wedge d\bar{z}^{\bar{\beta}_2}\wedge\cdots\wedge d\bar{z}^{\bar{\beta}_q}$$

$$= \frac{1}{(p-1)!(q-1)!} ih^{\bar{\beta}\alpha}(-1)^{p-1} f_{\alpha\alpha_2\cdots\alpha_p\bar{\beta}\bar{\beta}_2\cdots\bar{\beta}_q} dz^{\alpha_2}\wedge\cdots\wedge dz^{\alpha_p}\wedge d\bar{z}^{\bar{\beta}_2}\wedge\cdots\wedge d\bar{z}^{\bar{\beta}_q}$$

Since $f_{\alpha_1\cdots\alpha_p\bar{\beta}_1\cdots\bar{\beta}_q}$ is anti-symmetric with respect to its subscripts, a factor of $(-1)^{p-1}$ appears due to the $(p-1)$ times of shifts of the subscript $\bar{\beta}$.

Notice that the action of Λ does not affect the indices except α_1 and $\bar{\beta}_1$ Moreover, an i is introduced by Λ, and hence comparing to ΛF, $\bar{\Lambda} F$ should be written as

$$\bar{\Lambda} F = \frac{1}{(p-1)!(q-1)!}(-i)\overline{h^{\bar{\beta}\alpha}} f_{\bar{\alpha}\beta\alpha_2\cdots\alpha_p\bar{\beta}_2\cdots\bar{\beta}_q} dz^{\alpha_2}\wedge\cdots\wedge dz^{\alpha_p}\wedge d\bar{z}^{\bar{\beta}_2}\wedge\cdots\wedge d\bar{z}^{\bar{\beta}_q}$$

$$=\frac{1}{(p-1)!(q-1)!}(-i)h^{\bar{\alpha}\beta}(-1)^p f_{\beta\alpha_2\cdots\alpha_p\bar{\alpha}\bar{\beta}_2\cdots\bar{\beta}_q}dz^{\alpha_2}\wedge\cdots\wedge dz^{\alpha_p}\wedge d\bar{z}^{\bar{\beta}_2}\wedge\cdots\wedge d\bar{z}^{\bar{\beta}_q}$$

$$=\frac{1}{(p-1)!(q-1)!}ih^{\bar{\alpha}\beta}(-1)^{p+1} f_{\beta\alpha_2\cdots\alpha_p\bar{\alpha}\bar{\beta}_2\cdots\bar{\beta}_q}dz^{\alpha_2}\wedge\cdots\wedge dz^{\alpha_p}\wedge d\bar{z}^{\bar{\beta}_2}\wedge\cdots\wedge d\bar{z}^{\bar{\beta}_q}$$

$$=\Lambda F \tag{5.4.27}$$

Therefore Λ is a real operator.

Now we intend to prove the following relations:

$$\bar{\partial}\Lambda-\Lambda\bar{\partial}=-i\vartheta\,,\qquad \partial\Lambda-\Lambda\partial=i\bar{\vartheta} \tag{5.4.28}$$

Since Λ is a real operator, ∂ and $\bar{\partial}$ are complex conjugate to each other; and since that * is a real operator as well (see Eq.5.4.21), thus ϑ and $\bar{\vartheta}$ are also complex conjugate to each other (see Eq.5.4.18). Therefore we can see that the two equations in Eq.5.4.28 are complex conjugate to each other, and we only need to prove the first one.

Here we have to derive another feature of the Kähler 's metric:

Multiply $h^{\bar{j}m}$ to the first equation of Eq.5.4.6, and sum up with respect to $\bar{j}$,

$$h^{\bar{j}m}\frac{\partial h_{i\bar{j}}}{\partial z^k}=-h_{i\bar{j}}\frac{\partial h^{\bar{j}m}}{\partial z^k}=h^{\bar{j}m}h_{l\bar{j}}\Gamma^l_{ki}=\Gamma^m_{ki}$$

We now know that

$$\nabla_k h^{\bar{l}m}=\frac{\partial h^{\bar{l}m}}{\partial z^k}+h^{\bar{l}i}\Gamma^m_{ki}=0 \tag{5.4.29$_1$}$$

Similarly,

$$\nabla_{\bar{k}} h^{\bar{l}n}=\frac{\partial h^{\bar{l}n}}{\partial \bar{z}^{\bar{k}}}+h^{\bar{j}n}\Gamma^{\bar{l}}_{\bar{k}\bar{j}}=0 \tag{5.4.29$_2$}$$

This feature can be used to prove Eq.5.4.28.

First, according to Eq.5.4.27 we can write (use Eq.5.4.29):

$$\bar{\partial}\Lambda F=\frac{i}{(p-1)!(q-1)!}(-1)^{p-1}\frac{\partial}{\partial\bar{z}^{\bar{\beta}_1}}h^{\bar{\beta}\alpha}f_{\alpha\alpha_2\cdots\alpha_p\bar{\beta}\bar{\beta}_2\cdots\bar{\beta}_q}(-1)^{p-1}dz^{\alpha_2}\wedge\cdots\wedge dz^{\alpha_p}\wedge d\bar{z}^{\bar{\beta}_1}\wedge d\bar{z}^{\bar{\beta}_2}\wedge\cdots\wedge d\bar{z}^{\bar{\beta}_q}$$

$$=\frac{i}{(p-1)!(q-1)!}h^{\bar{\beta}\alpha}\nabla_{\bar{\beta}_1}f_{\alpha\alpha_2\cdots\alpha_p\bar{\beta}\bar{\beta}_2\cdots\bar{\beta}_q}dz^{\alpha_2}\wedge\cdots\wedge dz^{\alpha_p}\wedge d\bar{z}^{\bar{\beta}_1}\wedge d\bar{z}^{\bar{\beta}_2}\wedge\cdots\wedge d\bar{z}^{\bar{\beta}_q} \tag{5.4.30$_1$}$$

In Eq.5.4.30$_1$, we switch the indices $(\bar{\beta}_1\leftrightarrow\bar{\beta}_2)$, $(\bar{\beta}_2\leftrightarrow\bar{\beta}_3)$, $\cdots$, $(\bar{\beta}_{q-1}\leftrightarrow\bar{\beta}_q)$ respectively, and obtain q equivalent terms. Sum them up then divide by q, we have

$$\bar{\partial}\Lambda F=\frac{i}{(p-1)!q!}h^{\bar{\beta}\alpha}(\nabla_{\bar{\beta}_1}f_{\alpha\alpha_2\cdots\alpha_p\bar{\beta}\bar{\beta}_2\bar{\beta}_3\cdots\bar{\beta}_q}-\nabla_{\bar{\beta}_2}f_{\alpha\alpha_2\cdots\alpha_p\bar{\beta}\bar{\beta}_1\bar{\beta}_3\cdots\bar{\beta}_q}$$
$$+\nabla_{\bar{\beta}_3}f_{\alpha\alpha_2\cdots\alpha_p\bar{\beta}\bar{\beta}_1\bar{\beta}_2\cdots\bar{\beta}_q}-\cdots+(-1)^{q-1}\nabla_{\bar{\beta}_q}f_{\alpha\alpha_2\cdots\alpha_p\bar{\beta}\bar{\beta}_1\bar{\beta}_2\cdots\bar{\beta}_{q-1}})$$

$$\cdot dz^{\alpha_2}\wedge\cdots\wedge dz^{\alpha_p}\wedge d\bar{z}^{\bar{\beta}_1}\wedge d\bar{z}^{\bar{\beta}_2}\wedge\cdots\wedge d\bar{z}^{\bar{\beta}_q} \qquad (5.4.30)_2$$

Then we are going to find $\Lambda\bar{\partial}F$. An example of F and ΛF has been given by Eq.5.4.26. The main point is that $f_{\alpha\beta\alpha_2\cdots\alpha_p\bar{\beta}_2\cdots\bar{\beta}_q}$ must be completely anti-symmetric. Now take

$$\bar{\partial}F=\frac{(-1)^p}{p!q!}\frac{\partial}{\partial\bar{z}^{\bar{\beta}_0}}f_{\alpha_1\cdots\alpha_p\bar{\beta}_1\cdots\bar{\beta}_q}dz^{\alpha_1}\wedge\cdots\wedge dz^{\alpha_p}\wedge d\bar{z}^{\bar{\beta}_0}\wedge d\bar{z}^{\bar{\beta}_1}\wedge\cdots\wedge d\bar{z}^{\bar{\beta}_q}$$

Apply Λ to the equation. Since it is completely anti-symmetric with respect to $\bar{\beta}_0,\bar{\beta}_1,\cdots,\bar{\beta}_q$, thus $\bar{\beta}_0,\bar{\beta}_1,\bar{\beta}_2\cdots,\bar{\beta}_q$ of $\bar{\partial}F$ must be completely anti-symmetrized:

$$\Lambda\bar{\partial}F=\frac{(-1)^p(-1)^{p-1}}{(p-1)!(q-1)!}\frac{ih^{\bar{\beta}\alpha}}{q}\Big(\frac{\partial}{\partial\bar{z}^{\bar{\beta}_0}}f_{\alpha\alpha_2\cdots\alpha_p\bar{\beta}\bar{\beta}_2\bar{\beta}_3\bar{\beta}_4\cdots\bar{\beta}_q}-\frac{\partial}{\partial\bar{z}^{\bar{\beta}_1}}f_{\alpha\alpha_2\cdots\alpha_p\bar{\beta}\bar{\beta}_2\bar{\beta}_3\bar{\beta}_4\cdots\bar{\beta}_q}$$
$$+\frac{\partial}{\partial\bar{z}^{\bar{\beta}_2}}f_{\alpha\alpha_2\cdots\alpha_p\bar{\beta}\bar{\beta}_1\bar{\beta}_3\bar{\beta}_4\cdots\bar{\beta}_q}-\cdots\cdots+(-1)\frac{\partial}{\partial\bar{z}^{\bar{\beta}_q}}f_{\alpha\alpha_2\cdots\alpha_p\bar{\beta}\bar{\beta}_1\bar{\beta}_2\bar{\beta}_3\cdots\bar{\beta}_{q-1}}\Big)$$
$$\cdot dz^{\alpha_2}\wedge\cdots\wedge dz^{\alpha_p}\wedge d\bar{z}^{\bar{\beta}_1}\wedge d\bar{z}^{\bar{\beta}_2}\wedge\cdots\wedge d\bar{z}^{\bar{\beta}_q} \qquad (5.4.30)_3$$

$\dfrac{\partial}{\partial\bar{z}^{\bar{\beta}_i}}$'s in Eq.5.4.30$_3$ can all be replaced by $\nabla_{\bar{\beta}_i}$'s. Then subtract Eq.5.4.30$_3$ from Eq.5.4.30$_2$, we obtain

$$\left(\bar{\partial}\Lambda-\Lambda\bar{\partial}\right)F=\frac{i}{(p-1)!q!}h^{\bar{\beta}\alpha}\nabla_{\bar{\beta}_0}f_{\alpha\alpha_2\cdots\alpha_p\bar{\beta}\bar{\beta}_2\bar{\beta}_3\bar{\beta}_4\cdots\bar{\beta}_q}$$
$$\cdot dz^{\alpha_2}\wedge\cdots\wedge dz^{\alpha_p}\wedge d\bar{z}^{\bar{\beta}_1}\wedge d\bar{z}^{\bar{\beta}_2}\wedge\cdots\wedge d\bar{z}^{\bar{\beta}_q} \qquad (5.4.31)$$

At the mean time, we can compute directly that

$$-i\vartheta F=i*\bar{\partial}*F$$
$$=\frac{i}{(p-1)!q!}h^{\bar{\beta}\alpha}\nabla_{\bar{\beta}_0}f_{\alpha\alpha_2\cdots\alpha_p\bar{\beta}\bar{\beta}_2\bar{\beta}_3\bar{\beta}_4\cdots\bar{\beta}_q}\cdot dz^{\alpha_2}\cdots\wedge dz^{\alpha_p}\wedge d\bar{z}^{\bar{\beta}_1}\wedge\cdots\wedge d\bar{z}^{\bar{\beta}_q} \qquad (5.4.32)$$

Compare Eq.5.4.31 to Eq.5.4.32, we have that

$$\left(\bar{\partial}\Lambda-\Lambda\bar{\partial}\right)F=-i\vartheta F,$$

and hence proved Eq.5.4.28.

According to Eq.5.4.24 and Eq.5.4.28:

$$\Box=\partial\vartheta+\vartheta\partial=i\partial\left(\bar{\partial}\Lambda-\Lambda\bar{\partial}\right)+i\left(\bar{\partial}\Lambda-\Lambda\bar{\partial}\right)\partial$$
$$=i\left(\partial\bar{\partial}\Lambda-\partial\Lambda\bar{\partial}+\bar{\partial}\Lambda\partial-\Lambda\bar{\partial}\partial\right)$$
$$\overline{\Box}=\bar{\partial}\bar{\vartheta}+\bar{\vartheta}\bar{\partial}=-i\bar{\partial}\left(\partial\Lambda-\Lambda\partial\right)-i\left(\partial\Lambda-\Lambda\partial\right)\bar{\partial}$$
$$=-i\left(\bar{\partial}\partial\Lambda-\bar{\partial}\Lambda\partial+\partial\Lambda\bar{\partial}-\Lambda\partial\bar{\partial}\right)$$
$$=i\left(\partial\bar{\partial}\Lambda+\bar{\partial}\Lambda\partial-\partial\Lambda\bar{\partial}-\Lambda\bar{\partial}\partial\right)$$

Thus $\square = \overline{\square}$.

Then according to Eq.5.4.24, Eq.5.4.14, Eq.5.4.17 and Eq.5.4.28:

$$\begin{aligned}\Delta = d\delta + \delta d &= \left(\partial + \overline{\partial}\right)\left(\vartheta + \overline{\vartheta}\right) + \left(\vartheta + \overline{\vartheta}\right)\left(\partial + \overline{\partial}\right)\\ &= \partial\vartheta + \vartheta\partial + \overline{\partial\vartheta} + \overline{\vartheta\partial} + \partial\overline{\vartheta} + \overline{\vartheta}\partial + \overline{\partial}\vartheta + \vartheta\overline{\partial}\\ &= \square + \overline{\square} + \partial\overline{\vartheta} + \overline{\vartheta}\partial + \overline{\partial}\vartheta + \vartheta\overline{\partial}\end{aligned}$$

where

$$\partial\overline{\vartheta} + \overline{\vartheta}\partial = \partial\left(-i\partial\Lambda + i\Lambda\partial\right) + \left(-i\partial\Lambda + i\Lambda\partial\right)\partial = 0$$

$$\overline{\partial}\vartheta + \vartheta\overline{\partial} = \overline{\partial}\left(i\overline{\partial}\Lambda - i\Lambda\overline{\partial}\right) + \left(i\overline{\partial}\Lambda - i\Lambda\overline{\partial}\right)\overline{\partial} = 0$$

Therefore $\Delta = \square + \overline{\square} = 2\square = 2\overline{\square}$ in Eq.5.4.25 holds.

Part II. Global Topological Properties

Chapter 6. Homotopy Equivalence and Homotopy Groups of Manifolds

§ 6.1 Homotopic Mapping

Definition 6.1.1: A topological manifold is a Hausdorff space, of which each point has a neighborhood being homeomorphic to R^n.

Now through the continuous mappings between two topological manifolds, we will investigate the homotopy relationship between distinct continuous mappings.

Suppose that there are two topological manifolds *X* and *Y*. Also suppose that there are two continuous mappings *f* and *g* where *f* (or *g*) maps the points of *X* (or *Y*) injectively to the corresponding points of *Y* (or *X*):

$$f: \quad X \to Y$$
$$g: \quad Y \to X \tag{6.1.1}$$

Here *f* and *g* are both continuously deformable, and hence we have the following definition.

Definition 6.1.2: If a mapping f_a can be deformed continuously to f_b (or f_b can be deformed continuously to f_a), then f_a and f_b are called homotopic mappings.

The continuous deformation between f_a and f_b can be expressed by the function $F(x,\lambda)$:

$$F(x,a) = f_a(x) \quad \text{when } \lambda = a,$$
$$F(x,b) = f_b(x) \quad \text{when } \lambda = b. \tag{6.1.2}$$

where *x* represents the points of the topological manifold *X*, $f_a(x)$ and $f_b(x)$ are both points of *Y*. $F(x,\lambda)$ varies continuously from $f_a(x)$ to $f_b(x)$ when λ changes its value from *a* to *b*. In this case, we say that $f_a(x)$ and $f_b(x)$ are homotopic mappings. The homotopy property of $f_a(x)$ and $f_b(x)$ can be expressed as the following:

$$f_a \cong f_b \tag{6.1.3}$$

Remark: In the definition of homotopic mappings, f_a and f_b are only required to be 1-1 continuous mappings. However, they are not necessarily invertible.

Definition 6.1.3: The topological manifolds *X* and *Y* are *homotopically equivalent* if there exists *continuous mappings f* and *g* between them:

$$f: \quad X \to Y, \qquad g: \quad Y \to X. \tag{6.1.4}$$

where $g \cdot f$ is homotopic to the identical mapping on *X*, and $f \cdot g$ is homotopic to the identical mapping on *Y*.

Remark: Being homotopic to an identical mapping does not mean that the mapping itself is an identical one.

homotopy equivalence is also said to have the identical homotopy types.

Recall Definition 1.1.4, if two manifolds *X* and *Y* are homeomorphic to each other, i.e. there is a 1-1 homeomorphic mapping between *X* and *Y*, then *X* and *Y* must be *homotopically equivalent (or say to have the identical homotopy types).* But not vice versa, that is, even if *X and Y are homotopically equivalent, they are not necessarily homeomorphic manifolds.* For example, X and Y may have different dimensions. Therefore we can see that the dimension of a manifold is topologically invariant while its homotopy type may not have this property. But here we will not discuss those cases without topologically invariance.

Definition 6.1.4: Suppose that there are two manifolds $X \supset Y$. If there exists a continuous mapping

$$r: \; X \to Y$$
$$r(q) = q, \qquad q \in Y \tag{6.1.5}$$

Then $X \to Y$ is called a contraction of X, and r a contraction mapping.

Definition 6.1.5: Suppose that there is a contraction mapping r, and the following homotopy mapping

$$\begin{array}{ccc} X \otimes [0,1] & \longrightarrow & X \\ p \otimes \lambda & \longrightarrow & F(p,\lambda), \qquad \lambda \in [0,1]. \end{array}$$

where

i. $F(p,0) = p \subset X$ (The identity mapping on X)
ii. $F(p,1) = r(p) \subset Y$ (The contraction mapping)
iii. $F(q,\lambda) = q$ $(\forall q \subset Y)$

Then this mapping is called a deformation retraction of the manifold X, where

i. X is not contracted when λ=0;
ii. X is contracted to Y when λ=1;
iii. Y can be picked up from X.

Definition 6.1.6: A manifold M which have the same homotopy type as a single point $\{0\}$ is called a contractible manifold. Contractible manifolds are the simplest manifolds which are called the topologically trivial manifolds, e.g. R^n.

Example: Take a 2-dimensional disk D^2 in E^2 space (D^2 being the section of a 3-dimensional ball in E^3 space). Take away the central point of D^2. Then $D^2 - \{0\}$ can be deformed and contracted to a 1-dimensional S^1 (a circle).

$$D^2 - \{0\} \cong S^1 \tag*{(6.1.6)$_1$}$$

Similarly, take an n-dimensional disk D^n from E^n space (D^n being the section of an $(n+1)$-dimensional ball in E^{n+1} space). Take away the central point of D^n. Then $D^n - \{0\}$ can be deformed and contracted to an $(n-1)$-dimensional sphere S^{n-1}.

$$D^n - \{0\} \cong S^{n-1} \tag*{(6.1.6)$_2$}$$

§ 6.2 Fundamental Group $\Pi_1(M, x_0)$

Take a line segment I=[0, 1] from the real line and a continuous mapping f which maps each point of I injectively and continuously to the manifold M:

$$f:\ I \to M$$
$$x \to f(x) \tag{6.2.1}$$

$f(x)$ is called a path, which represents a curve segment on M mapped from I.

Definition 6.2.1: Suppose that there are two distinct paths f_0 and f_1. If one path can be changed continuously into the other one, then these two paths are said to be homotopic:

$$f_0 \sim f_1 \tag{6.2.2}$$

If both $f_0 \sim f_1$ and $f_1 \sim f_2$ are true, then it is easy to verify that $f_0 \sim f_2$. Hence all the mappings which are homotopic to f form an *equivalent homotopy class*, denoted by $\{f\}$.

If the starting point and the ending point of f coincide at a point x_0 on M, then f is called a circuit with a base point x_0 :

$$f(0) = f(1) = x_0 \tag{6.2.3}$$

If one circuit can be changed continuously into another circuit with the base point x_0 preserved, then these two circuits are said to be homotopically equivalent. The set of all the circuits homotopically equivalent to some circuit f forms an equivalent homotopy class, also denoted by $\{f\}$. Take the following figure as an example, where D_1 and D_2 are regions in R^2 space.

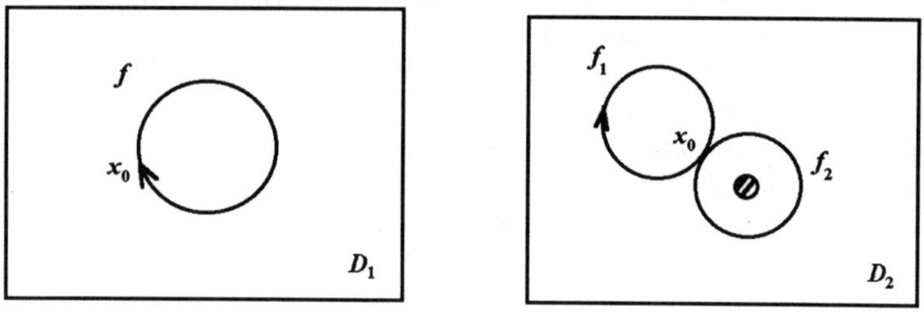

(a) (b)

Figure 6.2.1

All the circuits in D_1 with the base point x_0 (represented by f) can be contracted to the point x_0. That is, they all belong to the same homotopy equivalent class as the identity mapping (which contracts the whole circle to the point x_0) does.

The circuits in D_2 with the base point x_0 can be divided into two classes (represented by f_1 and f_2 respectively): one (represented by f_1) can be contracted to x_0 (similar to the f in part (a) of the figure); the other (represented by f_2) cannot be contracted to x_0 since there is a hole in D_2. In fact not only the f_2 in (b), but also those infinitely many circuits which circle the hole 2, 3, 4, $\cdots$ times in either directions, all cannot be contracted to the point x_0, and hence are not homotopic to f_1. Further more, the circuits with different numbers of circling times and different directions are not homotopic to each other as well.

In the language of homotopy class, all the close curves passing through the base point x_0 can be divided into several homotopy classes $\{f_i\}$ (i=1, 2, ⋯). The set of the homotopy classes $\{f_i\}$ is denoted by $\Pi_1(M, x_0)$.

Definition 6.2.2: $\Pi_1(M, x_0)$ is called the homotopy group, or the *homotopy group* with the base point x_0 on the manifold M. It is also called the *fundamental group* of manifold.

Definition 6.2.3: On a path connected M, $\Pi_1(M, x_0)$ depends only on M in case of isomorphism, and is irrelevant to the choice of the base point x_0. Therefore the notation of the fundamental group of the manifold M can be simplified as $\Pi_1(M)$.

$\Pi_1(M, x_0)$ must contain an identity element $\{e\}$, which is the equivalent homotopy class of the constant mapping. The constant mapping is defined as:

$$f(\lambda) = x, \quad \text{for all } \lambda \in [0,1].$$

All the curves in M which can be contracted continuously into one point x_0 belong to this equivalent homotopy class $\{e\}$, e.g., the f in part (a) and the f_1 in part (b) of Figure 6.2.1.

The rest of the elements of $\Pi_1(M, x_0)$ except $\{e\}$ are decided by the topological properties of M, e.g. in Figure 6.2.1, the number of holes in D_1 and D_2.

Example: In general, The following homotopic mapping R^n can be introduced into the real n-dimensional linear space $F(x,\lambda)$:

$$\vec{x} \in R^n, \quad \lambda \in [0,1], \quad \text{and the base point } \vec{x}_0 \in R^n$$

The homotopic mapping is :

$$F(\vec{x},\lambda) \to \lambda\vec{x}_0 + (1-\lambda)f(\vec{x}) \tag{6.2.4}$$

$$\lambda = 0: \quad F(\vec{x},0) = f(\vec{x})$$

$$\lambda = 1: \quad F(\vec{x},1) = \vec{x}_0 \qquad (6.2.4)_1$$

Since $f(\vec{x})$ is an arbitrary function of $\vec{x}$ and R^n has no hole, thus take any circuit in R^n with the base point $\vec{x}_0$, it will contract to x_0 after a deformation retraction. Therefore the set $\{f_i\}$ in R^n can only have one unit element or identity element $\{e\}$.

Now let us investigate the multiplication of the elements of homotopy groups.

Definition 6.2.4: The product of two elements $\{f_i\}$ and $\{f_j\}$ of the fundamental group $\Pi_1(M, x_0)$ is defined as the mapping which starts from x_0, circles along the path $f_i(\lambda)$ once and returns to x_0, then circles along the path $f_j(\lambda)$ once and returns to x_0. The homotopy class of the close curve formed by this product is denoted as:

$$\{f_i\}*\{f_j\} = \{f_i \cdot f_j\} \tag{6.2.5}$$

That is

$$\{f_i \cdot f_j\}(\lambda) = \begin{cases} f_i(2\lambda) & 0 \le \lambda \le \frac{1}{2} \\ f_j(2\lambda - 1) & \frac{1}{2} \le \lambda \le 1 \end{cases}$$

$$0 \le \lambda \le 1 \qquad (6.2.5)_1$$

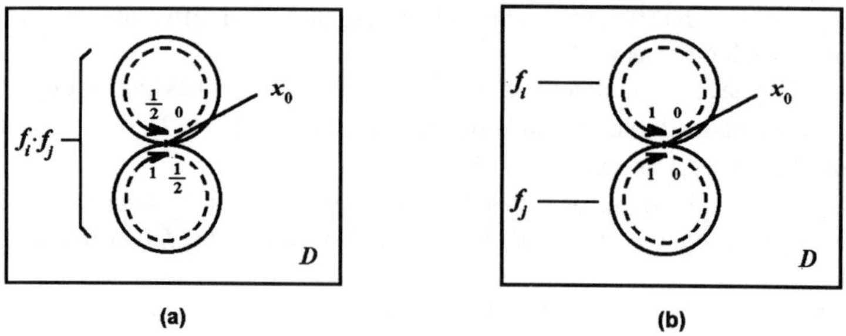

Figure 6.2.2

In part (a), $f_i \cdot f_j$ is the circuit of a continuous mapping with the base point x_0, the circuit of multiplication with $0 \le \lambda \le 1$. In part (b), f_i $(0 \le \lambda \le 1)$ and f_j $(0 \le \lambda \le 1)$ are two circuits which are independent to each other. They are not the circuit of multiplication.

We may define the circuit of f^{-1} as well. The circuit of $f^{-1}(\lambda)$ has the opposite direction of the circuit of $f(\lambda)$.

$f^{-1}(\lambda) = f(1-\lambda)$: $\qquad\qquad f(\lambda)$:

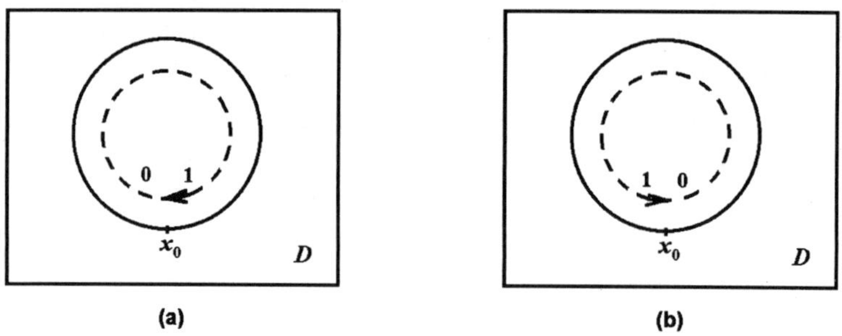

Figure 6.2.3

Definition 6.2.5: The homotopy class $\{f\}^{-1} \equiv \{f^{-1}\}$. (6.2.6)

According to the definition of multiplication, the circuit of $(f \cdot f^{-1})(\lambda)$ happens to start from x_0, come back to x_0 after completing a full circle, then travel backward along the same path and return to x_0. Hence it is homotopic to the identity mapping, that is,

$$\{f\} * \{f^{-1}\} = \{e\} \tag{6.2.7}$$

It can be proven as well that

$$\{f_i\} * (\{f_j\} * \{f_k\}) = (\{f_i\} * \{f_j\}) * \{f_k\} \tag{6.2.8}$$

i.e., $\{f_i\}$ satisfies the law of association.

Summing up we see that the homotopy class $\{f_i\}$ has the multiplication, the inverse element $\{f_i^{-1}\}$, the identity element $\{e\}$, and the law of association. Therefore the set of the homotopy class $\{f_i\}$ does form a group, that is, the *homotopy group* (see Definition 6.2.2) $\Pi_1(M, x_0)$ or $\Pi_1(M)$.

Example 1: If any two points of the manifold M can be connected by a curve on M, then this manifold M is called a *path connected manifold.*

If the fundamental group $\Pi_1(M, x_0)$ of M is trivial (i.e., having the identity element $\{e\}$ as the only group element), then M is called a *simply connected manifold.*

The fundamental groups $\Pi_1(M, x_0)$ of E^n (the n-dimensional Euclidean space) and S^n (the n-dimensional sphere with $n \geq 2$) are both trivial (the identity element $\{e\}$ is the only element), therefore E^n and S^n (the n-dimensional sphere with $n \geq 2$) are both simply connected manifolds of which any circuit can be contracted to a single point. But E^n and S^n are not topologically homeomorphic.

Example 2: Any two points on the torus T^2 can also be connected by a curve on the torus. Thus the torus T^2 is a path connected manifold. But not all of the circuits on T^2 can be contracted to a single point, therefore the fundamental group $\Pi_1(T^2, x_0)$ must have not only the identity element $\{e\}$, but also some other group elements (the homotopy class) of which the paths cannot be contracted to a single point. Hence T^2 is not a simply connected manifold.

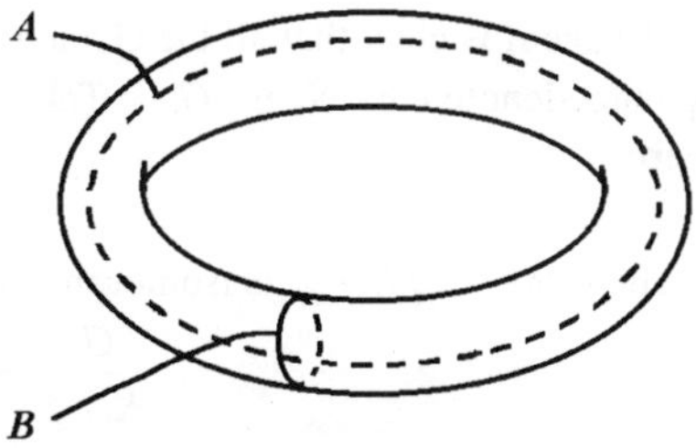

Figure 6.2.4

Example 3: The case of S^1 is similar to that of T^2 where the identity element $\{e\}$ is not the only element of the fundamental group $\Pi_1(S^1, x_0)$.

§ 6.3 Homomorphic Sequence and the Structure of Homotopy Groups

If there exists a fixed corresponding relation between the structures of G_1 group and G_2 group, then the two groups are said to be homomorphic to each other:

$$f: \ G_1 \to G_2 \tag{6.3.1}$$

If there is a group of elements in G_1 which can be mapped to the unit element e of G_2 by f, then this group of elements is called the kernel of f, denoted by *Ker f*:

$$Ker\ f = \{g \in G_1 \mid f(g) = e \in G_2\} \tag{6.3.2}$$

The image of a homomorphic mapping is denoted by *Im f* which is defined as the following:

$$Im\, f = \{g' \in G_2 \mid g' = f(g),\ g \in G_1\} \tag{6.3.3}$$

Note that in general, *Im f* does not contain the entire G_2.

Discussion 1: Take any two distinct elements g_0 and g'_0 from the subset *Ker f* of G_1. Then take another element *g* from G_1 (where *g* is not necessarily in *Ker f*). After a mapping of *f*, we must have the following conclusions:

$$f:$$
$$G_1 \longrightarrow Im\,f \in G_2$$
$$g_0 \longrightarrow e \in Im\,f$$
$$g'_0 \longrightarrow e \in Im\,f$$
$$g \longrightarrow \underline{g} \in Im\,f$$
$$g^{-1} \longrightarrow \underline{g}^{-1} \in Im\,f$$
$$g g \circ g^{-1} \longrightarrow \underline{g} e \underline{g}^{-1} = e \in Im\,f$$
$$\therefore \quad g g \circ g^{-1} \in Ker\,f.$$

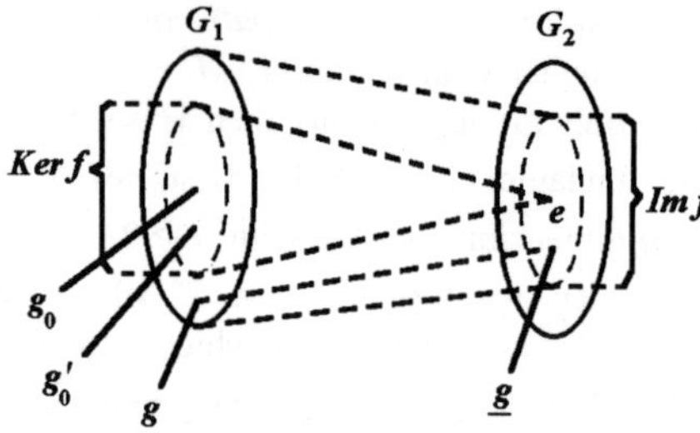

Figure 6.3.1

Hence it is clear that all the elements g_0, g'_0, g''_0, ⋯ of *Ker f* form an invariant subgroup, denoted by *S*, in G_1. Take any element *g* in G_1, *S* has the following property:

$$g S g^{-1} = S$$

Therefore $G_1/Ker\,f$ is isomorphic to $Im\,f/e$:

$$\frac{G_1}{Ker\,f} \sim \frac{Im\,f}{e} \sim Im\,f \tag{6.3.4}$$

Discussion 2: Suppose that G_k (k=1, 2, ⋯) is a sequence of groups, and there exists a homomorphic mapping f_k between the neighboring G_k and G_{k+1}, that is,

$$f_k:\ G_k \longrightarrow G_{k+1}.$$

The connection of the elements of this sequence can be illustrated as the following:

$$G_1 \xrightarrow{f_1} G_2 \xrightarrow{f_2} G_3 \xrightarrow{f_3} \cdots\cdots$$

In general,

$$Im\,f_1 \le G_2,\ Im\,f_2 \le G_3, \cdots,\ Im\,f_k \le G_{k+1}, \cdots\cdots$$

In the special case of $Im\,f_k = Ker\,f_{k+1}$, this *homomorphic sequence* G_k is called the *exact sequence*.

Here we prove two theorems of the exact sequence.

Theorem 6.3.1: Suppose there is an exact sequence of four groups *e*, G_1, G_2, *e* which is illustrated in Figure 6.3.2, where *O* is a trivial group with only the identity element *e*. Then the correspondence between the elements of G_1 and the elements of G_2 must be 1-1 (i.e., isomorphic).

Proof: Since $Im\,f_0 = Ker\,f_1 = e$, therefore the only element of G_1 which is mapped to the element *e* of G_2 is *e*, that is, $f_1 \cdot e = e$. Thus we know that f_1 itself is equivalent to $f_1 \cong e$. Therefore the correspondence between G_1 and G_2 is 1-1 (isomorphic). □

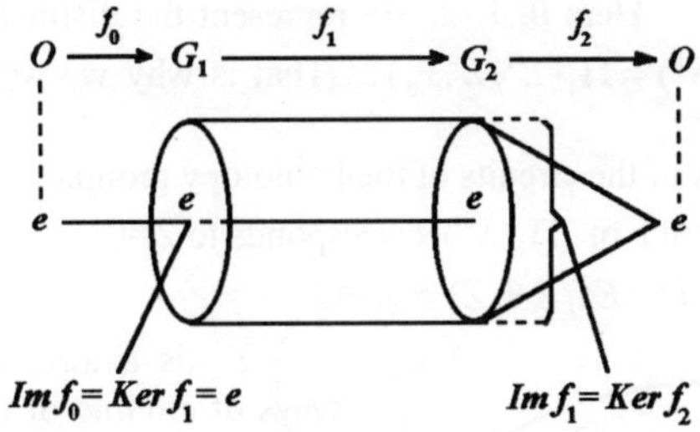

Figure 6.3.2

Theorem 6.3.2: Suppose there is an exact sequence of the following five groups:

$$O \xrightarrow{f_0} G_1 \xrightarrow{f_1} G_2 \xrightarrow{f_2} G_3 \xrightarrow{f_3} O$$

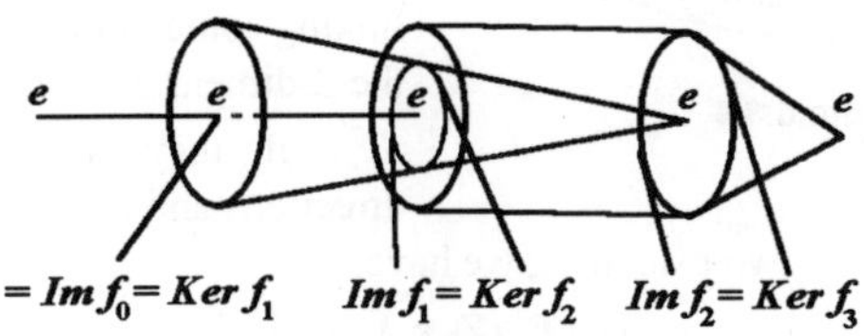

Figure 6.3.3

Then G_3 is isomorphic to G_2/G_1. That is, G_1 must be an invariant subgroup of G_2, and the quotient group G_2/G_1 is isomorphic to G_3.

Proof: According to Theorem 6.3.1, $e \xrightarrow{f_0} G_1 \xrightarrow{f_1} G_2$. In the case of $Ker\, f_1 = Im\, f_0 = e$, the homomorphic mapping $G_1 \xrightarrow{f_1} G_2$ (the part where $Im\, f_1 = Ker\, f_2$) is 1-1, i.e., G_1 is a subgroup of G_2: $G_1 \subset G_2$.

Then since $G_1 \sim Im\, f_1 = Ker\, f_2$, according to the analysis in Discussion 1, all the elements in $Ker\, f_2$ form an invariant subgroup of G_2. That is, *the set of* G_1 *is an invariant subgroup of* G_2.

At the mean time, $Im\, f_2 = Ker\, f_3 = G_3$, and thus $G_2 \xrightarrow{f_2} G_3$ is surjective.

Since that G_1 is an invariant subgroup of G_2, which corresponds to the identity element of G_3, it follows immediately from Figure 6.3.3 that

$$G_2/G_1 \sim \text{(isomorphism) } G_3/e \sim \text{(isomorphism) } G_3. \quad \square$$

The exact sequence can be used to demonstrate some properties and structures of homotopy groups. Here we first give two examples of homotopy groups.

Example 1: $\Pi_1(S^1) = \Pi_1(E^1/Z, x_0) \cong Z$ (6.3.5)

$\Pi_1(S^1)$ has only one circuit S^1. Z represents integers, means circling integer times (since S^1 is a circle). E^1 is the 1-dimensional Euclidean space. The paths which circle Z times will become part of E^1 if they are stretched straight. Thus E^1

contains Z=0, 1, 2, ⋯. Here 0, 1, 2, ⋯ represent the distinct group elements of the homotopy group $\Pi_1(S^1)=\Pi_1(E^1/Z,x_0)$. That is why we write $\cong Z$ to the right of Eq.6.3.5.

x_0 is the base point of the circuits of the homotopy group.

The identity element e of $\Pi_1(S^1)$ corresponds to Z=0.

Example 2: $\Pi_1(T^2)=\Pi_1(E^2/Z\oplus Z)\cong Z\oplus Z$ (6.3.6)

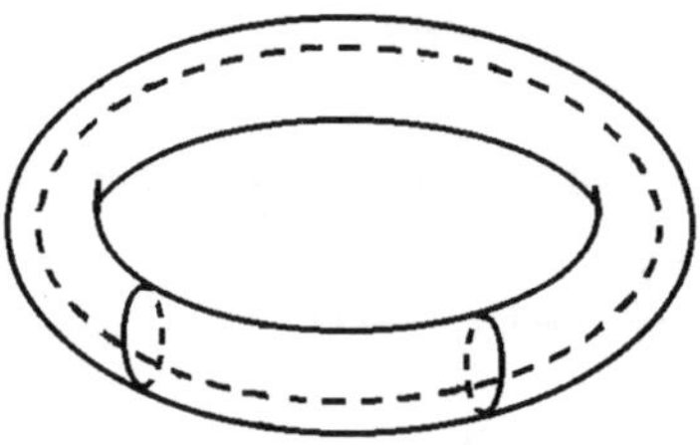

Figure 6.3.4

T^2 is a torus on which the different ways of circling of two distinct circuits can be represented by $Z\oplus Z$. The number of times of these two ways of circling are both integer Z. E^2 is the 2- dimensional Euclidean space. The two circling paths of $Z\oplus Z$ will become a path formed by linking some horizontal and vertical lines on the 2-dimensional Euclidean space (a plane) E^2, if they are flattened in different directions and linked in the order of circling.

Summing up these two examples, we have:

$$\left.\begin{array}{r}\Pi_1(S^1)=\Pi_1(E^1/Z,x_0)\\ \Pi_1(T^2)=\Pi_1(T^2/Z\oplus Z,x_0)\end{array}\right\}\ \Pi_1(M/Z,x_0) \qquad (6.3.7)$$

Taking $\Pi_1(S^1)=\Pi_1(E^1/Z,x_0)$:

When Z=0, $\Pi_1(E^1/0,x_0)$ ~ the identity element e, corresponding to the point x_0.

When Z=1, $\Pi_1(E^1/1,x_0)$ ~ circling S^1 once.

⋯ ⋯

When Z=k, $\Pi_1(E^1/k,x_0)$ ~ circling S^1 k times.

Taking $\Pi_1(T^2)=\Pi_1(T^2/Z\oplus Z,x_0)$:

When $Z\oplus Z=0\oplus 0$, $\Pi_1(E^2/0\oplus 0,x_0)$ ~ the identity element e, corresponding to the point x_0.

When $Z\oplus Z=m\oplus n$, $\Pi_1(E^2/m\oplus n,x_0)$ ~ a path of which the first circuit circles m times and the second circuit circles n times.

Hence we can define a group G_{x_0} acting on the integer groups Z, $Z\oplus Z$, $Z\oplus Z\oplus Z$, ⋯. The features of G_{x_0}'s action are similar to those of an isotropic subgroup.

$$\left.\begin{array}{l} G_{x_0} Z\ (Z \Rightarrow 0) \Rightarrow \text{the identity element } e, \text{ corresponding to } x_0 \\ G_{x_0} Z \oplus Z\ (Z \oplus Z \Rightarrow 0 \oplus 0) \Rightarrow \text{the identity element } e, \text{ corresponding to } x_0 \\ \cdots\cdots \\ G_{x_0} Z\ (Z \neq 0) = 0 \\ G_{x_0} Z \oplus Z\ (Z \oplus Z \neq 0 \oplus 0) = 0 \\ \cdots\cdots \end{array}\right\} \tag{6.3.8}$$

Since G_{x_0} has the features shown in Eq.6.3.8, the whole G_{x_0} group corresponds to the identity element *e* (corresponding to the point x_0) of $\Pi_1(M/G, x_0)$, and thus gives the following exact sequence:

$$O \xrightarrow{f_0} G_{x_0} \xrightarrow{f'} \Pi_1(M/G, x_0)$$

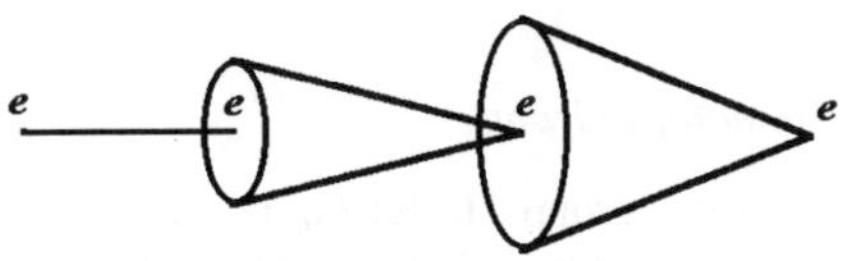

Figure 6.3.5

At the mean time, since *G* (which includes the group elements Z, $Z \oplus Z$, $\cdots$) is surjective to $\Pi_1(M/G, x_0)$ (which also includes Z, $Z \oplus Z$, $\cdots$), and since $Z = 0$, $Z \oplus Z = 0 \oplus 0$, $\cdots$ in *G* are all mapped to *e* (which corresponds to x_0) of $\Pi_1(M/G, x_0)$, therefore we have the following sequence:

$$G \xrightarrow{f_2} \Pi_1(M/G, x_0)$$

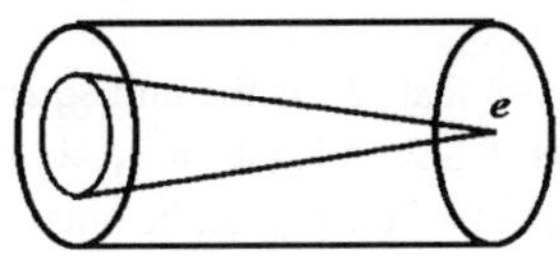

Figure 6.3.6

Put Figure 6.3.5 on top of Figure 6.3.6, we obtain Figure 6.3.7 which is of the similar type to that of Figure 6.3.3.

Similar to Figure 6.3.3, this is also an exact sequence which indicates that

i. G_{x_0} and $Im f_1$ are in injective correspondence. The group elements are $Z = 0$, $Z \oplus Z = 0 \oplus 0$, $Z \oplus Z \oplus Z = 0 \oplus 0 \oplus 0$, $\cdots$;

ii. G_{x_0} is homomorphic to the identity element $\{e\}$ of $\Pi_1(M/G,x_0)$, and $Im f_1 = Ker f_2$;

iii. the fundamental group $\Pi_1(M/G,x_0)$ and the G group (Z, $Z\oplus Z$, $Z\oplus Z\oplus Z$, ⋯) are in surjective correspondence;

iv. $\Pi_1(M/G,x_0) \sim G/G_{x_0}$ (according to Theorem 6.3.2).

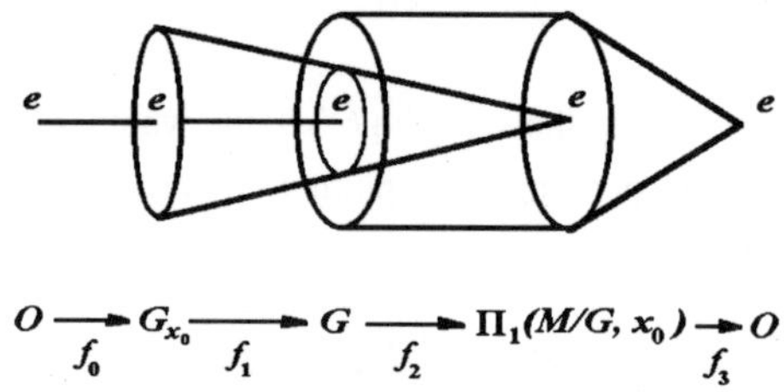

Figure 6.3.7

§ 6.4 Higher Order Homotopy Group

The first order homotopy group $\Pi_1(M)$ (x_0 is omitted here since its position is irrelevant to the properties of $\Pi_1(M, x_0)$) is proper for analyzing the homotopy properties of 2-dimensional surfaces. If the fundamental group is not trivial, then there must be a hole (or holes) on the 2-dimensional surface. But for higher order manifolds (n>2), the concept of higher order homotopy group must be introduced, otherwise we would not be able to distinguish the n-dimensional Euclidean space E^n from the n-dimensional sphere $S^n \cong D^{n+1} - \{0\}$.

The homotopy group of order k (k>1) is denoted by Π_k (M, x_0) (x_0 can be omitted here for the same reason as that of the first order homotopy group), which is formed by the set of equivalent homotopy classes of the k-dimensional circuits (with the base point x_0).

<u>2-dimensional circuit</u> Suppose that I^2 is the unit square in the 2-dimensional real space which can be expressed as $0 \le \lambda_i \le 1$ ($\lambda_i = \lambda_1, \lambda_2$). A 2-dimensional circuit (which is topologically equivalent to the sphere) can be obtained by the following steps:

Start from $0 \le \lambda_i \le 1$ (i=1, 2):

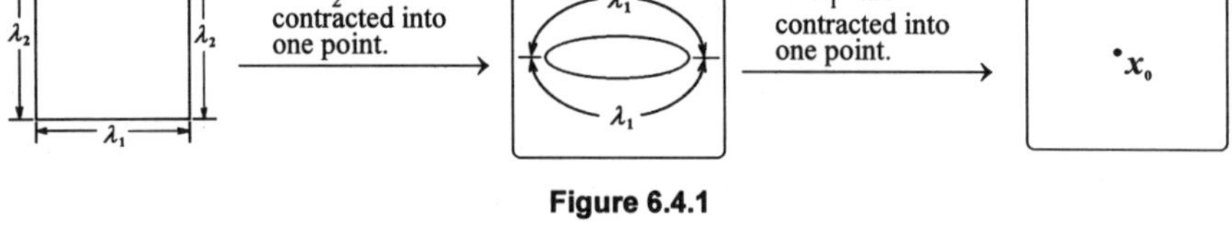

Figure 6.4.1

Obviously the 2-dimensional circuit is a closed bag which has the same topological properties as S^2.

n-dimensional circuit is a generalization of 2-dimensional circuit to the *n*-dimensional space: Suppose that I^n is the unit square in the *n*-dimensional real space, written as $0 \le \lambda_i \le 1$ (*i*=1, 2, ⋯, *n*). Contract the boundaries of λ_1, λ_2, ⋯, λ_n one by one into a single point, we may obtain an *n*-dimensional circuit — an *n*-dimensional closed bag which has the same topological properties as S^n. There is a point x_0 on this *n*-dimensional circuit as well, i.e.

$$f(\lambda_1,\lambda_2,\cdots,\lambda_n) = x_0 \qquad \text{when } \lambda_i = 0 \text{ and } \lambda_i = 1 \ (i=1, 2, \cdots, n)$$

Multiplication of n-dimensional circuits If $f(\lambda_1,\cdots,\lambda_n)$ and $g(\lambda_1,\cdots,\lambda_n)$ are two mappings of *n*-dimensional circuits, then $h = f \cdot g$ is defined as:

$$h(\lambda_1,\lambda_2,\cdots,\lambda_n) \equiv \begin{cases} f(2\lambda_1,\lambda_2,\cdots,\lambda_n) & 0 \le \lambda_1 \le \frac{1}{2} \quad (0 \le 2\lambda_1 \le 1) \\ g(2\lambda_1 - 1,\lambda_2,\cdots,\lambda_n) & \frac{1}{2} \le \lambda_1 \le 1 \quad (0 \le 2\lambda_1 - 1 \le 1) \end{cases} \tag{6.4.1}$$

Here we give an example of *n*=2:

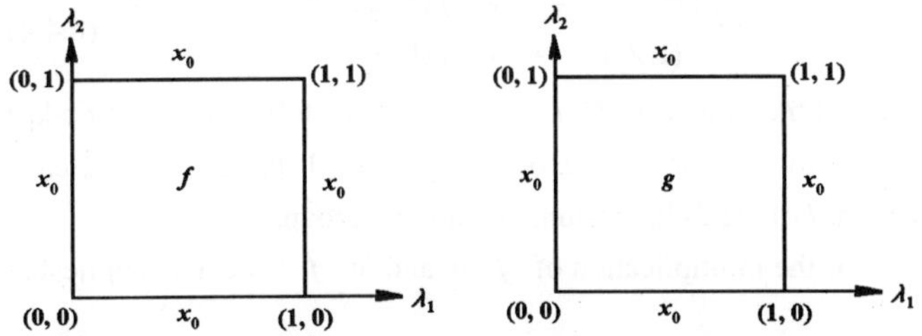

$$(i, j) = (\lambda_1, \lambda_2)$$
$$0 \le \lambda_1 \le 1$$
$$0 \le \lambda_2 \le 1$$

Figure 6.4.2 (a)

The sides marked with x_0 in the figures will all be contracted into a single point x_0. (i, j)'s are the coordinates of the vertices of I^2 (the 2-dimensional square).

According to the rule of multiplication in Eq.6.4.1, $h = f \cdot g$ can be illustrated by the following figure:

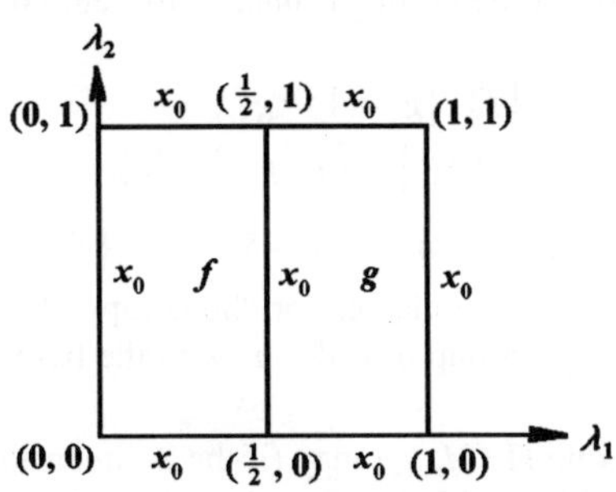

Figure 6.4.2 (b)

$(i, j) = (\lambda_1, \lambda_2)$ in the figure are the coordinates of the vertices.

$$\begin{array}{cc} f: & g: \\ 0 \le \lambda_1 \le \frac{1}{2} & \frac{1}{2} \le \lambda_1 \le 1 \end{array}$$

$$h(\lambda_1,\lambda_2) \equiv f \cdot g(\lambda_1,\lambda_2) \equiv \begin{cases} f(2\lambda_1,\lambda_2) & (0 \le \lambda_1 \le \frac{1}{2}) \\ g(2\lambda_1 - 1,\lambda_2) & (\frac{1}{2} \le \lambda_1 \le 1) \end{cases} \tag{6.4.2}$$

According to the above rule of multiplication, we can define f^{-1} as

$$f^{-1}(\lambda_1,\lambda_2) = f(1-\lambda_1,\lambda_2) \qquad (0 \le \lambda_1 \le 1) \tag{6.4.3}$$

Thus (similar to Eq.6.4.2):

$$f \cdot f^{-1}(\lambda_1,\lambda_2) \equiv \begin{cases} f(2\lambda_1,\lambda_2) & (0 \le \lambda_1 \le \frac{1}{2}) \\ f^{-1}(2\lambda_1 - 1,\lambda_2) & (\frac{1}{2} \le \lambda_1 \le 1) \end{cases}$$

(according to the definition of f^{-1} in Eq.6.4.3)

$$\equiv \begin{cases} f(2\lambda_1,\lambda_2) & (0 \le \lambda_1 \le \frac{1}{2}) \\ f^{-1}(2 - 2\lambda_1,\lambda_2) & (\frac{1}{2} \le \lambda_1 \le 1) \end{cases}$$

$$= \begin{cases} f(0,\lambda_2) & \longrightarrow & f(1,\lambda_2) \\ f(1,\lambda_2) & \longrightarrow & f(0,\lambda_2) \end{cases} \tag{6.4.4}$$

Therefore we can see that the action of $f \cdot f^{-1}(\lambda_1,\lambda_2)$ is the following: first f takes effect to move λ_1 from 0 to 1; then f^{-1} moves λ_1 from 1 back to 0. Hence $f \cdot f^{-1} =$ the identity element I of the 2-dimensional homotopy group.

The above definitions of the multiplication of $f \cdot g$ and of f^{-1} are also applicable to the case of $n>2$.

Two n-dimensional circuits are said to be homotopically equivalent if one can be deformed continuously into the other (with x_0 preserved). Same as in the case of 1-dimensional circuits, the equivalent homotopy class of the n-dimensional circuit f with $n>1$ can be represented by $\{f\}$ as well.

In the set of the equivalent homotopy class of n-dimensional circuits with the base point x_0, the multiplication of the n-dimensional equivalent homotopy classes can also be defined:

$$\{f\} \cdot \{g\} = \{f \cdot g\} \tag{6.4.5$_1$}$$

$$\{f\} \cdot \{f^{-1}\} = \{f \cdot f^{-1}\} = \{e\} \tag{6.4.5$_2$}$$

Since we have the multiplication $(6.4.5)_1$ and $\{f\}^{-1} = \{f^{-1}\}$, therefore the n-dimensional equivalent homotopy classes (or "homotopy classes" for short) also form a group, namely the homotopy group of order n (with the base point x_0) which is denoted by $\Pi_n(M,x_0)$.

The fundamental group $\Pi_1(M, x_0)$ may (if the 2-dimensional M has no hole) or may not (if the 2-dimensional M has 2-dimensional holes) be an Abelian group.

The higher order homotopy group $\Pi_k(M,x_0)$ with $k \ge 2$ must be an Abelian group, since the boundaries of a k-dimensional circuit are connected when $k \geq 2$, and $\{f\}\cdot\{g\}$ can be permuted to $\{g\}\cdot\{f\}$ by a continuous deformation without the blocking of any $(k+1)$-dimensional hole. (The detailed proof is omitted.)

§ 6.5 Homotopy Groups of the n-Dimensional Ball S^n

Here we sketch some interesting properties of the homotopy groups $\Pi_k(S^n)$:

1. $\Pi_n(S^n) = Z$ (6.5.1)

   An n-dimensional circuit can cover S^n for any integer number of times. Thus each homotopy class corresponds to an integer. The homotopy group $\Pi_n(S^n)$ should be an integer group Z.

2. $\Pi_k(S^n) = 0 \quad (k<n)$ (6.5.2)

   S^n has only one $(n+1)$-dimensional hole of which the boundary is the n-dimensional S^n. If $k<n$, then it is impossible for a k-dimensional circuit to cover an $(n+1)$-dimensional hole (that is, it is impossible to cover S^n). Thus the k-dimensional circuit on S^n can be continuously contracted into one point. Therefore Eq.6.5.2 holds.

   Example: $SU(2) \sim S^3$. According to the above properties 1 and 2, $\Pi_3(SU(2)) = Z$, $\Pi_2(SU(2)) = 0$, $\Pi_1(SU(2)) = 0$.

3. $\Pi_k(S^1) = 0 \quad (k>1)$ (6.5.3)

   What S^1 can cover is a 2-dimensional hole. What S^k $(k>1)$ can cover is a $(k+1)$-dimensional hole. S^k is homotopic to I^k (see Figure 6.4.1 and Figure 6.4.2, an example of $k=2$).

   But since I^k $(k \neq 1)$ is not homotopic to S^1, therefore I^k can not cover a 2-dimensional hole (of which the boundary is S^1).

   If we must map I^k to S^1 which is not homotopic to I^k, then the only possibility would be to map the whole I^k to some point x_0 on S^1, and at the mean time, to contract the whole S^1 to this point x_0. Then S^1 no longer exists, hence $\Pi_k(S^1) = 0$ (0 represents a trivial group with only the identity element).

Chapter 7. Homology and de Rham Cohomology

§ 7.1 *Integral Homology Group* $H_k(M) = Z_k(M) / B_k(M)$

Suppose that there is an *n*-dimensional Euclidean space R^n, of which the coordinates of each point can be written as $(x_1, x_2, \cdots, x_n)$. Take a mapping:

$$(x_1, x_2, \cdots, x_n) \mapsto (x_1, x_2, \cdots, x_n, 0) \tag{7.1.1}$$

We can see that R^n can be treated as a subspace of R^{n+1}.

Here we give some simple examples.

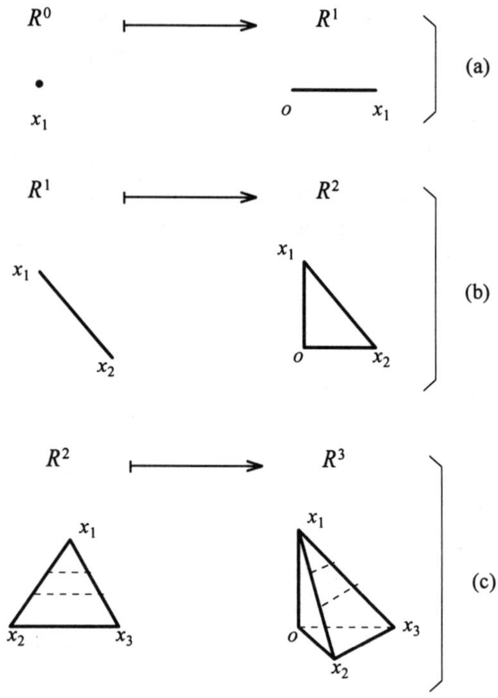

Figure 7.1.1

Generalize to R^k, we have *k* coordinates $(x_1, x_2, x_3, \cdots, x_k)$. Take *k* vectors:

$$\vec{x}_1 = (1,0,0,\cdots,0), \ \vec{x}_2 = (0,1,0,\cdots,0), \ \cdots, \ \vec{x}_k = (0,0,0,\cdots,1) \tag{7.1.2}$$

According to Figure 7.1.1 (a), (b) and (c), when *k*=0, 1, 2, it can be proven that the result of the manifolds on R^k (*k*=0, 1, 2) mapped to R^{k+1} by Eq.7.1.1 can be expressed as (take $\vec{x}_0 = (0,0,0,\cdots,0)$):

$$R^k \quad \mapsto \quad R^{k+1}$$

$$\{\vec{x}_1, \vec{x}_2, \cdots, \vec{x}_k\} \quad \mapsto \quad \left\{ \sum_{j=0}^{k} t_j \vec{x}_j \in R^{k+1} \;\middle|\; \sum_{j=0}^{k} t_j = 1,\ t_j \geq 0 \right\}$$

Compare to Figure 7.1.1 (c), we can see that $t_0 = 1$ represents the origin of R^{k+1}, $t_0 = 0$ represents the plane of (x_1, x_2, x_3), and $0 < t_0 < 1$ represents the plane which is parallel to the (x_1, x_2, x_3) plane but is closer to the origin.

Definition 7.1.1: The following Δ_k is called the canonical *k*-simplex (in *k*-dimensional spaces):

$$\Delta_k = \left\{ \sum_{j=0}^{k} t_j \vec{x}_j \in R^{k+1} \;\middle|\; \sum_{j=0}^{k} t_j = 1,\ t_j \geq 0 \right\} \tag{7.1.3}$$

Here *k* of R^{k+1} is no longer restricted to 0, 1, and 2, but can take 0 and all positive integers. The *k* points in Eq.7.1.2 together with $\vec{x}_0 = (0,0,0,\cdots,0)$ are the (*k*+1) vertices of the canonical *k*-simplex Δ_k.

Δ_k can be represented by its vertices $\{v_i\}$ as well:

$$\Delta_k = [v_0, v_1, \cdots, v_k] = \text{sign}(P)[v_{P_0}, v_{P_1}, \cdots, v_{P_k}] \tag{7.1.4}$$

where *P* in sign(*P*) represents the order of $v_{P_0}, v_{P_1}, \cdots, v_{P_k}$. $P \in P(k)$ where $P(k)$ is the permutation group of order *k*:

$$\text{sign}(P) = \begin{cases} +1 & \text{if } P \text{ is an even permutation} \\ -1 & \text{if } P \text{ is an odd permutation} \end{cases}$$

For example, $\text{sign}(P) = (-1)^{\frac{1}{2}k(k+1)}$ for $v_{P_0}, v_{P_1}, \cdots, v_{P_k} \to v_{P_k}, v_{P_{k-1}}, \cdots, v_{P_1}, v_{P_0}$.

Suppose that there is a manifold *M* of which the dimension is greater than *k*. A *k*-dimensional oriented submanifold of *M* can be obtained by a homeomorphism from Δ_k to *M*. The image σ_k of the homeomorphism can be written as:

$$\Phi:\ \Delta_k \longrightarrow \sigma_k \subset M \tag{7.1.5}$$

σ_k is called the *continuous k-simplex* of the manifold *M*, which can also be represented by its *k*+1 vertices $\{v_i\}$ as the following:

$$\sigma_k = [v_0, v_1, \cdots, v_k] \subset M \tag{7.1.6}$$

Continuous *k*-chain: The continuous *k*-chain of the manifold *M* is the linear combination of the continuous *k*-simplexes of *M* with integral coefficients:

$$c_k = \sum_i a_i \sigma_k^i \qquad a_i \in Z \tag{7.1.7}$$

The set of the continuous *k*-chains of the manifold *M* is denoted by

$$C_k(M) = \{c_k\} \tag{7.1.8}$$

There are two directions of each *k*-simplex, denoted as $\pm\sigma_k$

$$\sigma_k + (-\sigma_k) = 0$$
$$a\sigma_k + b\sigma_k = (a+b)\sigma_k$$
$$a(-\sigma_k) = -a\sigma_k \qquad a,b \in Z \tag{7.1.9}$$

The set $\{c_k\}$ combined with these additive operations forms the Abelian group.

Boundary operator ∂: ∂ maps each *k*-simplex to the (*k* − 1)-simplex of its boundary, e.g. according to (7.1.6):

$$\partial\sigma_k = \partial[v_0, v_1, \cdots, v_k] = \sum_{i=0}^{k} (-1)^i [v_0, \cdots, \hat{v}_i, \cdots, v_n] \tag{7.1.10}$$
$$\partial\sigma_k \in C_{k-1}(M) \qquad \text{(see Eq.7.1.8)}$$

where the symbol $\wedge$ in $\hat{v}_i$ means the absence of the vertex v_i. For example:

$\partial\sigma_0 = 0$ (The 0-chain has only the v_0),

$\partial\sigma_1 = v_1 - v_0$

$\partial\sigma_2 = [v_1, v_2] - [v_0, v_2] + [v_0, v_1]$

……

The actions of ∂ to any *k*-chain will preserve the additive operation of the chain group:

$$\partial C_k = \partial(\sum_i a_i \sigma_k^i) = \sum_i a_i (\partial\sigma_k^i) \in C_{k-1}(M) \tag{7.1.11}$$

From Eq.7.1.11 we can see that ∂ is the homomorphic operator between the chain groups $C_k(M)$ and $C_{k-1}(M)$.

$$\partial\colon \; C_k(M) \longrightarrow C_{k-1}(M) \tag{7.1.12}$$

Two subgroups of $C_k(M)$:

1. The kernel of the boundary homomorphism is a subgroup which is defined as
$$Z_k(M) = \{\, c_k;\; \partial c_k = 0 \,\} \tag{7.1.13}$$
c_k satisfying $\partial c_k = 0$ is called a closed *k*-chain. Z_k is a closed *k*-chain group.
2. The image of the boundary homomorphism is also a subgroup defined as
$$B_k(M) = \{\, b_k = \partial c_{k+1} \mid c_{k+1} \in C_{k+1}(M) \,\} \tag{7.1.14}$$
$b_k = \partial c_{k+1}$ is called the boundary chain and B_k is a boundary *k*-chain group.

Now we prove $\partial \cdot \partial = 0$:

Take an arbitrary *k*-simplex $\sigma_k = [v_0, v_1, v_2, \cdots, v_k]$, we have

$$\partial\cdot\partial\sigma_k = \partial\left(\sum_{i=0}^{k}(-1)^i[v_0,\cdots,\hat{v}_i,\cdots,v_k]\right) = \sum_{i=0}^{k}(-1)^i\partial[v_0,\cdots,\hat{v}_i,\cdots,v_k]$$
$$= \sum_{j<i}(-1)^{i+j}[v_0,\cdots,\hat{v}_j,\cdots,\hat{v}_i,\cdots,v_k]$$
$$+ \sum_{j>i}(-1)^{i+j-1}[v_0,\cdots,\hat{v}_i,\cdots,\hat{v}_j,\cdots,v_k] \overset{i,j\ \text{symmetric}}{=} 0$$

Therefore a boundary chain must be closed, and hence B_k must be a subgroup of Z_k:

$$B_k \subset Z_k . \tag{7.1.15}$$

And B_k and Z_k are both Abelian subgroups.

Invariant subgroup: Suppose that b_l is a group element of $B_k(M)$, c_k is a group element of $Z_k(M)$, and each $b_l \in B_k(M)$ satisfies

$$c_k \cdot b_l \cdot c_k^{-1} \in B_k(M), \tag{7.1.16}$$

Then $B_k(M)$ is called an invariant subgroup of $Z_k(M)$.

In case that both $B_k(M)$ and $Z_k(M)$ are Abelian groups, the times sign $(\cdot)$ should be replaced by the plus sign (+), and $c_k^{-1} = -c_k$. Thus if $b_l \in B_k(M)$, then we must have $c_k + b_l - c_k = b_l \in B_k(M)$ which satisfies Eq.7.1.16. This indicates that the Abelian group $B_k(M)$ is an invariant subgroup of the Abelian group $Z_k(M)$.

Quotient group: If $B_k(M)$ is an invariant subgroup of $Z_k(M)$, and if $c_i \in Z_k(M)$, $c_j \in Z_k(M)$, and

$$\left(c_i \cdot B_k(M)\right) \cdot \left(c_j \cdot B_k(M)\right) = (c_i \cdot c_j) \cdot Z_k(M) \tag{7.1.17}$$

Then $Z_k(M)/B_k(M)$ is a group as well, called the quotient group. In case that $B_k(M)$ and $C_k(M)$ are both Abelian groups, the times sign should be replaced by the plus sign, and Eq.7.1.17 should be modified as

$$\left(c_i + B_k(M)\right) + \left(c_j + B_k(M)\right) = (c_i + c_j) + Z_k(M) \tag{7.1.17$_1$}$$

Since that $Z_k(M)$ defined in Eq.7.1.13 and $B_k(M)$ defined in Eq.7.1.14 satisfy Eq.7.1.17$_1$, therefore we can use the Abelian groups $Z_k(M)$ and $B_k(M)$ to define a quotient group $H_k(M)$:

$$H_k(M) = Z_k(M)/B_k(M) \tag{7.1.18}$$

$H_k(M)$ is called the *k*-homology group of the manifold *M*.

The homology group H_k is the set of the equivalent classes (differed by a boundary chain) of the closed *k*-chain. That is, if the *k*-chains c_k^1 and c_k^2 are differed by a boundary chain,

$$c_k^1 - c_k^2 \in B_k(M) \tag{7.1.19}$$

then c_k^1 and c_k^2 are said to be equivalent.

$$c_k^1 \sim c_k^2 \tag{7.1.19$_1$}$$

This kind of equivalence is called homology, i.e. c_k^1 and c_k^2 are homologous to each other, and belong to the same homology class, denoted by $< Z_k >$.

In the homology group H_k, non-trivial closed *k*-chains (that is, a closed chain c_k satisfies that $\partial c_k = 0$, but *it is not the boundary chain* ∂c_{k+1}) are *k*-submanifolds circling the (*k*+1)-dimensional holes. What is wrapped up by a closed *k*-chain which does not circle the holes is a solid (k+1)-manifold N. Hence this kind of closed k-chain can be written as ∂c_{k+1} and must be *a boundary chain.*

Contained in $Z_k(M)$ are the non-trivial closed k-chain which is exactly along the boundary of the (k+1)-dimensional hole, and the non-trivial sum of this closed k-chain and all kinds of boundary chains. This sum and the closed k-chain which is exactly along the boundary of the (k+1)-dimensional hole belong to the same equivalent homology class $Z_k(M)$ (they are homotopic to each other as well).

If *all the k-chains* in a equivalent homology class $Z_k(M)$ are of the type ∂c_{k+1}, then ∂c_{k+1} must belong to the same equivalent homology class as the 0-chain since they are differed by a boundary chain ∂c_{k+1}. In this case, there is no (k+1)-dimensional hole in M, and $H_k(M)$ has only the 0 element.

Suppose that the manifolds M and N are homeomorphic, and there exists a homeomorphic mapping f. Then

$$f: \ M \longrightarrow N$$

Since the homeomorphic mapping operator f is exchangeable with the boundary operator ∂, therefore we obtain

$$f\,\partial M = \partial f\ M = \partial N$$

Thus we can see that the homeomorphic manifolds M and N have the boundary chain groups which are equivalent under the homeomorphic mapping. Hence the homology groups of M and N are isomorphic as well. Therefore we say that *the homeomorphic manifolds have the same homology group.*

Each generator of the homology group $H_k(M)$ on the manifold M is an equivalent homotopy class of the k-submanifold circling the (k+1)-dimensional hole, and is invariant under deformation retractions. Hence the homology groups of any two manifolds with the same homotopy types are isomorphic to each other. This fact illustrates the inner connections between homotopy groups and homology groups.

Example 1: R^n is a manifold without holes and it can be contracted into a single point. Thus it has the same homotopy type as the single point v_0. The corresponding homology group is:

$$H_k(R^n) = H_k(v_0) = \begin{cases} Z & k=0 \\ 0 & k=1,2,3,\cdots \end{cases} \tag{7.1.20}$$

Example 2: In a path connected manifold M, any two points v_0 and v_1 can be connected by a 1-submanifold (a simplex of k=1) $\sigma_1 = [v_0, v_1]$. v_0 and v_1 are two 0-submanifolds between which the difference $v_1 - v_0 = \partial[v_0, v_1]$ happens to be a 0-dimensional boundary chain. Hence according to the definitions of Eq.7.1.19 and Eq.7.1.19$_1$, v_0 and v_1 are homologous to each other, and belong to the same homology class (the 0-dimensional homology class). Further more, any other point v_1' in this path connected manifold M is homologous to v_0 as well. Therefore the 0-dimensional homology group of the path connected manifold M is

$$H_0(v_0) = H_0(v_1) = H_0(v_1') = H_0(M) = Z \tag{7.1.21}$$

(where Z represents the integers. See Eq.7.1.7 where all the a_i's are positive integers, i.e. $a_i \in Z$).

Definition 7.1.1: The rank of the k-dimensional homology group $H_k(M)$ of the manifold M (i.e. the dimension of the homology group $H_k(M)$) is called the *Betti number* of order k of the manifold M, denoted by

$$b_k = \dim(H_k(M)) \tag{7.1.22}$$

As we have previously mentioned that under the mapping of the homeomorphism f,

$$f: \ H_k(M) \longrightarrow H_k(N)$$

is isomorphic (for all $k \geq 0$). Therefore the Betti number is the topological invariant of the manifold (i.e. the invariant under the homeomorphism).

Definition 7.1.2: The alternating sum of the Betti numbers of various orders on the manifold M is called the Euler's number $\chi(M)$.

$$\chi(M) = \sum_{k=0}^{n} (-1)^k b_k \tag{7.1.23}$$

Example 1: The Euler's number of a tetrahedron is $\chi(M)=2$.

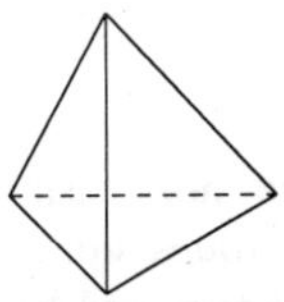

Figure 7.1.2

A tetrahedron has

4 independent vertices: $b_0 = 4 = \dim(H_0(M))$

6 independent edges: $b_1 = 6 = \dim(H_1(M))$

4 independent side faces: $b_2 = 4 = \dim(H_2(M))$

$\therefore \quad \chi(M) = 4 - 6 + 4 = 2$

Example 2: The Euler's number of the sphere S^2 is $\chi(M)=2$.

Using topological invariance, a tetrahedron can be deformed into a sphere S^2 by topological deformations:

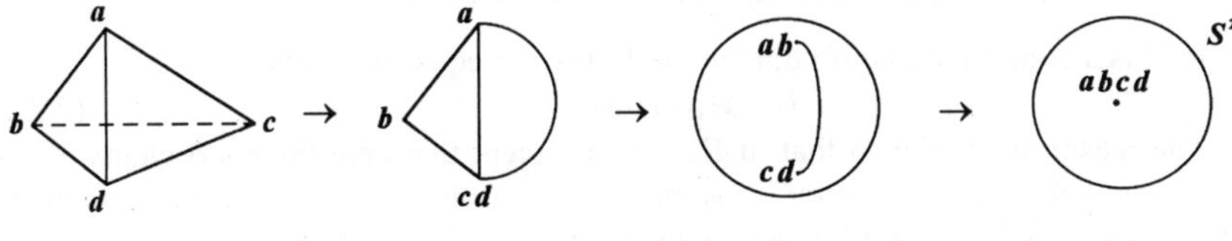

Figure 7.1.3

The sphere S^2 has only a 0-simplex (the point *abcd*) and a 2-simplex (the sphere), but no 1-simplex. Therefore $\chi(M)=1+0+1=2$ is the same as $\chi(M)=2$ of the tetrahedron. In fact $\chi(M)$ of any polyhedron which is topologically homeomorphic to the sphere is 2.

In general, suppose there is an arbitrary connected n-manifold M. After the triangulation of M (triangulation is possible for any polygon), the Euler's number of M can be defined as the alternating sum of the simplex numbers a_k:

$$\chi(M) = \sum_{k=0}^{n}(-1)^k a_k \left(= \sum_{k=0}^{n}(-1)^k b_k \right) \tag{7.1.24}$$

The point is that $\chi(M)$ is the topological invariant of the manifold *M*. The detailed proof of Eq.7.1.24 is omitted here.

The chain groups discussed above are all the linear combinations with integral coefficients. Hence we give the following definition:

Definition 7.1.3: The homology group given by the linear combination of the chain groups on the manifold *M* with integral coefficients is called the *integral homology group*, or *continuous integral homology group* $H_k(M,Z)$, where *Z* represents the integral additive group and hence is an Abelian group, and *k* represents the continuous *k*-simplex on the manifold *M* (see Eq.7.1.6 and Eq.7.1.7).

The *Z* group can be replaced by other Abelian group *G*. Thus we obtain the corresponding chain group $C_k(M,G)$, closed chain group $Z_k(M,G)$, boundary chain group $B_k(M,G)$, and homology group $H_k(M,G)$, etc.. Here we list some typical groups as the following:

Definition 7.1.4: $H_k(M,Z_2)$ — homology group of module 2.
$H_k(M,R)$ — real homology group.
$H_k(M,Q)$ — rational homology group.
$H_k(M,C)$ — complex homology group.

Among all the homology groups, integral homology group is the most basic one which contains more information than the rest of the homology groups do. The properties of other homology groups $H_k(M,G)$ can be completely determined by the properties of the integral homology group $H_k(M,Z)$ and the Abelian group *G*. That is in fact the *Universal Coefficient Theorem* (the details are omitted).

§ 7.2 Universal Coefficient Theorem of Homology Groups

If *M* is a connected manifold, then the following equation holds:

$$H_0(M,G) = G \tag{7.2.1}$$

The reason is similar to that of Eq.7.1.21 except that here the coefficients of the chain group belong to *G* (where *G* is some Abelian group) while the coefficients of the chain group of Eq.7.1.21 are integers (which belong to the *Z* group).

If *M* can be decomposed into *l* connected branches, then it is obvious that

$$H_0(M,G) = G \oplus \cdots \oplus G \tag{7.2.2}$$

where at the right hand side is the direct sum of the *l G* groups.

Theorem 7.2.1: If the *n*-manifold *M* is a compact connected oriented manifold, then

$$H_n(M,G) = G \tag{7.2.3}$$

Proof: Since the boundary chain of the highest order of the *n*-manifold *M* can only be the $(n-1)$-dimensional $B_{n-1}(M,G)$, therefore $B_n(M,G) = 0$. Then in the case that

$H_k(M,G)$, $Z_k(M,G)$, and $B_k(M,G)$ are all Abelian groups, Eq.7.1.18 gives

$$H_n(M,G) = Z_n(M,G) - B_n(M,G) = Z_n(M,G).$$

Since the n-manifold M is both an oriented region and a compact connected n-dimensional closed chain (which contains an $(n+1)$-dimensional hole), therefore the unique generator $Z_k(M,G)$ may have is this n-dimensional compact connected oriented manifold M itself. The corresponding homology group $H_n(M,G)$ is thus the group elements of G multiplying this unique generator. Hence we have Eq.7.2.3. □

Examples: 1-dimensional A straight line:  can be extended to infinities and is not compact.

S^1:  is compact, oriented, and contains a 2-dimensional hole.

2-dimensional A plane: 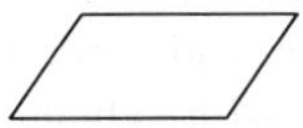 can be extended to infinities and is not compact.

S^2: 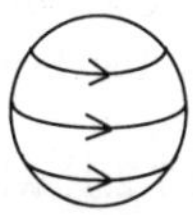 is compact, oriented, and contains a 3-dimensional hole.

Theorem 7.2.2: On a noncompact connected n-manifold M, we should have

$$H_n(M,G) = 0 \tag{7.2.4}$$

Proof: Since that the n-dimensional closed chain does not exist on a noncompact connected n-manifold M, hence $H_n(M,G)$ does not exist either, and therefore Eq.7.2.4 holds. The reader may refer to the above two simplest examples of manifolds which are not compact. □

If M is a non-orientable manifold, then we can subdivide it into simplexes and hence obtain an oriented region Σ. There must be an $(n-1)$-submanifold which is closely related to the non-orientable property on the original manifold. M is not oriented before the subdivision of this submanifold; however, after the subdivision there will be two $(n-1)$-dimensional boundary chains which are both the boundaries of the oriented region Σ. Obviously these two $(n-1)$-dimensional boundary chains are both equivalent to that original $(n-1)$-submanifold, thus the $(n-1)$-dimensional boundary chains on the oriented region Σ after the subdivision must double the $(n-1)$-submanifold on M before the subdivision.

Again we take the Möbius strip as an example, and make the subdivision as the following:

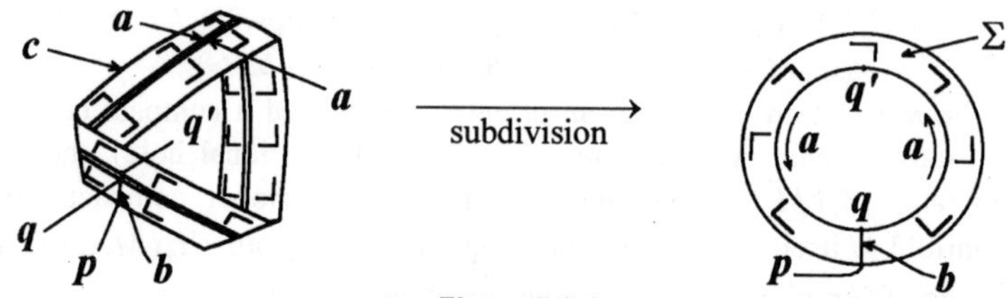

Figure 7.2.1

On the left of the figure is the Möbius strip. From the subdivision we can see that if we fold along the diameter, stick q and q' together, cut the strip along the line segment b which connects the points p and q, and stick the boundary chains a of both sides together, then the Möbius strip on the left will be recovered from the Σ region on the right of the figure.

The symbols └,└,··· illustrate the orientation. The Σ region on the right is clearly oriented while the Möbius strip on the left is not, since after sticking the boundaries together, we can see two different orientations └ and ┌ appear simultaneously in a local region.

From the figure we can also see that the edge a of the Möbius strip is a 1-submanifold which is formed by gluing the two 1-dimensional boundary chains of Σ together. Hence there must be a 1-dimensional boundary chain on the oriented region Σ, and this chain must double the 1-submanifold, i.e. the edge a, of the Möbius strip before the subdivision.

According to the Figure 7.2.1, when stick q and q' together, and a and a together after folding along the diameter, we will see that the submanifolds of the Möbius strip include:

2 vertices (There should be 3 vertices q, q', and p after the subdivision. The q and q' become one vertex after being glued together), 3 edges (There should be 4 edges a, a, b, and c. The a and a become one edge after being glued together), and 1 surface Σ.

Therefore according to Eq.7.1.24, the Euler's number of the Möbius strip should be

$$\chi(M) = 2 - 3 + 1 = 0$$

Here we give a brief summary of the basic concepts of homotopy and homology:

(i) Homotopy group reflects the topological barrier of the continuous mapping from the k-dimensional sphere S^k to the manifold M. $\Pi_k(M)$ is the homotopy group corresponding to S^k. $\Pi_k(M) \neq 0$ indicates the existence of a (k+1)-dimensional hole in M. This hole is a topological barrier since it prevents the k-dimensional circuits in the homotopy group $\Pi_k(M)$ from being contracted into one point.

(ii) Homology group reflects the amount of the independent topologically non-trivial k-submanifolds on an n-manifold M ($k<n$). $H_k(M)$ is called the homology group of order k of the manifold M.

§ 7.3 *Introducing Dual Chain Group and Cohomology Group through Dual Homomorphism*

Let A be an n-dimensional Abelian group, in which there exists a group of basis

$$X = \{x_1, x_2, \cdots, x_n\}.$$

Each group element a of A is then a linear combination of the base elements in X with integer coefficients:

$$a = \sum_{i=1}^{n} z_i x_i \quad (z_i \in Z) \quad \text{where } Z \text{ is the additive group of integers.} \qquad (7.3.1)$$

Assume that $Hom(A,Z)$ is the collection of the homomorphic mappings from the group A to the additive group of integers Z (it is also called the "*homomorphism group*"). Then we may pick an arbitrary homomorphic mapping f from this set ($f \in Hom(A,Z)$), and require the following features for $Hom(A,Z)$ and f:

$$Hom(A,Z): \; A \longrightarrow Z \qquad (7.3.2)_1$$

$$f: \; a \longrightarrow \langle f, a \rangle \in Z \qquad (7.3.2)_2$$

To achieve Eq.7.3.2, we need to take a group of basis which is dual to X:

$$f = \{\xi_1, \xi_2, \cdots, \xi_n\}, \quad \text{satisfying}$$

$$\langle \xi_j, x_i \rangle = \delta_{ij}, \qquad (7.3.3)$$

and hence obtain

$$\langle \xi_j, a \rangle = \langle \xi_j, \sum_{i=1}^{n} z_i x_i \rangle = z_j \in Z \qquad (7.3.4)$$

which can also be written as

$$\langle f, a \rangle \longrightarrow \left\{ \begin{array}{c} \langle \xi_1, a \rangle = z_1 \\ \langle \xi_2, a \rangle = z_2 \\ \cdots\cdots \\ \langle \xi_n, a \rangle = z_n \end{array} \right\} \in Z \qquad (7.3.4)_1$$

By the way, $\langle \xi_j$ and $x_i \rangle$ have the features of the Dirac notations *bra* and *ket*.

Suppose that we have two Abelian groups A and B. Then we can explain dual homomorphism using the above notations of *bra* and *ket* together with Figure 7.3.1.

Remarks: (i) The arrows of the homomorphism φ and $\tilde{\varphi}$ are in opposite directions. The action of φ to a is in the direction of $A \to B$.

(ii) b is chosen from $Hom(B, Z)$, the collection of the homomorphic mappings. The action of $\tilde{\varphi}$ to b is in the direction of $B \to A$.

(iii) From $Hom(A, Z)$, the collection of the homomorphisms, we have chosen

$$f = \tilde{\varphi}(b) = \tilde{\varphi}_{ij} b_j = b_j \varphi_{ji} \qquad (7.3.5)$$

(iv) Hence we have:

$$< \tilde{\varphi}(b), a >= b_j \varphi_{ji} a_i$$

$$< b, \varphi(a) >= b_i \varphi_{ij} a_j$$

$$\therefore \quad < \tilde{\varphi}(b), a >=< b, \varphi(a) >\in Z \tag{7.3.6}$$

Eq.7.3.6 is an essential property of dual homomorphism.

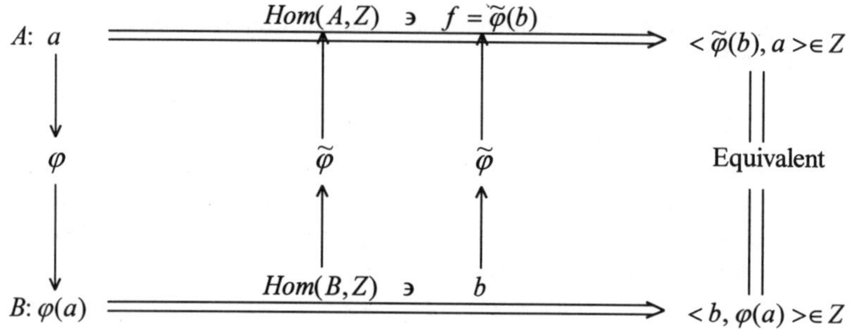

Figure 7.3.1

Similar to Figure 7.3.1, starting from $Hom(C_k(M), Z)$, the set of all the homomorphic mappings from the *k*-chain $C_k(M)$ (an Abelian group) to the additive group of integers *Z*, we can define $C^k(M)$, the dual chain group of $C_k(M)$. $C^k(M)$ and $C_k(M)$ belong to the cochain and chain group series respectively.

Suppose that a chain group has the following *boundary homomorphic relation*:

$$\partial: \; C_k(M) \longrightarrow C_{k-1}(M) \qquad (7.3.7)_1$$

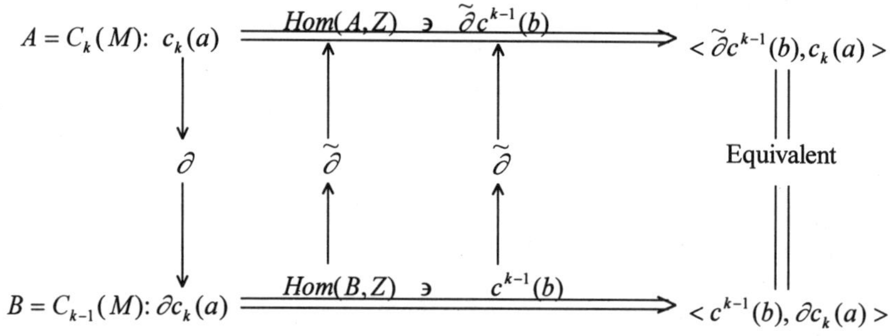

Figure 7.3.2

Remarks: (i) The *coboundary homomorphism* (compare to Eq.7.3.7$_1$) can be obtained from Figure 7.3.2:

$$\tilde{\partial}: \; C^{k-1}(M) \longrightarrow C^k(M) \qquad (7.3.7)_2$$

Similar to Eq.7.3.7$_1$, this is the dual homomorphic relation between the cochain groups.

(ii) $< c^{k-1}(b), \partial c_k(a) >= c^{k-1}(b) \partial_{ba} c_k(a)$

$$<\widetilde{\partial}c^{k-1}(b),c_k(a)>=\widetilde{\partial}_{ab}c^{k-1}(b)c_k(a)=c^{k-1}(b)\partial_{ba}c_k(a)$$

$$\therefore\ <c^{k-1}(b),\partial c_k(a)>=<\widetilde{\partial}c^{k-1}(b),c_k(a)> \tag{7.3.8}$$

(iii) $\partial\cdot\partial=0$ (see Section 7.1) $\longrightarrow \partial_{k-1}\cdot\partial_k=0$ (7.3.9)

The dual operator of $\partial_{k-1}\cdot\partial_k$ is $\widetilde{(\partial_{k-1}\partial_k)}=\widetilde{\partial}_k\cdot\widetilde{\partial}_{k-1}$.

$$\therefore\ \widetilde{\partial}_k\cdot\widetilde{\partial}_{k-1}=\widetilde{(\partial_{k-1}\partial_k)}=0\longrightarrow\widetilde{\partial}\widetilde{\partial}=0 \tag{7.3.10}$$

(iv) Thus we can define the subgroup of the cochain group $C^k(M)$ (refer to Eq.7.1.13):

Coclosed chain group: $Z^k(M)$ is $Ker\ \widetilde{\partial}$,

$$Z^k(M)=\left\{c^k;\widetilde{\partial}c^k=0\right\} \tag{7.3.11}$$

Coboundary group: $B^k(M)$ is $Image\ \widetilde{\partial}_{k-1}$

$$B^k(M)=\left\{b^k=\widetilde{\partial}c^{k-1}\right\} \tag{7.3.12}$$

Hence the cohomology group can be defined as (compare to Eq.7.1.18):

$$H^k(M)=Z^k(M)/B^k(M) \tag{7.3.13}$$

$H^k(M)$ is a collection of the equivalent classes of k-coclosed chains (mod coboundary), and forms an Abelian group.

From Figure 7.3.2 we can see as well that cohomology groups are isomorphic to homology groups, and hence the following theorem exists:

$$b^k=\dim H^k(M)=b_k=\dim H_k(M) \tag{7.3.14}$$

where $b^k=\dim H^k(M)$ is the co-Betti number and $b_k=\dim H_k(M)$ is the Betti number.

§ 7.4 de Rham Cohomology Theory

Write the Stokes' Theorem according to Eq.1.5.14 of Chapter 1:

$$\int_c d\omega_{n-1}=\int_{\partial c}\omega_{n-1}$$

and compare it to Eq.7.3.8 and Eq.7.3.6, we can see that Eq.1.5.14 is a kind of scalar product as well, similar to Eq.7.3.8 and Eq.7.3.6:

$$<c,d\omega_{n-1}>=<\partial c,\omega_{n-1}> \tag{7.4.1}$$

If c is replaced by ∂c, using the Stokes' Theorem, then Eq.7.4.1 gives the following equation:

$$<\partial c,d\omega_{n-1}>=<\partial\partial c,\omega_{n-1}>=0 \tag{7.4.2}$$

Here we have applied the fact that the boundary of a boundary chain must be zero, i.e. $\partial\partial=0$ (see Section 7.1).

Then replace ω by $d\omega$, Eq.7.4.1 also gives

$$<c,dd\omega_{n-1}>=<\partial c,d\omega_{n-1}>=0 \tag{7.4.3}$$

where $dd\omega=0$ is applied (see Eq.1.4.17).

Combine Eq.7.4.2 and Eq.7.4.3 we obtain

$$0=<\partial^2 c,\omega_{n-1}>=<\partial c,d\omega_{n-1}>=<c,d^2\omega_{n-1}> \tag{7.4.4}$$

where c, ∂c and $\partial^2 c$ are the integral regions; ω_{n-1}, $d\omega_{n-1}$ and $d^2\omega_{n-1}$ are the differential forms.

Compare with the homomorphisms Eq.7.3.7$_1$ and Eq.7.3.7$_2$ of Section 7.3, we may obtain the following sequence from Eq.7.4.2, Eq.7.4.3 and Eq.7.4.4:

$$\partial:\ c_{k+1}\xrightarrow{\partial}c_k\xrightarrow{\partial}c_{k-1} \tag{7.4.5$_1$}$$

Here the arrows point to the direction of decreasing k. c_{k+1}, c_k, and c_{k-1} belong to the chain group series. Eq.7.4.5$_1$ has the property of homology.

From Eq.7.4.2, Eq.7.4.3 and Eq.7.4.4 we obtain the following sequence as well:

$$d:\ \omega_{p+1}\xleftarrow{d}\omega_p\xleftarrow{d}\omega_{p-1} \tag{7.4.5$_2$}$$

Here the arrows point to the direction of increasing p. ω_{p+1}, ω_p, and ω_{p-1} belong to the cochain group series. Eq.7.4.5$_2$ has the property of cohomology.

It was de Rham who first discovered and investigated the cochain group series from the Stokes' Theorem, and hence the cohomology theory about this cochain group series is called *de Rham cohomology theory*.

Similar to Eq.7.3.11, Eq.7.3.12 and Eq.7.3.13, and refer to Eq.7.1.13, Eq.7.1.14 and Eq.7.1.18, the following definitions can be given successively:

Definition 7.4.1: $Z^p_{dR}=\left\{\ \omega_p, d\omega_p=0\ \right\}$, i.e. $Ker\ d_p$, is called a *closed form.*

Definition 7.4.2: $B^p_{dR}=\left\{\ \omega_p,\omega_p=d\alpha_{p-1}\ \right\}$, i.e. $Img\ d_{p-1}$, is called an *exact form.*

Since

$$\int_{c_p}\omega_p+d\alpha_{p-1}=\int_{c_p}\omega_p+\int_{\partial c_p}\alpha_{p-1}=\int_{c_p}\omega_p\quad(\because\ \ c_p\in Z_p\to\partial c_p=0),$$

$$\int_{c_p+\partial a_{p+1}}\omega_p=\int_{c_p}\omega_p+\int_{a_{p+1}}d\omega_p=\int_{c_p}\omega_p\quad(\because\ \ \omega_p\in Z^p_{dR}\to d\omega_p=0),$$

therefore $<c_p,\omega_p>=\int_{c_p}\omega_p$ depends only on the equivalent classes of c_p and ω_p, not on the choice of c_p or $c_p'=c_p+\partial a_{p+1}$ from the equivalent class of $c_p\in Z_p$, and not on the choice of ω_p or $\omega_p'=\omega_p+d\alpha_{p-1}$ from the equivalent class of $\omega_p\in Z^p_{dR}$.

The group of the equivalent classes of c_p is the homology group (see Eq.7.1.18)

$$H_p(M,R)=Z_p(M,R)/B_p(M,R)=Ker\ \partial_p/Img\ \partial_{p+1} \tag{7.4.6}$$

where $H_p(M,G)$ is taken as $H_p(M,R)$, and $G=R$ is the set of real numbers.

The group of the equivalent classes of ω_p is

$$H^p_{dR}(M,R) = Z^p_{dR}(M,R)/B^p_{dR}(M,R) = Ker\, d_p / Img\, d_{p-1} \tag{7.4.7}$$

where $H^p_{dR}(M,R)$ is the de Rham cohomology group.

For the sake of convenience, the right hand side of the equation Eq.7.4.7 can be considered as subtracting the set $B^p_{dR} = \{ \omega_p, \omega_p = d\alpha_{p-1} \}$ from the set $Z^p_{dR} = \{ \omega_p, d\omega_p = 0 \}$, according to the Definitions 7.4.1 and 7.4.2 ("Division" of Abelian groups is just the same as "subtraction").

To indicate that Eq.7.4.7 is a cohomology group, we must also give the dual relationship between a group of its basis $\{\omega_i\}$ (closed forms, i=1, 2, ⋯, $\dim H^p_{dR}(M,R)$) and a group of basis $\{c_i\}$ (c_i are closed chains, i=1, 2, ⋯, $\dim H_p(M,R)$) of the homology group (i.e. $H_p(M,R)$). de Rham has proven the following two theorems (The prerequisite of his proofs is that M must be a compact unbounded manifold. The proofs are omitted.):

Theorem 7.4.1: Given a group of v_i, i=1, 2, 3, ⋯, $\dim H_p$. There must exist a closed p-form ω which satisfies

$$v_i = \pi(c_i,\omega) = \int_{c_i} \omega, \quad i=1, 2, 3, \cdots, \dim H_p \tag{7.4.8}$$

Theorem 7.4.2: If there is a p-form α_p such that all $\pi(c_i,\alpha_p)$ are equal to zero, i.e.

$$0 = \pi(c_i,\alpha_p) = \int_{c_i} \alpha_p, \quad i=1, 2, 3, \cdots, \dim H_p \tag{7.4.9}$$

Then α_p must be an exact form: $\alpha_p = d\beta_{p-1}$.

According to Theorem 7.4.1, suppose that the dimension of the homology group of the manifold M is $\dim H_p = r$, then there must be r independent closed chains (the submanifolds of M) in this manifold M: c_1, c_2, ⋯, c_r. A closed p-form ω_i ($d\omega_i = 0$) can be defined on each closed chain c_i. Hence we obtain (here i is *not* a dummy index of summation):

$$\pi_{ii} = \pi(c_i,\omega_i) = \int_{c_i} \omega_i \neq 0$$

$$\pi_{ij} = \pi(c_i,\omega_j) = \int_{c_i} \omega_j = 0 \quad \text{(since that } \omega_j \text{ is not defined on } c_i\text{)}.$$

To make it brief, we have

$$\pi_{ij} = \pi(c_i,\omega_j) = \langle c_i,\omega_j \rangle = \delta_{ij} \tag{7.4.10}$$

According to Theorems 7.4.1 and 7.4.2, it is also clear that the value of $\langle c_i,\omega_j \rangle$ will not be effected by either the choice of c_p or $c_p' = c_p + \partial a_{p+1}$ from the equivalent class of $c_p \in Z_p$, or the choice of ω_p or $\omega_p' = \omega_p + d\alpha_{p-1}$ from the equivalent class of $\omega_p \in Z^p_{dR}$.

Thus the group of the equivalent classes ω_i (i=1, 2, 3, ⋯, r) found according to de Rham's Theorem 1 and Theorem 2 possesses the property of Eq.7.4.10. That indicates that this group of ω_i (i=1, 2, 3, ⋯, r) is dual to a group of basis of $H_p(M,R)$ (i.e. an equivalent class of c_i). Therefore we can consider the equivalent classes of ω_i (i=1, 2, 3, ⋯, r) as a linear basis, and hence obtain the $H^p_{dR}(M,R)$ group, which is a kind of cohomology group dual to the homology group $H_p(M,R)$, and is equivalent to the simple cohomology group we have mentioned in the preceding section:

$$H^p_{dR}(M,R) \cong H^p(M,R) \qquad (7.4.11)$$

where R indicates that the coefficients of an arbitrary linear combination of ω_j' and ω_j can be taken as any real numbers in R:

$$\pi(c_i,\omega_j') = \int_{c_i} \omega_j' \in R$$

Then ω_j' can become ω_j of Eq.7.4.10 after some linear transformation.

One thing that is worth noticing is that the space $H^k_{dR}(M,R)$ which depends on the differential structure is in fact topologically invariant. Therefore $H^k_{dR}(M,R)$ is the connection between the topological properties and the differential structure of a manifold.

de Rham Cohomology Ring

Consider the following features:

1. If $d\omega_i = 0$, and $d\omega_j = 0$, then

   $d(\omega_i \wedge \omega_j) = d\omega_i \wedge \omega_j + (-1)^p \omega_i \wedge d\omega_j = 0$ (where ω_i's are the p-forms).

   Hence we have

   $$\textit{closed form} \wedge \textit{closed form} = \textit{closed form}. \qquad (7.4.12)$$

2. If $d\omega_i = 0$, and $\omega_j = d\alpha_{j-1}$, then

   $\omega_i \wedge \omega_j = \omega_i \wedge d\alpha_{j-1} = (-1)^p d(\omega_i \wedge \alpha_{j-1})$ (where ω_i's are the p-forms).

   Hence we have

   $$\textit{closed form} \wedge \textit{exact form} = \textit{exact form}. \qquad (7.4.12)$$

With the above two features, we can define the coproduct of de Rham cohomology group through the exterior product $\wedge$:

$$H^p_{dR}(M,R) \otimes H^q_{dR}(M,R) = H^{p+q}_{dR}(M,R) \qquad (7.4.14)$$

since the homology equivalent classes in $H^{p+q}_{dR}(M,R)$ happen to be the product (exterior product) of the homology equivalent classes of $H^p_{dR}(M,R)$ and $H^q_{dR}(M,R)$.

Suppose that the dimension of M is n. Then a collection of $H^0_{dR}(M,R)$, $H^1_{dR}(M,R)$, $H^2_{dR}(M,R)$, ⋯, $H^n_{dR}(M,R)$ can be given by Eq.7.4.14. This set can also be written as:

$$\oplus H^p_{dR}(M,R) \qquad (p=0, 1, 2, \cdots, n) \qquad (7.4.15)$$

which possesses the following *properties of a ring*:

1. This set has both addition and multiplication.
2. This group is Abelian under the addition.
3. The multiplication satisfies the law of association: $(ab)c=a(bc)$, and the law of distribution: $a(b+c)=ab+ac$, $(a+b)c=ac+bc$.

   The set $\oplus H^p_{dR}(M,R)$ (p=0, 1, 2, ⋯, n) of Eq.7.4.15 is called the *de Rham cohomology ring.* A general cohomology has its *cohomology ring* as well, which is isomorphic to the de Rham cohomology ring (the details are omitted).
4. If $M = M_1 \times M_2$ is a product manifold, in a similar way of Eq.7.4.14 we may define the following:

$$H^k(M,R) = \bigoplus_{i+j=k} H^i(M,R) \otimes H^j(M,R) \tag{7.4.16}$$

   which is called the Künneth 's formula.

§ 7.5 *Harmonic Forms Harm^k(M, R)*

When a metric tensor field exists on a manifold, we may introduce the Hodge* operation, and define the elliptic differential operator Δ (see Eq.4.1.26$_1$ in Chapter 4). The kernel φ_p ($\Delta\varphi_p = 0$) of the operator Δ is called the harmonic p-form. The set of the harmonic p-forms forms the linear space $Harm^p(M,R)$ defined on the real numbers.

Theorem 7.5.1: Suppose that the manifold M is compact, unbounded, and oriented. Then $Harm^p(M,R)$ is isomorphic to $H^p(M,R)$.

Proof: An arbitrary p-form on the compact unbounded oriented manifold M can be uniquely decomposed into (see Eq.3.4.21)

$$\omega_p = d\alpha_{p-1} + \delta\beta_{p+1} + \gamma_p \qquad \text{(where } \gamma_p \text{ is a harmonic form).}$$

If $\omega_p \in Z_p(M)$, i.e. $d\omega_p = 0$, then since $dd\alpha_{p-1} = 0$ and $d\gamma_p = 0$ (see the features of the harmonic form γ_p in Section 3.4), we must have

$$d\delta\beta_{p+1} = 0$$

Then use Eq.4.1.26$_1$ we obtain $(\beta_{p+1}, d\delta\beta_{p+1}) = (\delta\beta_{p+1}, \delta\beta_{p+1}) = 0 \Rightarrow \delta\beta_{p+1} = 0$.

$$\therefore \quad \omega_p = d\alpha_{p-1} + \gamma_p$$

where $d\alpha_{p-1}$ is an exact form, and hence each cohomology class $\{\omega_p\}$ corresponds to only one γ_p. Therefore $H^p_{dR}(M,R)$ (the linear space formed by the set of the cohomology classes $\{\omega_p\}$'s) and $Harm^p(M,R)$ (the linear space formed by the set of the harmonic p-forms defined on the real numbers) are isomorphic.

Corollary 1: The dimension of $Harm^p(M,R)$ is finite on a compact unbounded manifold. Therefore the dimension of $H^p_{dR}(M,R)$ is finite as well.

Corollary 2: Since Hodge* and Δ are commutative, if γ is a harmonic form, then so is

$*\gamma$ (since $*\Delta\gamma = 0 = \Delta * \gamma$). Hence $Harm^p(M,R)$ is isomorphic to $*Harm^p(M) = Harm^{n-p}(M,R)$, and thus $H^p_{dR}(M,R)$ is isomorphic to $H^{n-p}_{dR}(M,R)$:

$$b^p = b^{n-p} \tag{7.5.1}$$

which is called the *Poincaré duality*.

Corollary 3: The manifold M with the Poincaré duality must be oriented (the proof is omitted).

Corollary 4: The manifold M with the Poincaré duality must be compact and unbounded (the proof is omitted).

Chapter 8. Fibre Bundles and Their Topological Structures

§ 8.1 *What are Fibre Bundles*

Bundle consists of three contents: the base space *B*, the entire space *E*, and the projection of the bundle Π. Therefore a bundle can be represented by (*E*, *B*, Π) where *E*, *B*, and Π have the following relation:

$$\Pi:\ E \to B\,. \tag{8.1.1}$$

The simplest bundle is the product bundle ($B_1 \times B_2$, B_1, Π), where $E = B_1 \times B_2$ is the entire space, B_1 is the base space, and the action of the projection Π is

$$\Pi:\ B_1 \times B_2 \to B_1\,. \tag{8.1.2}$$

To be more specific, if x_1 is the coordinate in B_1, and x_2 is the coordinate in B_2, then (x_1, x_2) is the coordinate in $E = B_1 \times B_2$. Thus the action of Π in Eq.8.1.2 can also be written as

$$\Pi:\quad (x_1, x_2) \to x_1. \qquad (8.1.2)_1$$

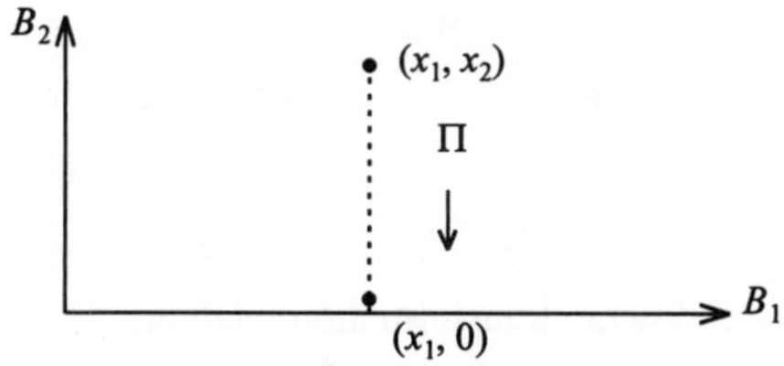

Figure 8.1.1

Use a small piece of paper to make a cylinder, we can obtain the simplest product bundle $\left(S^1 \times I, S^1, \Pi\right)$, where S^1 is a circle which corresponds to the base space B_1, and *I* corresponds to the height of the cylinder which belongs to the B_2 space. See Figure 8.1.2 below:

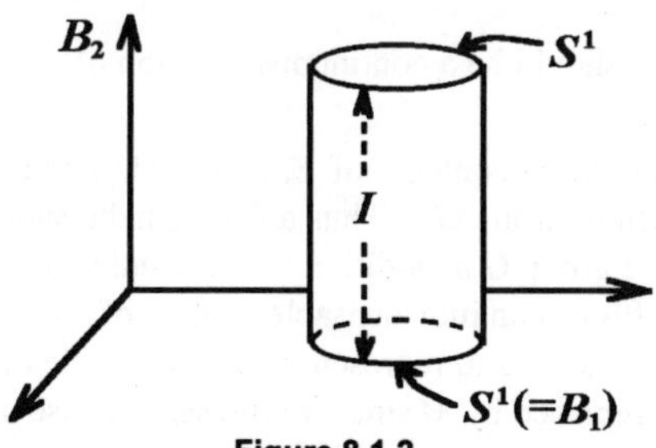

Figure 8.1.2

In those more general cases, such as the Möbius strip for example, *E* is no longer a simple product of two spaces at a global point of view. However, *E* can be considered as the product of two local spaces $U \times I$, where $U\left(\subset S^1\right)$ is a small subset of S^1 (i.e. an open cover of S^1). But to twist the Möbius strip at somewhere there has to be another part involved. Hence sometimes a fibre bundle is also called "the product of the twisted spaces".

The topological space $\Pi^{-1}(x)$ (for all $x \in B_1$) is also required to be homeomorphic to some space F (which is equivalent to some subspace of B_2 in the previously mentioned product space $B_1 \times B_2$), i.e.

$$\Pi^{-1}(x) = F_x, \quad \text{where } F_x \text{ is homeomorphic to } F \tag{8.1.3}$$

Then F_x is called the fibre of the base manifold B_1 at x; and F is called the standard fibre.

Fibre Bundle If the base space B of a fibre (E, B, Π) can be covered by a family of open sets $\{U_j\ ;\ j \in J \leq N\}$, and the fibre $\Pi^{-1}(x) = F_x$ of each U_j is homeomorphic to a standard fibre F, then the entire space E can be considered as a collection of a family of fibres $\Pi^{-1}(x)$ (for all $x \in B_1$). Furthermore, let U_α and U_β be two intersecting open covers of the base space B. Then we can take a point $p \in U_\alpha \cap U_\beta$. Use the two homeomorphisms φ_α and φ_β of U_α and U_β, respectively, we obtain:

$$\varphi_\alpha : \begin{array}{l} \Pi^{-1}(U_\alpha) \to U_\alpha \times F \\ \quad x_p \quad \to (p, \xi(p)) \end{array} \qquad \varphi_\beta : \begin{array}{l} \Pi^{-1}(U_\beta) \to U_\beta \times F \\ \quad x_p \quad \to (p, \xi'(p)) \end{array} \tag{8.1.4}$$

where $\xi(p) = \xi_p^1, \cdots, \xi_p^k$ and $\xi'(p) = \xi'^1_p, \cdots, \xi'^k_p$ are two vectors in the (k-dimensional) fibre space F and have the following transformation relation:

$$\xi^\alpha(p) = a_{\alpha\beta}(p)\xi'^\beta(p) \tag{8.1.5}$$

where $a_{\alpha\beta}(p)$ is a $k \times k$ matrix (α, β=1, 2, ⋯, k).

The set $\{a_{\alpha\beta}(p)\}$ forms a group $G = \{a_{\alpha\beta}(p)\}$. This group ($G$) is called the *structure group* which gives the structure of the fibre bundle, e.g. the twisting structure of the Möbius strip.

Notice that $a_{\alpha\beta}(p)$ should be a continuous function of x_p as it varies according to p.

In short, besides the three contents of E, B and Π, a fibre bundle has yet another content, that is, the structure group G. Thus a fibre bundle should be represented by (E, B, Π, G), with a structure group G in addition to the bundle represented by (E, B, Π).

Since the standard fibre is an indispensable content of a fibre bundle as well, another notation, (E, B, F, Π, G), is used to represent the fibre bundle of which F is the standard fibre. Sometimes B is replaced by M since it represents a base manifold.

Now take the tangent bundle $T(M) = \bigcup_p T_p(M)$ as an example, we state the features of the fibre bundles in more details:

Let M be the (m-dimensional) base manifold of the tangent bundle. Then the fibre F_p of the point p on M is $T_p(M)$. $T_p(M)$ is an m-dimensional vector space which is isomorphic to the standard fibre $F = R^m$.

Then choose a group of base vectors $\{e_a(p)\}$ in *M*. An arbitrary tangent vector X_p in $T_p(M)$ can be written as

$$X_p = \xi^a(p)e_a(p), \text{ where } a=1, 2, \cdots, m \text{ and } \xi^a(p) \in R. \tag{8.1.6}$$

The projection Π is defined in the tangent bundle $T(M)$ as well. As we have previously mentioned, the action of Π is

$$\begin{aligned} \Pi:\ & T(M) \to M \\ & X_p \to p \end{aligned} \tag{8.1.7}$$

Since the tangent bundle is the product of the base manifold U_α and the standard fibre $F = R^m$ in a local region, thus we have the homeomorphism

$$\begin{aligned} \varphi_\alpha:\ & \Pi^{-1}(U_\alpha) \to U_\alpha \times R^m \\ & X_p \to (p, \xi(p)) \quad (\text{where } p \text{ is a point on } U_\alpha) \end{aligned} \tag{8.1.8}$$

If $p \in U_\alpha \cap U_\beta$, there exists another homeomorphism

$$\begin{aligned} \varphi_\beta:\ & \Pi^{-1}(U_\beta) \to U_\beta \times R^m \\ & X_p \to (p, \xi'(p)) \quad (\text{where } p \text{ is a point on } U_\beta) \end{aligned} \tag{8.1.8$_1$}$$

where $\xi(p)$ and $\xi'(p)$ are related by Eq.8.1.5.

A local coordinate $x = (x^1, x^2, \cdots, x^m)$ can be chosen from the neighborhood of the point *p* on the base manifold *M*. Then the base vectors $e_a(p)$ (a=1, 2, $\cdots$, m) of the tangent space can be written by the natural frame

$$e_a(p) = e_a^i(p)\frac{\partial}{\partial x^i} \tag{8.1.9}$$

Taking a transformation of the coordinates for the manifold will not vary the point *p* and the base vectors $e_a(p)$ of the tangent space, and therefore will not effect *the vector* $\xi^a(p)$ *in the fibre space F* (see the remarks of Eq.8.1.4), the transformation function $a_{\alpha\beta}(p)$ (see Eq.8.1.5), and $G = \{a_{\alpha\beta}(p)\}$.

The Consistent Condition: For each point x_i in the base manifold *M*, we can find a neighborhood U_i and an isomorphic mapping Φ_i. The action of Φ_i is to map $U_i \otimes F$ into the subset $\Pi^{-1}(U_i)$ of the fibre bundle *E*:

$$\Phi_i:\ (U_i \otimes F) \to \Pi^{-1}(U_i) \tag{8.1.10}$$

Since $(x_i \otimes f) \in U_i \otimes F$ ($x_i \in U_i$, $f \in F$), we obtain by Eq.8.1.10

$$\Phi_i:\ (x_i \otimes f) \to \Pi^{-1}(x_i) \tag{8.1.10$_1$}$$

$$\therefore\ \Pi(\Phi_i(x_i \otimes f)) \to \Pi\,\Pi^{-1}(x_i) = x_i \tag{8.1.10$_2$}$$

Therefore we see that it is a consistent condition which must be satisfied.

From the isomorphic mapping Φ_i, we can define the transformation matrix Φ_{ij} of two intersecting neighborhoods U_i and U_j (from U_j to U_i):

$$\Phi_{ij} = \Phi_i \Phi_j^{-1} \quad \text{when } x \in U_i \cap U_j \tag{8.1.11}$$

The action of Φ_{ij} for each fixed x ($x \in U_i \cap U_j$) is to transform the coordinates of x in U_j into the coordinates of x in U_i. Hence we have

$$\Phi_{ij} = I \tag{8.1.12$_1$}$$

$$\Phi_{ij}\Phi_{jk} = \Phi_{ik} \quad \text{when } x \in U_i \cap U_j \cap U_k \tag{8.1.12$_2$}$$

Obviously Eq.8.1.12$_1$ and Eq.8.1.12$_2$ are the necessary and sufficient conditions for making the bonding of all the neighborhoods U_i, U_j, U_k, $\cdots$ possible. The features of a fibre bundle are completely determined by the features of the transformation matrix.

§ 8.2 Fibre Bundles and Cross Sections

As we have commented in the previous section, the tangent bundle $T(M) = \bigcup_p T_p(M)$ is a kind of fibre bundle. A tangent vector field can be obtained on M if we choose a tangent vector from the tangent fibre bundle $T_p(M)$ for each point p (on the manifold M) by a certain given rule (which is related to p). Thus a tangent vector field on M can be considered as the cross section of a tangent bundle on M. Hence the concept of the "cross section" is introduced by the following:

Definition 8.2.1: Let p be a point on the base manifold M, and $\Pi^{-1}(p)$ be the fibre which belongs to the point p on M. Fix S_p on the fibre $\Pi^{-1}(p)$ for each point p. Then a cross section $S(p)$ of the fibre bundle is formed by the collection of S_p's for all of the points on the base manifold M.

Remark 1: Usually the dimension of a fibre is taken as k. But the dimension of the fibre of a tangent bundle $T(M)$ must be m which is the dimension of the base manifold M.

Remark 2: If the fibre is a collection of vectors, then S_p is one of the vectors; if the fibre is a collection of tensors, then S_p is one of the tensors, etc..

Definition 8.2.2: If the fibre bundle is a tangent bundle, then the cross section $S(p)$ is a tangent vector field, of which the dimension is m, the same as the dimension of the base manifold M.

Definition 8.2.3: If the fibre bundle is a k-dimensional vector bundle, then the cross section $S(p)$ is a k-dimensional vector field on the base manifold M.

The base vector $e = \{\, e_i(x) \,\}$ (i=1, 2, $\cdots$, k) of a vector bundle is called the frame field of the vector bundle. The rest of the k-dimensional vector fields are some linear combinations of the frame field $\{\, e_i(x) \,\}$ (i=1, 2, $\cdots$, k). The coefficients of the linear combinations are functions on the base manifold M. Therefore linear transformations of a frame field give other frame fields.

$$e'_j(x) = e_i(x)c^i_j(x)\,, \quad \det c^i_j \neq 0 \tag{8.2.1}$$

Definition 8.2.4: The set of all the frames of the vector bundle E on the manifold M, $L(E)$, is called the frame bundle of the vector bundle E:

$$L(E) = \{\ (x,e'), e'(x) \text{ represents all of the possible frames at } x\ \} \tag{8.2.2}$$

Let $e_i(x)$ be smooth vector-valued functions defined on the open set U_α which covers the base manifold M. All of the possible frames can be represented by $e' = ec$. Here $\{c^i_j(x)\}$ can be considered as a smooth matrix function on this open set U_α.

Fix a frame field $\{\ e_i(x), i = 1,\cdots,k\ \}$. A new frame field can be obtained with each different matrix function $C(x) = \left(c^i_j(x)\right)$. The set of all the $k \times k$ nonsingular matrix functions $\left(c^i_j(x)\right)$ has a 1-1 correspondence to the set of all frame bundles. Hence the set of the matrix functions $\left(c^i_j(x)\right)$ is equivalent to the frame bundle $L(E)$.

Since the dimensions of the base manifold M and the $k \times k$ nonsingular matrix function $\left(c^i_j(x)\right)$ are m and $k \times k = k^2$, respectively, the total degree of freedom of the frame bundle $L(E)$ (i.e. the dimension of $L(E)$ as a manifold) is given by

$$m + k^2 \tag{8.2.3}$$

where m represents the degree of freedom of the open set U_i on the base manifold M, and k^2 represents the degree of freedom of the transformation matrix for the k-dimensional frame field. $L(E)$ is locally homeomorphic to $U_i \times GL(k,R)$, where $GL(k,R)$ is the "kth ordered real general linear group". Thus we see that the transformation of the fibre of the frame bundle $L(E)$ possesses some features of the group $GL(k,R)$. Therefore the fibre of the frame bundle $L(E)$ is a group manifold. A bundle of which the fibre is a group manifold is called a *principal bundle* (see Definition 8.3.2 in the next section). According to this definition, the frame bundle $L(E)$ is a kind of principal bundle, which is called the adjoint principal bundle.

§ 8.3 Some Representative Fibre Bundles

A. The General Vector Bundle $E(M,F,\Pi,G)$:

The general vector bundle E consists of the base manifold M and the fibre space F, where M is an m-dimensional manifold and F is a k-dimensional linear vector space. Since E is a general vector bundle, not a tangent bundle, therefore k is usually *not* equal to m.

The projection operator between the vector bundle E and the base manifold M is still denoted by Π:

$$\Pi: \quad E \rightarrow M \tag{8.3.1}$$

This projection Π is a continuous surjective mapping. Hence we see that the basic contents of the general vector bundle E are M, F, Π, and G. We can denote E and its contents as $E(M,F,\Pi,G)$, and illustrate their relations in Figure 8.3.1.

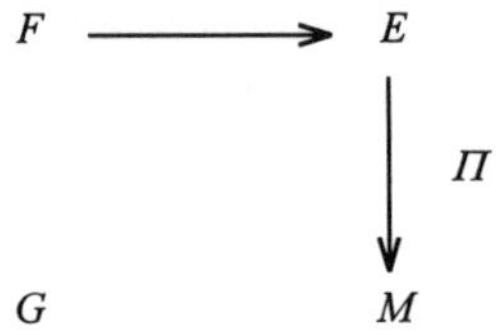

Figure 8.3.1

In Figure 8.3.1, $G=\left\{g_{\alpha\beta}(x)\right\}$ represents the collection of the transition matrices for the vector space F at each fixed point $x\in U_\alpha\cap U_\beta$ on the base manifold. G is a structure group.

Now we give some definitions which may be useful in the future.

Definition 8.3.1: The self-action of the group G has two forms:

(i) L_g: $G\to G$ $L_g(h)=gh$ is called the left (translation) action, and

(ii) R_g: $G\to G$ $R_g(h)=hg$ is called the right (translation) action.

where $g,h\in G$.

Definition 8.3.2: Suppose there is a fibre bundle P of which the group spaces of the standard fibre F and the structure group G are equivalent. Let G apply the left translation action onto F. Then this fibre bundle is called the principal bundle (The frame bundle is an example where G applies the left translation action onto the fibre of the frame bundle).

Definition 8.3.3: Let the transformation group G be functioning on some space S. Take $g\in G$, and $x\in S$. If $gx\neq x$ except when $g=e$, then it is said that G acts on S freely.

B. The Principal Bundle $P(MG\Pi)$:

The fibre of the principal bundle P is a group manifold G of Lie group. The transition function of the principal bundle P (i.e. the structure group) belongs to G group and applies the left translation action onto the fibre G.

Since the global structure of the bundle requires the smooth adhesion of the fibres on each intersecting region (i.e. $U_\alpha\cap U_\beta$) of the base manifold M. Hence we should find, with some help of Figure 8.3.2 and Figure 8.3.3, the relation between $\Pi^{-1}(U_\alpha)$ and $\Pi^{-1}(U_\beta)$ at the point $x\,(\in U_\alpha\cap U_\beta)$.

Figure 8.3.2 shows the fibre $\Pi^{-1}(x)=G_x$ of x on U_α, and the homeomorphism between the point u on $\Pi^{-1}(x)$ and the point $\varphi_{\alpha,x}(u)$ on the standard fibre G ($\varphi_{\alpha,x}(u)$ is related to U_α).

Figure 8.3.3 shows the fibre $\Pi^{-1}(x)=G_x$ of x on U_β, and the homeomorphism between the point u on $\Pi^{-1}(x)$ and the point $\varphi_{\beta,x}(u)$ on the standard fibre G ($\varphi_{\beta,x}(u)$ is related to U_β).

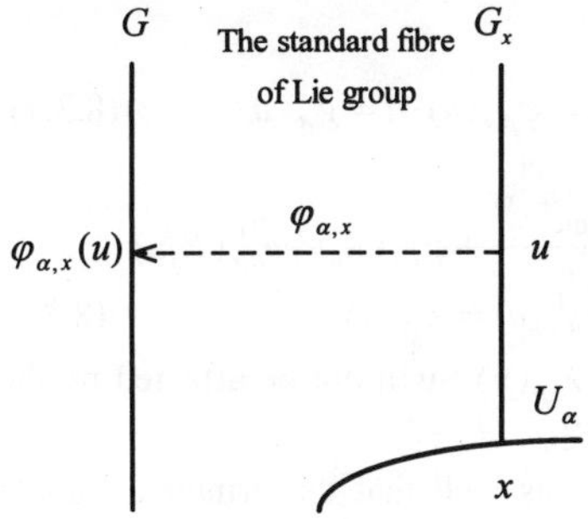

Figure 8.3.2

The principal bundle is locally trivial. Hence for the local open set U_α we have the following homeomorphism φ_α:

$$\begin{array}{lccc} \varphi_\alpha: & \Pi^{-1}(U_\alpha) & \to & U_\alpha \times G \\ \text{Homeomorphism} & u & \to & (x, \varphi_{\alpha,x}(u)) \end{array} \tag{8.3.2}$$

$$\left(u \in \Pi^{-1}(U_\alpha),\ x \in U_\alpha,\ \varphi_{\alpha,x}(u) \in G \right)$$

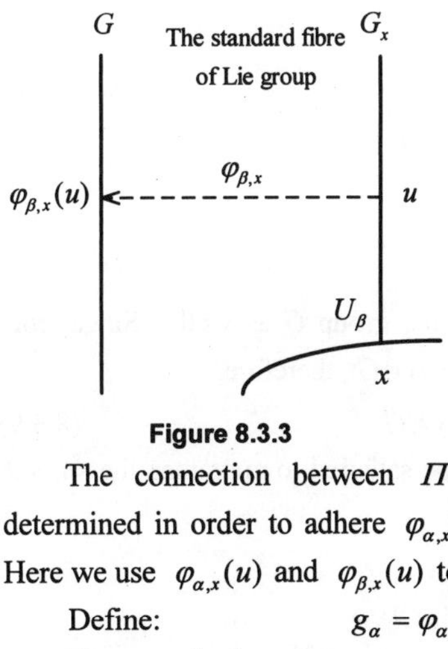

Figure 8.3.3

If $x \in U_\alpha \cap U_\beta$, then there is another homeomorphism φ_β:

$$\begin{array}{lccc} \varphi_\beta: & \Pi^{-1}(U_\beta) & \to & U_\beta \times G \\ \text{Homeomorphism} & u & \to & (x, \varphi_{\beta,x}(u)) \end{array} \tag{8.3.3}$$

$$\left(u \in \Pi^{-1}(U_\beta),\ x \in U_\beta,\ \varphi_{\beta,x}(u) \in G \right)$$

The connection between $\Pi^{-1}(U_\alpha)$ and $\Pi^{-1}(U_\beta)$ at $x \in U_\alpha \cap U_\beta$ must be determined in order to adhere $\varphi_{\alpha,x}(u)$ to $\varphi_{\beta,x}(u)$ smoothly in the neighborhood of x. Here we use $\varphi_{\alpha,x}(u)$ and $\varphi_{\beta,x}(u)$ to define $\lambda_{\alpha\beta}(x)$.

Define: $$g_\alpha = \varphi_{\alpha,x}(u), \qquad g_\beta = \varphi_{\beta,x}(u) \tag{8.3.4}$$

Then we obtain:

$$g_\alpha = \varphi_{\alpha,x}(u) = \varphi_{\alpha,x}(u)\varphi_{\beta,x}^{-1}(u)\varphi_{\beta,x}(u) = \varphi_{\alpha,x}(u)\varphi_{\beta,x}^{-1}(u)g_\beta = \lambda_{\alpha\beta}(x)g_\beta \tag{8.3.5}$$

$$\lambda_{\alpha\beta}(x) = \varphi_{\alpha,x}(u)\varphi_{\beta,x}^{-1}(u) = g_\alpha g_\beta^{-1} \tag{8.3.6}$$

where $\lambda_{\alpha\beta}(x)$ is called the transfer function.

To investigate the relations between the transfer function $\lambda_{\alpha\beta}$ and the right action of the group G onto the principal bundle, we replace each u in Eq.8.3.6 by $R_a u$ which follows the rules (see Definition 8.3.1 (ii)):

$$\varphi_{\alpha,x}(u) \xrightarrow[\text{onto the principal bundle}]{\substack{\text{Applying the right action of}\\ \text{the element } a \text{ of the group } G}} \varphi_{\alpha,x}(R_a u) = \varphi_{\alpha,x}(u)\cdot a = g_\alpha \cdot a \qquad (8.3.7)_1$$

$$\varphi_{\beta,x}(u) \xrightarrow[\text{onto the principal bundle}]{\substack{\text{Applying the right action of}\\ \text{the element } a \text{ of the group } G}} \varphi_{\beta,x}(R_a u) = \varphi_{\beta,x}(u)\cdot a = g_\beta \cdot a \qquad (8.3.7)_2$$

$$\therefore\ \lambda_{\alpha\beta}(x) = \varphi_{\alpha,x}(u)\varphi_{\beta,x}^{-1}(u) \xrightarrow[\text{onto the principal bundle}]{\substack{\text{Applying the right action of}\\ \text{the element } a \text{ of the group } G}} \varphi_{\alpha,x}(R_a u)\varphi_{\beta,x}^{-1}(R_a u)$$

$$= \varphi_{\alpha,x}(u)\cdot a \cdot a^{-1} \cdot \varphi_{\beta,x}^{-1}(u) = \varphi_{\alpha,x}(u)\varphi_{\beta,x}^{-1}(u) = \lambda_{\alpha\beta}(x) \qquad (8.3.8)$$

Thus it is illustrated that the "transfer function $\lambda_{\alpha\beta}(x)$" will not be affected by the "right action of the group *G* onto the principal bundle"!

From Figure 8.3.2 and Figure 8.3.3 we can see as well that the principal bundle consists of the base manifold *M*, the Lie group *G* (as the fibre), and the projection *Π*. Hence the principal bundle *P* can be denoted by *P*(*MGP*). The inner relations of the principal bundle are illustrated in the following figure:

$$\begin{array}{ccc} G & \longrightarrow & P \\ & & \big\downarrow \Pi \\ & & M \end{array}$$

Figure 8.3.4

From Eq.8.3.8 we see that $\lambda_{\alpha\beta}(x)$ belongs to the group *G* as well. Since from Eq.8.3.2 and Eq.8.3.3 we have $\varphi_{\alpha,x}(u) \in G$ and $\varphi_{\beta,x}(u) \in G$, therefore

$$\lambda_{\alpha\beta}(x) = \varphi_{\alpha,x}(u)\varphi_{\beta,x}^{-1}(u) \in G. \qquad (8.3.9)$$

The following conditions for consistency must be satisfied to guarantee the smooth adhesion of the open sets U_α, U_β, U_γ, $\cdots$:

$$\begin{array}{ll} \lambda_{\alpha\alpha}(x) = 1 & x \in U_\alpha \\ \lambda_{\alpha\alpha}(x) = \lambda_{\beta\beta}(x) & x \in U_\alpha \cap U_\beta \\ \lambda_{\alpha\beta}(x)\cdot\lambda_{\beta\gamma}(x) = \lambda_{\alpha\gamma}(x) & x \in U_\alpha \cap U_\beta \cap U_\gamma \end{array}$$

$$(8.3.10)$$

Thus the set of the transfer functions forms the structure group $\{\lambda_{\alpha\beta}(x)\} \subset G$ of the fibre bundle (the principal bundle).

When $x \in U_\alpha \cap U_\beta$, the point *u* on $\Pi^{-1}(x)$ can be mapped into g_α and g_β in *G* through the homeomorphisms $\varphi_{\alpha,x}$ and $\varphi_{\beta,x}$, referring to the equations Eq.8.3.4 and Eq.8.3.5.

Hence the right multiplication of the group element *a* of *G* to the principal bundle *P* can be defined by the right multiplication of *a* to the standard fibre *G*. To make this point clear, we use Eq.8.3.4:

$$\varphi_{\alpha,x}(R_a u) = \varphi_{\alpha,x}(u)\cdot a = g_\alpha \cdot a$$

$$\longrightarrow \quad R_a u = \varphi_{\alpha,x}^{-1}(u)\cdot g_\alpha \cdot a \qquad \text{(summing with respect to } \alpha\text{)}$$

$$\varphi_{\beta,x}(R_a u) = \varphi_{\beta,x}(u)\cdot a = g_\beta \cdot a$$

$$\longrightarrow \quad R_a u = \varphi_{\beta,x}^{-1}(u)\cdot g_\beta \cdot a \qquad \text{(summing with respect to } \beta\text{)} \qquad (8.3.11)$$

Thus $R_a u$ is equal to both $\varphi_{\alpha,x}^{-1}\, g_\alpha \cdot a$ and $\varphi_{\beta,x}^{-1}\, g_\beta \cdot a$, which means that the right multiplication (i.e. $R_a u$) of the group element *a* of *G* to the principal bundle *P* (which is represented by *u*) is acting on the whole *G*. Each group index in *G* is represented by *a* (or *b*) in $\varphi_{\alpha,x}^{-1}(u)\, g_\alpha$ (or $\varphi_{\beta,x}^{-1}(u)\, g_\beta$). That is, *a* (or *b*) represents the whole group *G*. If we treat each orbit on which the group *G* acts as a point, then

$$P/G = M \qquad (8.3.12)$$

where *M* is the base manifold of the principal bundle *P*. Thus *M* in Figure 8.3.4 should be taken as $M = P/G$.

In brief, $R_a u$ acts on the entire *G*, and hence can be defined globally. The right action of *G* onto the principal bundle *P* does not vary the structure group of the bundle, nor does it affect the global topological structure of the bundle. Since $\lambda_{\alpha\beta}(x)$'s are identical on the open sets U_α, U_β, U_γ, ⋯, the smooth adhesion of the fibres in the intersecting regions will not be affected either.

C. *The Adjoint Vector Bundle $E(M, F, G, \Pi, P)$:*

Let *P*(*MGP*) be a principal bundle, abbreviated as *P*. Let *G* represent the fibre of the principal bundle *P*. *G* is then the structure group of *P* as well.

Let *F* be a vector space on which *G* may act. Then we may define the direct product bundle $P \times F$ of which any point can be written as (p, f): $p \in P$, $f \in F$.

Here *p* is a point on the principal bundle *P*, it corresponds to a group element of *G*.

f, on which the group element *g* of *G* in the principal bundle may act, is a vector in the vector space *F*.

Therefore we define the action formulae of the group *G* of the principal bundle and the group element $g \in G$ onto the direct product bundle $P \times F$ as the following:

$$G\colon \quad P \times F \longrightarrow P \times F$$

$$g\colon \quad (p, f) \longrightarrow (pg, g^{-1} f) \qquad (8.3.13)$$

That is, the action of *G* onto *F* is induced by the action of *G* onto *P*. While $g \in G$ is applying the right action onto *P*, it gives the left action on *F* as well.

Treat the point (p, f) on $P \times F$ and the related point $(pg, g^{-1} f)$ through the group action equivalently, i.e. $(p, f) \sim (pg, g^{-1} f)$. Then we obtain the subdivision of $P \times F$ according to the equivalent classes: $\{\, (pG, G^{-1} f) \,\}$. Treat each equivalent class as a point, we have

$$\{ (pG, G^{-1}f) \} = \{ (\bar{p}, \bar{f}) \} = P \times F/G \tag{8.3.14}$$

where the quotient manifold obtained here is called the *adjoint vector bundle* E of the principal bundle P:

$$E = P \times F/G \tag{8.3.15}$$

§ 8.4 *A few Definitions, Principles and Examples of Fibre Bundles*

a) The Pullback Bundle

Let E be a fibre bundle on the base manifold M, and F be the fibre. Let f be a mapping which maps another manifold M' onto M, that is,

$$\begin{aligned} f: \quad & M' \rightarrow M \\ & x' \rightarrow x = f(x') \end{aligned} \tag{8.4.1}$$

where specifically $f(x')$ maps x' of M' onto $x = f(x')$ of M. Hence a fibre bundle can be established on the base manifold M'. Taking the fibre at the point x' of M' as the fibre at the point $x = f(x')$ of M, we obtain a fibre bundle and denote it as E'. Thus f^* can be defined by:

$$E' = f^* E$$

or

$$\begin{aligned} E' \quad & \xleftarrow{f^*} \quad E \\ F_{f(x')} \quad & \xleftarrow{f^*} \quad F_x \end{aligned} \tag{8.4.1$_1$}$$

$E' = f^* E$ is called the *pullback bundle.*

If $M = M'$, and $f(x) = x$, then $f^* E = E$.

b) The Equivalent Bundle and Homotopic Mapping

Let h and g be two mappings from M' to M ($h^* E$ and $g^* E$ are not the same in general). If there exists an $H(x', \alpha)$ with the parameter a varying smoothly from 0 to 1, such that $H(x', \alpha = 0) = h(x')$ and $H(x', \alpha = 1) = g(x')$, then the mappings h and g are homotopic to each other, or are said to be equivalent. $h^* E$ and $g^* E$ are equivalent bundles (i.e. isomorphic to each other).

Theorem 8.4.1: If the base manifold M is contractible, then the fibre bundle constructed on M must be trivial (i.e. it is isomorphic to $M \times F$)[1].

Proof: Since M can be contracted into arbitrary small M', including a single point $M' = x_0$. That means that there is no special topological structure on M'. Hence when M is contracted into M', the fibre bundle must be $E' = M' \times F$. Then we define the "*identity mapping*" h such that $h(x') = x'$ (x' is invariant), and the

---

1 The solid sphere in R^n and the manifold formed by S^n taking away one point are both manifolds which can be contracted into one point.

"*constant mapping*" g such that $g(x) = x_0$ (All the points on M are contracted to x_0). Then M being contractible means that the mappings h and g are homotopic. Since h^* and g^* are the inverses of h and g, respectively, then E is equivalent to $M \times F$, and hence E must be trivial.

$$E = h^* E \xleftrightarrow{\text{Homotopically equivalent}} M \times F = g^* E'$$

□

According to Theorem 8.4.1, a non-trivial fibre bundle can be obtained only if the base manifold is not contractible. The simplest noncontractible manifold is the n-dimensional sphere S^n (n=1, 2, 3, ···).

c) The Trivial Bundle

The global topological properties of the trivial bundle are equivalent to those of $E = M \times F$.

Theorem 8.4.2: The principal bundle P is trivial if and only if it has a globally smooth section. (The proof is omitted.)

Theorem 8.4.3: If the vector bundle E has global sections which are nonzero everywhere, and if the dimension of the maximum collection of the linearly independent sections are equal to the dimension k of the fibre. Then the vector bundle E is trivial. (The proof is omitted.)

d) The Line Bundle

The line bundle is a kind of vector bundle as well, of which the fibre is the 1-dimensional vector space.

If we replace the interval $[-1, 1]$ on the Möbius strip by the real line R, then we can obtain a non-trivial real line bundle with the circle as its base manifold.

If the interval $[-1, 1]$ is replaced by the complex set C, then the line bundle we obtain is isomorphic to $S^1 \times C$, which is a trivial complex line bundle.

e) The Tangent Bundle and the Cotangent Bundle

Let M be an n-dimensional real manifold. Suppose that a local coordinate system $x = (x_1, x_2, \cdots, x_n)$ is taken on a neighborhood U of M. Then the tangent bundle $T(M)$ on the base manifold M can be defined as the real vector bundle, of which the fibre at the point $x \in M$ is taken as the tangent space $T_x(M)$ at the point x. The cotangent bundle $T^*(M)$ on the base manifold M can be defined as the real vector bundle as well, of which the fibre at the point $x \in M$ is taken as the cotangent space $T_x^*(M)$ at the point x. The standard basis can be defined as the following:

The standard basis of the tangent bundle $T(M)$: $\left\{\dfrac{\partial}{\partial x_1}, \cdots, \dfrac{\partial}{\partial x_n}\right\}$.

The standard basis of the cotangent bundle $T^*(M)$: $\{dx_1, \cdots, dx_n\}$.

If U^a and U^b are two intersecting neighborhoods on M, of which the coordinates are written as x_i^a and x_j^b, respectively. Then the transformation of the standard basis in $U^a \cap U^b$ is:

$$\frac{\partial}{\partial x_i^a} = \frac{\partial x_j^b}{\partial x_i^a} \frac{\partial}{\partial x_j^b} \qquad \text{for the tangent bundle; and}$$

$$dx_i^a = \frac{\partial x_i^a}{\partial x_j^b} dx_j^b \qquad \text{for the cotangent bundle.}$$

§ 8.5 Universal Bundle and Classifying Space

How many fibre bundles which are not equivalent can be formed after the base manifold *M* and the fibre *F* are given? That is the topic of our discussion in this section (where the concepts of equivalence and equivalent bundles can be found in b) of Section 8.4).

We have known that the fibre of the frame bundle *L*(*E*) is a group manifold. Thus *L*(*E*) is in fact a kind of principal bundle, namely the adjoint principal bundle of the vector bundle *E* (see Section 8.2). In general, the transitions of the vector space of the vector bundle *E* require the introduction of the structure group *G* (see Part A of Section 8.3). The adjoint vector bundle *E*, which is called the adjoint principal bundle sometimes, is determined by the structure group *G* as well (see Part C of Section 8.3). Therefore we can see that after the base manifold *M* and the fibre space *F* are given, the structures of all the adjoint bundles (such as the frame bundle, the general vector bundle, and the adjoint principal bundle, etc.) are deeply connected to the principal bundle (especially the properties of the structure group of the principal bundle). Hence it is necessary to first investigate the classification of the principal bundles, that is, to analyze how many principal bundles are not equivalent after *M* and the group *G* are given.

The idea of classifying the principal bundles with the same structure group is mainly to use the analysis of the homotopic groups.

Definition 8.5.1: Suppose that *M* is the base manifold of a fibre bundle, on which the "*family of transition functions* $\{g_{ij}\}$" (or the "*family of structure functions*") covers the complete procedure of adhering the local fibre space to the base manifold *M* to obtain the entire fibre bundle. Take a continuous mapping *f*: $M \to M'$, from which we obtain another family of functions $\{g_{ij} \cdot f\}$, which covers the complete procedure of adhering the local fibre space to the base manifold M' to obtain another fibre bundle. Such a fibre bundle we obtained on M' is called the *induced bundle*. The homotopic mappings on a compact space induce the same bundle.

Definition 8.5.2: The space *X* is called a *compact space* if each of its open covers has a finite subset which is still a cover of *X*.

Now we begin to investigate the classification of the principal *G* bundle. There is a kind of special and essential principal G bundle, denoted by $\xi(G)$. The specialty is that every other principal bundle with G as its structure group is an induced bundle of this $\xi(G)$.

For $\xi(G)$ we have a figure similar to Figure 8.3.4 which illustrates the inner relations of the principal G bundle.

$$\begin{array}{ccc} G & \longrightarrow & \xi(G) \\ & & \Big\downarrow \Pi \\ & & B_G \end{array}$$

Figure 8.5.1

Here corresponding to the base manifold M in Figure 8.3.4 is the base manifold B_G. The bundle $\xi(G)$ is called the *universal bundle* which has the feature that its homotopy groups of all orders are zero.

$$\Pi_k(\xi) = 0, \qquad k \in Z^+ \tag{8.5.1}$$

There is another case where $\xi(G)$ satisfies only the following:

$$\Pi_k(\xi) = 0, \qquad k \le n+1 \tag{8.5.2}$$

In this case $\xi(G)$ is called the *n-universal bundle*, and is denoted by $\xi(n,G)$.

If a continuous mapping f can be defined from the manifold M to B_G,

$$f\colon \ M \to B_G \tag{8.5.3}$$

then the principal bundle $P(M,G)$ with M as its base manifold must be isomorphic to the pullback bundle $f^*\xi(G)$ of the universal bundle $\xi(G)$:

$$P(M,G) \cong f^*\xi(G) \tag{8.5.4}$$

Similar to Eq.8.5.3 we have

$$\begin{aligned} f^*\colon \quad & B_G \longrightarrow M \\ f^*\colon \quad & \xi(G) \longrightarrow P(M,G) \end{aligned} \tag{8.5.5}$$

Hence there must exist a 1-1 correspondence between the set of the equivalent classes of principal bundles with G as their structure group on the manifold M, such as $f^*\xi(G)$, $g^*\xi(G)$, $h^*\xi(G)$, $\cdots$, and the set of the homotopy classes represented by the continuous mappings f, g, h, $\cdots$ from M to B_G. This is the *Classification Theorem of Principal Bundles*. The manifold B_G is called the *classifying space*. An illustration is shown in Figure 8.5.2.

If the dimension of the manifold M is less than n, then by replacing $\xi(G)$ by $\xi(n,G)$ we may have the similar theorem as well, since those manifolds with dimensions greater than n are irrelevant.

The classification of vector bundles is completely determined by the classification of principal bundles. Thus we can find the universal vector bundle as well. (The details are omitted.)

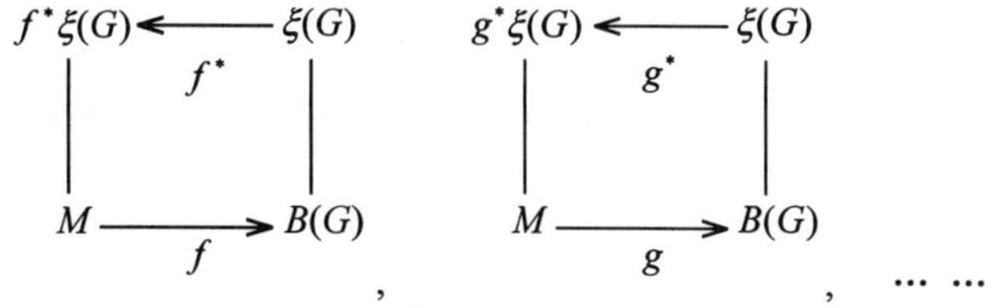

Figure 8.5.2

Another conclusion we can obtain from Theorem 8.4.1 is that the base manifold M of a non-trivial bundle must be noncontractible into a single point. The simplest noncontractible manifold is the n-dimensional sphere S^n. The fibre bundles with S^n as their base manifold are possibly non-trivial (n>0).

Here we give an example of Yang-Mills instanton where $M = S^4$. It is a principal bundle with $SU(2)$ group as its fibre and S^4 as its base manifold. It has the following features:

$M = S^4$ with four real coordinates r, q, j, and y.

$SU(2)$ group manifold is S^3 with three parameters a, b, and g of $SU(2)$ group.

S^4 can be divided into two semi-spheres H_+ and H_-, which are both contractible with their boundaries being S^3. The coordinates on H_+ and H_-, and the parameters of the fibre $SU(2)$ group are

$$\begin{aligned} &H_+ \times SU(2) \qquad (r,\theta,\varphi,\psi;\alpha_+,\beta_+,\gamma_+) \\ &H_- \times SU(2) \qquad (r,\theta,\varphi,\psi;\alpha_-,\beta_-,\gamma_-) \end{aligned} \tag{8.5.8}$$

respectively.

Extend H_+ and H_- a little to form the intersecting region $H_+ \cap H_-$:

$$H_+ \cap H_- = L \times S^3 \tag{8.5.9}$$

where L can be contracted into infinitesimal such that the intersecting region $H_+ \cap H_-$ is homotopic to S^3. The transition functions between the open covers on both H_+ and H_- are uniformly denoted as

$$g_{+,-} \in SU(2) \tag{8.5.10}$$

Notice that the group manifold of $SU(2)$ is 3-dimensional (since the group manifold is S^3). Hence $g_{+,-} \in SU(2)$ in Eq.8.5.10 is 3-dimensional as well, and $g_{+,-}$ form the 3-dimensional circuits on the intersecting region $H_+ \cap H_-$ or S^3. According to Eq.6.5.1, take n=3, we have

$$\Pi_3(S^3) = Z \longrightarrow \Pi_3(SU(2)) = Z \tag{8.5.11}$$

Therefore $g_{+,-}$ can classify the principal bundles of $SU(2)$ group on S^4 according to the values of the integer Z in the equation $\Pi_3(SU(2)) = Z$.

Generalize the conclusion to the principal G bundle on S^n, we can classify with the homotopy group $\Pi_{n-1}(G)$ of order $(n-1)$, in a way similar to the above example where $M = S^4$.

Chapter 9. Connections and Curvatures on Fibre Bundles

§ 9.1 Connections on a General Fibre Bundle

Let $E(M,F,\Pi,G)$ be a general vector bundle with a *k*-dimensional vector space. Then on each neighborhood *U* of the base manifold *M*, we may choose a local vector bundle frame $\{e_1,e_2,\cdots,e_k\}$ and write the section $S(x)$ of the vector bundle on $\Pi^{-1}(U)=U\times F$ as the following vector function:

$$S(x)=\sum_{i=1}^{k}e_i(x)z^i(x) \tag{9.1.1}$$

where $e_i(x)$ is the vector bundle frame at the point *x* of the fibre space, and $z^i(x)$ is the *i*th component of the vector bundle's section at *x*.

In the vector bundle *E*, the local base vectors of the tangent space $T(E)$ and the cotangent space $T^*(E)$ are $\left(\frac{\partial}{\partial x^\mu},\frac{\partial}{\partial z^i}\right)$ and $\left(dx^\mu,dz^i\right)$[1], respectively.

Here we look into some equivalent definitions of connection on the vector bundles.

1. Definition given by the parallel translation: Let $x(\tau)$ be a curve in *M*, and the vector space $S(x)$ be a section on the vector bundle *E*. If $S(x(\tau))$ satisfies the following differential equation

$$\nabla_{\frac{d}{d\tau}}(S)=0 \tag{9.1.2}$$

Then $S(x)$ is said to be a parallel translation along $x(\tau)$. $\frac{d}{d\tau}=\frac{dx^\mu}{d\tau}\frac{\partial}{\partial x^\mu}$ is the vector in the direction of $x(\tau)$.

According to the first equation in Eq.3.1.9 of Chapter 3, since that $\frac{\partial}{\partial x^\mu}$ and e_μ are vectors of the same type, we have

$$\nabla e_\mu(x)=\Gamma^\nu_\mu(x)e_\nu(x) \qquad (\Gamma^\nu_\mu(x)=\Gamma^\nu_{\lambda\mu}(x)dx^\lambda) \tag{9.1.3}$$

Then according to Eq.3.1.13 of Chapter 3, we obtain

$$\nabla_{\frac{\partial}{\partial x^\nu}}(e_i)=<\nabla e_i,\frac{\partial}{\partial x^\nu}>=<e_j\Gamma^j_{\mu i}dx^\mu,\frac{\partial}{\partial x^\nu}>=e_j\Gamma^j_{\nu i}(x) \tag{9.1.4}$$

Hence using Eq.9.1.1 we can write Eq.9.1.2 as

$$\nabla_{\frac{d}{d\tau}}(S)=\nabla_{\frac{d}{d\tau}}(e_iz^i)=\left(\nabla_{\frac{d}{d\tau}}e_i\right)z^i+e_i\left(\nabla_{\frac{d}{d\tau}}z^i\right)$$

---

[1] i,j=1, 2, $\cdots$, k are the indices of the vector space; μ, ν $\cdots$ are the indices of the coordinates on the base manifold *M*.

$$=<e_j\Gamma^j_{\mu i}dx^\mu, \frac{dx^\nu}{d\tau}\frac{\partial}{\partial x^\nu}>z^i + e_i < \frac{\partial z^i}{\partial x^\mu}dx^\mu, \frac{dx^\nu}{d\tau}\frac{\partial}{\partial x^\nu}>$$

$$= e_j\Gamma^j_{\nu i}\frac{dx^\nu}{d\tau}\cdot z^i + e_i\frac{\partial z^i}{\partial x^\nu}\frac{dx^\nu}{d\tau} = \frac{dx^\nu}{d\tau}e_j\left(\Gamma^j_{\nu i}z^i + \frac{\partial z^i}{\partial x^\nu}\right) = 0 \qquad (9.1.5)$$

Compare to Eq.3.1.7 and Eq.3.1.8 of Chapter 3, we see that $X = \xi^a\frac{\partial}{\partial x^a}$ being a parallel translation requires $\nabla X = \nabla\xi^a\frac{\partial}{\partial x^a} = \left(\frac{\partial\xi^a}{\partial x^i} + \Gamma^a_{ib}\xi^b\right)dx^i\frac{\partial}{\partial x^b} = 0$, which is exactly in the same form as that of the parallel translation in Eq.9.1.5, with difference only being that ξ^a and ξ^b are replaced by z^i and z^j, respectively. Therefore we define the connection on the vector bundle to be $\Gamma^j_{\mu i}$.

2. Definition in the tangent space's point of view: Let $x(\tau)$ be a curve on the base manifold *M*. We may promote it onto the vector bundle $E(M,F,\Pi,G)$, and denote the promoted curve as $c(\tau)$. The coordinates of each point on $c(\tau)$ should include both $x^\mu(\tau)$ and $z^i(\tau)$. Thus the curve promoted onto *E* should be written as

$$c(\tau) = (x^\mu(\tau), z^i(\tau)) \qquad (i=1, 2, \cdots, k) \qquad (9.1.6)$$

The differential along $c(\tau)$ should be defined as

$$\frac{d}{d\tau} = \frac{dx^\mu}{d\tau}\frac{\partial}{\partial x^\mu} + \frac{dz^i}{d\tau}\frac{\partial}{\partial z^i} \qquad (9.1.7)$$

According to the Levi-Civita's rule of parallel translation, $\frac{dz^i}{d\tau}$ must satisfy the following parallel transportation equation:

$$\frac{dz^i}{d\tau} + \Gamma^i_{\mu j}\frac{dx^\mu}{d\tau}z^j = 0 \qquad (9.1.8)$$

Plug into Eq.9.1.7 we obtain

$$\frac{d}{d\tau} = \frac{dx^\mu}{d\tau}\frac{\partial}{\partial x^\mu} - \Gamma^i_{\mu j}\frac{dx^\mu}{d\tau}z^j\frac{\partial}{\partial z^i}$$

$$= \frac{dx^\mu}{d\tau}\left(\frac{\partial}{\partial x^\mu} - \Gamma^i_{\mu j}z^j\frac{\partial}{\partial z^i}\right) = \frac{dx^\mu}{d\tau}D_\mu \qquad (9.1.9)$$

where

$$D_\mu = \frac{\partial}{\partial x^\mu} - \Gamma^i_{\mu j}z^j\frac{\partial}{\partial z^i} \qquad (9.1.10)$$

D_μ is a symbol of covariant differentiation, and $\Gamma^i_{\mu j}$ is the connection in the tangent space $T(E)$.

From Eq.9.1.10 we can see as well that $T(E)$ can be decomposed into two parts:

$$T(E) = V(E) \oplus H(E) \qquad (9.1.11)$$

where $V(E)$, with the base vectors $\frac{\partial}{\partial z^i}$, is called the vertical part of the tangent space; and $H(E)$, with the base vectors $D_\mu = \frac{\partial}{\partial x^\mu} - \Gamma^i_{\mu j} z^j \frac{\partial}{\partial z^i}$, is called the horizontal part of the tangent space. The base vectors D_μ in H may be considered as a horizontal promotion of $\frac{\partial}{\partial x^\mu}$ on the manifold M onto the manifold E.

3. Definition in the cotangent space's point of view: Consider an 1-form in the cotangent space $T^*(E)$:

$$\omega^i = dz^i + \Gamma^i_{\mu j} dx^\mu z^j \tag{9.1.12}$$

It has the following features:

(i) $$< \omega^i, D_\mu > = < dz^i + \Gamma^i_{\nu j} dx^\nu z^j, \frac{\partial}{\partial x^\mu} - \Gamma^k_{\mu l} z^l \frac{\partial}{\partial z^k} > = -\Gamma^i_{\mu l} z^l + \Gamma^i_{\mu j} z^j = 0 \tag{9.1.13}$$

(ii) $$< \omega^i, \frac{\partial}{\partial z^j} > = < dz^i + \Gamma^i_{\nu j} dx^\nu z^j, \frac{\partial}{\partial z^j} > = \delta^i_j \tag{9.1.14}$$

Eq.9.1.13 and Eq.9.1.14 indicate that the 1-form ω^i in the cotangent space $T^*(E)$ is orthogonal to the horizontal part ($D_\mu \in H(E)$) of the tangent space $T(E)$; the relation between ω^i and the vertical part ($\frac{\partial}{\partial z^j} \in V(E)$) of the tangent space $T(E)$ is given by Eq.9.1.14.

The 1-form $\Gamma^i_{\mu j} dx^\mu$ introduced by Eq.9.1.12 is the connection 1-form on a general vector bundle. For example, take $s = e_i z^i$, we have

$$\begin{aligned} \nabla(s) &= \nabla(e_i z^i) = e_i(\nabla z^i) + \nabla(e_i) z^i \\ &= e_i\, dz^i + \Gamma^j_i e_j z^i \qquad (\text{where } \Gamma^j_i = \Gamma^j_{\mu i} dx^\mu) \end{aligned} \tag{9.1.15}$$

4. Definition in the frame transformation's point of view: Take a frame transformation

$$e'_j = e_i (f^{-1})^i_j(x), \quad z'^i = f^i_j(x) z^j . \tag{9.1.16}$$

The section $S(x)$ is invariant, that is,

$$S(x) = e_i(x) z^i(x) = e'_i(x) z'^i(x) = S'(x) . \tag{9.1.17}$$

Since that the transformation properties of $e_i(x)$ and the tangent vector $\frac{\partial}{\partial x^i}$ are the same, therefore according to the first equation in Eq.3.1.9, $\nabla e'_j(x)$ must be equal to (using Eq.9.1.16 as well):

$$\begin{aligned} \nabla e'_j(x) &= \nabla\left(e_i (f^{-1})^i_j(x)\right) = \nabla e_i \cdot (f^{-1})^i_j + e_i \cdot \nabla (f^{-1})^i_j \\ &= e'_i(x) \Gamma'^i_j = e_k(x) \cdot (f^{-1})^k_i \Gamma'^i_j \end{aligned} \tag{9.1.18}$$

Apply the first equation in Eq.3.1.9 once more, we obtain from Eq.9.1.18

$$\nabla e_i \cdot (f^{-1})^i_j + e_i \cdot \nabla (f^{-1})^i_j = e_k \cdot (f^{-1})^k_i \Gamma'^i_j$$

$$\longrightarrow e_k \Gamma^k_i \cdot (f^{-1})^i_j + e_k \cdot \nabla (f^{-1})^k_j = e_k \cdot (f^{-1})^k_i \Gamma'^i_j \tag{9.1.19}$$

Since that e_k 's are independent to each other, Eq.9.1.19 gives that

$$\Gamma^k_i (f^{-1})^i_j + d(f^{-1})^k_j = (f^{-1})^k_i \Gamma'^i_j \tag{9.1.20}$$

$$\longrightarrow (f)^l_k \Gamma^k_i (f^{-1})^i_j + (f)^l_k d(f^{-1})^k_j = \Gamma'^l_j \tag{9.1.21}$$

where the transformation Eq.9.1.21 between Γ^k_i and Γ'^l_j is obviously not the transformation of the tensorial type.

Now let us take a look at the normal transformation of a normal field.

$$T^a A^a_\mu \longrightarrow T^a A^{a'}_\mu = e^{-i\theta^i T^i}\left[T^a A^a_\mu - \frac{i}{g} e^{i\theta^i T^i} \partial_\mu (e^{-i\theta^i T^i})\right] e^{i\theta^i T^i}$$

$$\to e^{i\theta^i T^i} T^a A^{a'}_\mu e^{-i\theta^i T^i} = T^a A^a_\mu - \frac{i}{g} e^{i\theta^i T^i} \partial_\mu e^{-i\theta^i T^i}$$

$$\to T^a A^{a'}_\mu = e^{-i\theta^i T^i} T^a A^a_\mu e^{i\theta^i T^i} - \frac{i}{g} \partial_\mu e^{-i\theta^i T^i} \cdot e^{i\theta^i T^i}$$

Take $U = e^{-i\theta^i T^i}$, and $U^{-1} = e^{i\theta^i T^i}$ (9.1.22)

$$\to T^a A^{a'}_\mu = U T^a A^a_\mu U^{-1} + \frac{i}{g} U \cdot \partial_\mu U^{-1} \tag{9.1.23}$$

Compare Eq.9.1.21 and Eq.9.1.23 we can see that $T^a A^{a'}_\mu$ and $T^a A^a_\mu$ correspond to Γ' and Γ, respectively; U and U^{-1} correspond to $e^{-i\theta^i T^i}$ and $e^{i\theta^i T^i}$, respectively. Thus the transformation from Γ^k_i to Γ'^l_j in Eq.9.1.21 happens to be similar to the transformation from $T^a A^a_\mu$ to $T^a A^{a'}_\mu$ in the normal field, but is definitely not a tensor transformation. Therefore the connection 1-form Γ^i_j can be defined according to the transformation in Eq.9.1.21 as well.

It is known that $S(x) = S'(x)$ in the intersecting region $U \cap U'$ (see Eq.9.1.17). It can be proved as well that $\nabla(S) = \nabla(S')$. The proof is the following:

According to Eq.9.1.15, Eq.9.1.16 and Eq.9.1.21,

$$\begin{aligned}
\nabla(S') &= e'_j dz'^j + e'_j \Gamma'^i_j z'^j \\
&= e_i \left(f^{-1}\right)^i_j d\left(f^j_k z^k\right) + e_l \left(f^{-1}\right)^l_i \left(f^i_k \Gamma^k_m \left(f^{-1}\right)^m_j + f^i_k d\left(f^{-1}\right)^k_j\right) f^j_k z^k \\
&= e_i \delta^i_k dz^k + e_i \left(f^{-1}\right)^i_j \left(df^j_k\right) z^k + e_l \delta^l_k \Gamma^k_m \delta^m_k z^k + e_l \delta^l_k d\left(f^{-1}\right)^k_j f^j_k z^k \\
&= e_k dz^k + e_i \left(f^{-1}\right)^i_j \left(df^j_k\right) z^k + e_k \Gamma^k_m z^m + e_i \left(df^{-1}\right)^i_j f^j_k z^k \\
&= e_k dz^k + e_k \Gamma^k_m z^m = \nabla(S)
\end{aligned} \tag{9.1.24}$$

Since Eq.9.1.17 and Eq.9.1.24 have proved that $S(x) = S'(x)$ and $\nabla(S) = \nabla(S')$, that is, both $S(x)$ and $\nabla(S)$ are invariant under the transformation of the coordinate system. Therefore consistent definitions which are independent of the choices of frames can be made with respect to $S=e_i z^i$ and $\nabla(S)=e_k dz^k + e_k\Gamma^k_m z^m$ in the two neighborhoods U and U', and in the intersecting area $U\cap U'$.

§ 9.2 Remarks on Curvature of Vector Bundles

1. Curvature tensor: According to Eq.9.1.10, we obtain $[D_\mu, D_\nu]$ as follows:

$$[D_\mu,D_\nu]=\left[\frac{\partial}{\partial x^\mu}-\Gamma^i_{\mu j}z^j\frac{\partial}{\partial z^i},\frac{\partial}{\partial x^\nu}-\Gamma^k_{\nu l}z^l\frac{\partial}{\partial z^k}\right]$$

$$=-\left(\frac{\partial}{\partial x^\mu}\Gamma^k_{\nu l}\right)z^l\frac{\partial}{\partial z^k}+\left(\frac{\partial}{\partial x^\nu}\Gamma^i_{\mu j}\right)z^j\frac{\partial}{\partial z^i}+\Gamma^i_{\mu j}\Gamma^k_{\nu i}z^j\frac{\partial}{\partial z^k}-\Gamma^k_{\nu l}\Gamma^i_{\mu k}z^l\frac{\partial}{\partial z^i}$$

$$=-\left(\frac{\partial}{\partial x^\mu}\Gamma^i_{\nu j}-\frac{\partial}{\partial x^\nu}\Gamma^i_{\mu j}+\Gamma^i_{\mu k}\Gamma^k_{\nu j}-\Gamma^i_{\nu k}\Gamma^k_{\mu j}\right)z^j\frac{\partial}{\partial z^i}=-R^i_{j\mu\nu}z^j\frac{\partial}{\partial z^i} \qquad (9.2.1)$$

$$R^i_{j\mu\nu}=\frac{\partial}{\partial x^\mu}\Gamma^i_{\nu j}-\frac{\partial}{\partial x^\nu}\Gamma^i_{\mu j}+\Gamma^i_{\mu k}\Gamma^k_{\nu j}-\Gamma^i_{\nu k}\Gamma^k_{\mu j} \qquad (9.2.2)$$

where $R^i_{j\mu\nu}$ is the curvature tensor of the vector bundle.

Here is something interesting: According to Eq.9.1.11, D_μ and D_ν are the base vectors of the horizontal part of the tangent space; $\frac{\partial}{\partial z^i}$ and $\frac{\partial}{\partial z^k}$ are the base vectors of the vertical part of the tangent space; and moreover, the base vectors of the vertical and of the horizontal parts are independent to each other. However, Eq.9.2.1 shows that commutant of the base vectors of the horizontal part, D_μ and D_ν, gives the base vectors $\frac{\partial}{\partial z^i}$ of the vertical part.

2. Γ^i_j can be written as the pure gauge when $R^i_{j\mu\nu}=0$.

Suppose that there exists a locally flat frame $e'_j=e_i\left(f^{-1}\right)^i_j$ satisfying $\nabla e'_j=0$. Then

$$\left(\nabla e_i\right)\left(f^{-1}\right)^i_j+e_i\left(\nabla f^{-1}\right)^i_j=0$$

$$\longrightarrow e_k\Gamma^k_i\left(f^{-1}\right)^i_j+e_i\left(df^{-1}\right)^i_j=0.$$

Since that e_i and e_k are independent to each other, hence

$$\Gamma^k_i\left(f^{-1}\right)^i_j+\left(df^{-1}\right)^k_j=0.$$

Multiply f^j_l to the right, we have

$$\Gamma_l^k = -\left(df^{-1}\right)_j^k f_l^j = \left(f^{-1}\right)_j^k \left(df\right)_l^j \tag{9.2.3}$$

$$\therefore \quad \Gamma_{\mu l}^k = \left(f^{-1}\right)_j^k \frac{\left(\partial f\right)_l^j}{\partial x^\mu} = -f_l^j \left(\frac{\partial f^{-1}}{\partial x^\mu}\right)_j^k \tag{9.2.4}$$

where the right-hand-sides of Eq.9.2.3 and Eq.9.2.4 are both pure gauge forms[1].

Then plug Eq.9.2.4 into Eq.9.2.2 we have

$$\begin{aligned}
R_{j\mu\nu}^i &= \frac{\partial}{\partial x^\mu}\left((f^{-1})_l^i \frac{(\partial f)_j^l}{\partial x^\nu}\right) - \frac{\partial}{\partial x^\nu}\left((f^{-1})_l^i \frac{(\partial f)_j^l}{\partial x^\mu}\right) \\
&\quad + \left((f^{-1})_l^i \frac{(\partial f)_k^l}{\partial x^\mu}\right)\left((f^{-1})_m^k \frac{(\partial f)_j^m}{\partial x^\nu}\right) - \left((f^{-1})_l^i \frac{(\partial f)_k^l}{\partial x^\nu}\right)\left((f^{-1})_m^k \frac{(\partial f)_j^m}{\partial x^\mu}\right) \\
&= \frac{\partial (f^{-1})_l^i}{\partial x^\mu}\frac{\partial (f)_j^l}{\partial x^\nu} + (f^{-1})_l^i \frac{(\partial^2 f)_j^l}{\partial x^\mu \partial x^\nu} - \frac{\partial (f^{-1})_l^i}{\partial x^\nu}\frac{\partial (f)_j^l}{\partial x^\mu} - (f^{-1})_l^i \frac{(\partial^2 f)_j^l}{\partial x^\nu \partial x^\mu} \\
&\quad - \left(\frac{\partial (f^{-1})_l^i}{\partial x^\mu} f_k^l\right)\left((f^{-1})_m^k \frac{\partial f_j^m}{\partial x^\nu}\right) + \left(\frac{\partial (f^{-1})_l^i}{\partial x^\nu} f_k^l\right)\left((f^{-1})_m^k \frac{\partial f_j^m}{\partial x^\mu}\right) \\
&= \frac{\partial (f^{-1})_l^i}{\partial x^\mu}\frac{\partial (f)_j^l}{\partial x^\nu} - \frac{\partial (f^{-1})_l^i}{\partial x^\nu}\frac{\partial (f)_j^l}{\partial x^\mu} \\
&\quad - \frac{\partial (f^{-1})_l^i}{\partial x^\mu}\frac{\partial (f)_j^l}{\partial x^\nu} + \frac{\partial (f^{-1})_l^i}{\partial x^\nu}\frac{\partial (f)_j^l}{\partial x^\mu} = 0
\end{aligned} \tag{9.2.5}$$

Thus we see that if the connection Γ_k^i of the vector bundle is taken as a pure gauge, then the curvature tensor $R_{j\mu\nu}^i = 0$.

The inverse proposition holds as well. That is, if $R_{j\mu\nu}^i = 0$ then Γ_j^i can be written in a pure gauge form as in Eq.9.2.3. It can be proved by Frobenius Theorem. (The proof is omitted.)

3. The transformation properties of the curvature 2-form Ω_j^i on vector bundles.

From Eq.9.1.24: $\nabla(s) = e_i\left(dz^i + \Gamma_j^i z^j\right) = e_i \omega^i$.

Then from Eq.9.1.12: $\omega^i = dz^i + \Gamma_{\mu j}^i z^j dx^\mu$,

$$\begin{aligned}
\therefore \quad \nabla^2(s) &= \nabla\left(e_i \omega^i\right) = \nabla e_i \wedge \omega^i + e_i \nabla \omega^i \\
&= e_j \Gamma_i^j \wedge \omega^i + e_i d\omega^i \\
&= e_j \Gamma_i^j \wedge \left(dz^i + \Gamma_k^i z^k\right) + e_i\left(d\Gamma_j^i z^j - \Gamma_j^i \wedge dz^j\right) \\
&= e_j\left(d\Gamma_k^j + \Gamma_i^j \wedge \Gamma_k^i\right) z^k
\end{aligned} \tag{9.2.6}$$

Use Eq.9.2.2 we obtain the curvature 2-form:

[1] The form as $g^{-1}dg$ is called the pure gauge form.

$$\Omega^i_j = \frac{1}{2} R^i_{j\mu\nu} dx^\mu \wedge dx^\nu = \frac{1}{2}\left(\frac{\partial}{\partial x^\mu}\Gamma^i_{\nu j} - \frac{\partial}{\partial x^\nu}\Gamma^i_{\mu j} + \Gamma^i_{\mu k}\Gamma^k_{\nu j} - \Gamma^i_{\nu k}\Gamma^k_{\mu j}\right) dx^\mu \wedge dx^\nu \tag{9.2.7}$$

Eq.9.2.7 can also be written as:

$$\Omega^i_j = \frac{1}{2} R^i_{j\mu\nu} dx^\mu \wedge dx^\nu = d\Gamma^i_j + \Gamma^i_k \wedge \Gamma^k_j \tag{9.2.8}$$

If $R^i_{j\mu\nu} = 0$, then $\Omega^i_j = 0$ and thus

$$d\Gamma^i_j = -\Gamma^i_k \wedge \Gamma^k_j \tag{9.2.9}$$

Plug Eq.9.2.8 into Eq.9.2.6 we have

$$\nabla^2(s) = e_i \Omega^i_k z^k \tag{9.2.10}$$

where Ω^i_k varies according to the rules of a tensor, e_i and z^k vary according to the rules of covariance and contravariance, respectively. Therefore we can see that $\nabla^2(s) = e_i \Omega^i_k z^k$ is invariant under frame transformations.

§ 9.3 Connections on Principal Bundles

The fibre of the principal bundle $P(MG\Pi)$ is the group *G*. If local coordinates are taken, then the local coordinates on the principal bundle *P* should include (x, g) where *x* is a point on the base manifold *M*; and *g* is some group element in the group *G* space with g_{ij} being its matrix elements.

Now suppose that there is a curve $x(\tau)$ on the base manifold *M*. Promote the curve to the principal bundle *P*. Then the matrix elements of the group element *g* of the group *G* after promotion should be written as $g_{ij}(\tau)$ which is a function of *t*.

Starting from this point, we can define the connection on the principal bundle $P(MG\Pi)$ and observe its behavior in various situations.

1. Connections on Principal Bundles and Parallel Translation: Let $x(\tau)$ be a curve on *M*, and let the section $g_{ik}(\tau)$ be a parallel translation along $x(\tau)$. Similar to Levi-Civita's rule of parallel translation in Eq.9.1.8, the parallel translation of $g_{ik}(\tau)$ on the principal bundle is defined as:

$$\frac{dg_{ik}}{d\tau} + A^a_\mu(x)\frac{(\lambda_a)_{ij}}{2!}\frac{dx^\mu}{d\tau} g_{jk} = 0 \tag{9.3.1}$$

Compare to Eq.9.1.8, we see that z^i in Eq.9.1.8 has only one index *i*, whereas g_{ik} in Eq.9.3.1 has two indices *i* and *k*. Thus $A^a_\mu(x)(\lambda_a)_{ij}\frac{dx^\mu}{d\tau} g_{ik}$ has to be introduced to replace $\Gamma^i_{\mu j}\frac{dx^\mu}{d\tau} z^j$ in Eq.9.1.8. (λ_a) is the generator of the fibre group *G* of the principal bundle. (λ_a) and (λ_b) are distinct and independent generators when $a \neq b$.

Hence similar to that Eq.9.1.8 having only the index i left after contraction, $A^a_\mu(x)\cdot(\lambda_a)_{ij}dx^\mu g_{ik}$ has the indices i and k left.

Compare Eq.9.3.1 to Eq.9.1.8 we see that $\Gamma^i_{\mu j}$ in Eq.9.1.8 and $A^a_\mu(x)\frac{(\lambda_a)_{ij}}{2!}$ in Eq.9.3.1 play the same role in both formulae. Therefore we may call

$$A^a_\mu(x)\frac{(\lambda_a)_{ij}}{2!} \tag{9.3.2}$$

the connection on the principal bundle.

Since that $g_{ik}(\tau)$, $x(\tau)$ and $(\lambda_a)_{ij}$ in Eq.9.3.1 are all given, $A^a_\mu(x)$ can be solved.

Take $a \in G$ (where a is an arbitrary element of the group G). Then we have the following transformation of right multiplication:

$$g'_{kj} = g_{kl}a_{lj}, \qquad g_{mn} = g'_{ml}a^{-1}_{ln} \tag{9.3.3}$$

$$\frac{\partial}{\partial g'_{ij}} = \frac{\partial g_{mn}}{\partial g'_{ij}}\frac{\partial}{\partial g_{mn}} = \frac{\partial g'_{ml}}{\partial g'_{ij}}a^{-1}_{ln}\frac{\partial}{\partial g_{mn}} = \delta^m_i\delta^l_j a^{-1}_{ln}\frac{\partial}{\partial g_{mn}} = a^{-1}_{jn}\frac{\partial}{\partial g_{in}}$$

$$\therefore\quad g'_{ln}\frac{\partial}{\partial g'_{mn}} = g_{lj}a_{jn}a^{-1}_{nk}\frac{\partial}{\partial g_{mk}} = g_{lk}\frac{\partial}{\partial g_{mk}} \tag{9.3.4}$$

Notice that Eq.9.3.3 are right multiplications of a_{lj} and a^{-1}_{ln}. Thus the property of Eq.9.3.4

$$g'_{ln}\frac{\partial}{\partial g'_{mn}} = g_{lk}\frac{\partial}{\partial g_{mk}}$$

is called the property of right invariance.

2. Connections on Principal Bundles and Tangent Spaces: Let $x(\tau)$ be a curve on the base manifold M. Promote it to the principal bundle $P(MG\Pi)$, we obtain a curve $c(\tau)$ on P.

Similar to Eq.9.1.6, $c(\tau)$ can be written as

$$c(\tau) = \left(x^\mu(\tau), g_{ij}(\tau)\right), \qquad g \in G \tag{9.3.5}$$

$$\Pi\left(c(\tau)\right) = x^\mu(\tau) \tag{9.3.6}$$

The complete differential along this promoted curve $c(\tau)$ is

$$\frac{d}{d\tau} = \frac{dx^\mu}{d\tau}\frac{\partial}{\partial x^\mu} + \frac{dg_{ij}}{d\tau}\frac{\partial}{\partial g_{ij}} \tag{9.3.7}$$

When $g_{ik}(\tau)$ of the principal bundle makes a parallel translation, we obtain from the definition in Eq.9.3.1 that

$$\frac{dg_{ij}}{d\tau} = -A^a_\mu(x)\frac{(\lambda_a)_{ik}}{2!}\frac{dx^\mu}{d\tau}g_{kj} = \dot{g}_{ij}. \tag{9.3.8}$$

Plug Eq.9.3.8 into Eq.9.3.7 we have

$$\frac{d}{d\tau} = \frac{dx^\mu}{d\tau}\left(\frac{\partial}{\partial x^\mu} - A^a_\mu(x)\frac{(\lambda_a)_{ik}}{2!}g_{kj}\frac{\partial}{\partial g_{ij}}\right)$$

$$= \dot{x}^\mu\left(\frac{\partial}{\partial x^\mu} - A^a_\mu(x)\overline{L}_a\right) = \dot{x}^\mu D_\mu \tag{9.3.9}$$

where

$$\overline{L}_a = \frac{(\lambda_a)_{ik}}{2!}g_{kj}\frac{\partial}{\partial g_{ij}} \tag{9.3.10}$$

D_μ in Eq.9.3.9 is the covariant differential on principal bundle which is defined as

$$D_\mu = \frac{\partial}{\partial x^\mu} - A^a_\mu(x)\overline{L}_a \tag{9.3.11}$$

Recall what has been proven in Section 9.1 that the tangent space $T(E)$ of a general vector bundle can be decomposed into two parts: $T(E) = V(E) \oplus H(E)$ (see Eq.9.1.11), where $V(E)$ is the vertical part and $H(E)$ is the horizontal part. Similar to $T(E)$, the tangent space $T(P)$ on the principal bundle can be decomposed into two parts as well:

$$T(P) = V(P) + H(P) \tag{9.3.12}$$

where $V(P)$ with its base vectors $\frac{\partial}{\partial g^{ij}}$ is called the vertical part of the tangent space $T(P)$; and $H(P)$ with its base vectors $D_\mu = \frac{\partial}{\partial x^\mu} - A^a_\mu(x)\overline{L}_a$ is called the horizontal part of the tangent space $T(P)$.

The base vector D_μ can be considered as a horizontal promotion of the vector $\frac{\partial}{\partial x^\mu}$ on the manifold M to the principal bundle P. $A^a_\mu(x)$ in D_μ is thus the connection on the principal bundle P.

According to Eq.9.3.4 and Eq.9.3.10, $\overline{L}_a = \frac{(\lambda_a)_{ik}}{2!}g_{kj}\frac{\partial}{\partial (g^T)_{ji}}$ is right-invariant, and hence D_μ in Eq.9.3.11 is right-invariant as well. Therefore we know that the decomposition of $T(P) = V(P) + H(P)$ in Eq.9.3.12 is right-invariant as well.

The reason we emphasize the right-invariance of the group is that the horizontal part $H(P)$ of the tangent space $T(P)$ is invariant under the G group transformation (since $\overline{L}_a$ in D_μ is right-invariant under the G group transformation. See Eq.9.3.10 and Eq.9.3.4.). Therefore we know that the difference between $H(P)$ and $V(P)$ is unique and is invariant under the G group transformation.

Now let us take a look at the relationship between the connection $A^a_\mu(x)$ and the curvature $F^a_{\mu\nu}$ on the principal bundle.

In $D_\mu = \frac{\partial}{\partial x^\mu} - A_\mu^a(x)\overline{L}_a$ of Eq.9.3.11, x^μ is not contained in $\overline{L}_a$ (see Eq.9.3.10). Therefore

$$\left[D_\mu , D_\nu\right] = -\left(\frac{\partial}{\partial x^\mu} A_\nu^b(x)\right)\overline{L}_b + \left(\frac{\partial}{\partial x^\nu} A_\mu^a\right)\overline{L}_a$$
$$+\left[A_\mu^a\left(\frac{\lambda_a}{2!}\right)_{ml} g_{ln}\frac{\partial}{\partial g_{mn}}, A_\nu^b\left(\frac{\lambda_b}{2!}\right)_{ij} g_{jk}\frac{\partial}{\partial g_{ik}}\right]_- \tag{9.3.13}$$

where g_{ij} is the element of the matrix g at the ith row and the jth column. When a partial differentiation is taken,

$$\frac{\partial}{\partial g_{mn}} g_{jk} = \delta_j^m \ \delta_k^n . \tag{9.3.14}$$

Thus the third term $[\ ]_-$ of Eq.9.3.13 is equal to

$$\left[A_\mu^a\left(\frac{\lambda_a}{2i}\right)_{ml} g_{ln} A_\nu^b\left(\frac{\lambda_b}{2i}\right)_{il}\left(\frac{\partial}{\partial g_{mn}} g_{jk}\right)\frac{\partial}{\partial g_{ik}} - A_\nu^b\left(\frac{\lambda_b}{2i}\right)_{ij} g_{jk} A_\mu^a\left(\frac{\lambda_a}{2i}\right)_{ml}\left(\frac{\partial}{\partial g_{ik}} g_{ln}\right)\frac{\partial}{\partial g_{mn}}\right]$$
$$= A_\mu^a A_\nu^b\left(\frac{\lambda_a}{2i}\right)_{ml}\left(\frac{\lambda_b}{2i}\right)_{ij} g_{ln}\ \delta_j^m\ \delta_k^n \frac{\partial}{\partial g_{ik}} - A_\nu^b A_\mu^a\left(\frac{\lambda_b}{2i}\right)_{ij}\left(\frac{\lambda_a}{2i}\right)_{ml} g_{jk}\ \delta_l^i\ \delta_n^k \frac{\partial}{\partial g_{mn}}$$
$$= A_\mu^a A_\nu^b\left(\left(\frac{\lambda_b}{2i}\right)_{im}\left(\frac{\lambda_a}{2i}\right)_{ml} g_{lk}\frac{\partial}{\partial g_{ik}} - \left(\frac{\lambda_a}{2i}\right)_{ml}\left(\frac{\lambda_b}{2i}\right)_{lj} g_{jn}\frac{\partial}{\partial g_{mn}}\right)$$
$$= A_\mu^a A_\nu^b\left(\left(\frac{\lambda_b}{2i}\right)_{im}\left(\frac{\lambda_a}{2i}\right)_{ml} g_{lk}\frac{\partial}{\partial g_{ik}} - \left(\frac{\lambda_a}{2i}\right)_{im}\left(\frac{\lambda_b}{2i}\right)_{ml} g_{lk}\frac{\partial}{\partial g_{ik}}\right)$$
$$= A_\mu^a A_\nu^b\left[\frac{\lambda_b}{2i}, \frac{\lambda_a}{2i}\right]_{-il} g_{lk}\frac{\partial}{\partial g_{ik}} = A_\mu^a A_\nu^b f_{bac}\left(\frac{\lambda_c}{2i}\right)_{il} g_{lk}\frac{\partial}{\partial g_{ik}}$$
$$= A_\mu^a A_\nu^b f_{bac}\overline{L}_c \qquad \text{(Use Eq.9.3.10)} \tag{9.3.15}$$

Plug it into Eq.9.3.13, we obtain

$$\left[D_\mu , D_\nu\right]_- = -\left[\frac{\partial}{\partial x^\mu} A_\nu^a(x) - \frac{\partial}{\partial x^\nu} A_\mu^a(x) + f_{abc} A_\mu^b(x) A_\nu^a(x)\right]\overline{L}_a = -F_{\mu\nu}^a \overline{L}_a \tag{9.3.16}$$

where

$$F_{\mu\nu}^a = \frac{\partial}{\partial x^\mu} A_\nu^a - \frac{\partial}{\partial x^\nu} A_\mu^a + f_{abc} A_\mu^b A_\nu^a \tag{9.3.17}$$

Following it we have Jacobi's identity:

$$\left[D_\mu , \left[D_\nu , D_\lambda\right]\right] + \left[D_\nu , \left[D_\lambda , D_\mu\right]\right] + \left[D_\lambda , \left[D_\mu , D_\nu\right]\right] = 0 \tag{9.3.18}$$

When the base manifold M is 4-dimensional, $\mu , \nu = 1,2,3,4$. Eq.9.3.18 can be written as

$$\varepsilon_{\mu\nu\lambda\rho}\left[D_{\mu}, \left[D_{\nu}, D_{\lambda}\right]\right] = 0 \tag{9.3.19}$$

Plug Eq.9.3.16 into the above equation, we have

$$\varepsilon_{\mu\nu\lambda\rho}\left[D_{\mu}, F^{a}_{\nu\lambda}\overline{L}_{a}\right] = 0 \tag{9.3.20}$$

Furthermore, since that $\overline{L}_a$ $(a = 1,2,\cdots,k)$ are independent to each other, from Eq.9.3.20 we obtain

$$\varepsilon_{\mu\nu\lambda\rho}\left[D_{\mu}, F^{a}_{\nu\lambda}\right] = 0 \tag{9.3.21}$$

Definition 9.3.1: $*F^{a}_{\mu\nu} = \frac{1}{2}\varepsilon_{\mu\nu\lambda\rho}F^{a}_{\lambda\rho}$ (where $*F^{a}_{\mu\nu}$ is the duel tensor of $F^{a}_{\lambda\rho}$). Then we have

$$\left[D_{\mu}, *F^{a}_{\mu\nu}\right] = 0 \tag{9.3.22}$$

The equations Eq.9.3.21 and Eq.9.3.22 are the Bianchi's equations which must be satisfied by the curvature tensors and the connections on the principal bundles.

3. The Principal Bundle's Connection and the Cotangent Space: Denote the cotangent space on the base manifold M as $T^{*}(M)$. Then every 1-form in $T^{*}(M)$ can be written as $A = A_{\mu}(x)dx^{\mu}$.

Let λ_a be the generator of the G group of the principal bundle (λ_a is a constant matrix). Then A can be taken as

$$A = A_{\mu}(x)dx^{\mu} = A^{a}_{\mu}(x)\frac{\lambda_a}{2!}dx^{\mu} \tag{9.3.23}$$

The variables g_{ij} of the principal bundle's fibre $F(\cong G)$ are not included in Eq.9.3.23. Hence A of Eq.9.3.23 is still a connection 1-form which belongs to $T^{*}(M)$.

If the connection 1-form in $T^{*}(M)$ is to be extended into a connection 1-form in $T^{*}(P)$, then Eq.9.3.23 has to be extended to a 1-form which contains both dx^{μ} and dg_{ij}.

Therefore we can take the left-invariant 1-form $g^{-1}dg$ in Eq.2.6.6 of Chapter 2, and then generalize Eq.9.3.23 which belongs to $T^{*}(M)$ into the following 1-form in $T^{*}(P)$:

$$\omega = g^{-1}Ag + g^{-1}dg = A^{a}_{\mu}(x)\left(g^{-1}\frac{\lambda_a}{2i}g\right)dx^{\mu} + g^{-1}dg \tag{9.3.24}$$

where $g^{-1}Ag = A^{a}_{\mu}(x)\left(g^{-1}\frac{\lambda_a}{2i}g\right)dx^{\mu}$ corresponds to Eq.9.3.23; and the extra term of $g^{-1}dg$ is the left-invariant Maurer-Cartan form which is linearly independent to $T^{*}(M)$ and hence is the part in ω which is "orthogonal" to $T^{*}(M)$.

With the definition in Section 3.1: $< dx^{a}, \frac{\partial}{\partial x^{b}} >= \delta^{a}_{b}$, we may prove that a cotangent space has the following properties:

(i) Use the rule in Section 3.1, namely $< dx^a , \frac{\partial}{\partial x^b} >_- = \delta^a_b$, it can be proven that

$$< \omega , D_\mu >_- = 0 \tag{9.3.25}$$

Proof: According to Eq.9.3.24 and Eq.9.3.11, we have

$$< \omega_{ij} , D_\mu >_- = < g^{-1}_{im} A^a_\nu(x) \frac{(\lambda_a)_{mn}}{2i} dx^\nu g_{nj} + g^{-1}_{im} dg_{mj} , \frac{\partial}{\partial x^\mu} - A^a_\mu \frac{(\lambda_a)_{pq}}{2i} g_{qr} \frac{\partial}{\partial g_{pr}} >_-$$

$$= g^{-1}_{im} A^a_\mu(x) \frac{(\lambda_a)_{mn}}{2i} g_{nj} - g^{-1}_{im} A^a_\mu \frac{(\lambda_a)_{pq}}{2i} g_{qr} \delta^p_m \delta^r_j = 0$$

□

(ii) Take $\overline{L}_a$ from Eq.9.3.10 we can prove that

$$< \omega_{ij} , \overline{L}_a >_- = \left(\frac{\lambda_a}{2i} \right)_{ij} \tag{9.3.26}$$

Proof: Again, according to Eq.9.3.24 and Eq.9.3.10 we have

$$< \omega_{ij} , \overline{L}_a >_- = < g^{-1}_{im} A^a_\nu(x) \frac{(\lambda_a)_{mn}}{2i} dx^\nu g_{nj} + g^{-1}_{im} dg_{mj} , \left(\frac{\lambda_a}{2i} \right)_{ik} g_{kl} \frac{\partial}{\partial g_{il}} >_-$$

$$= g^{-1}_{im} \left(\frac{\lambda_a}{2i} \right)_{ik} g_{kl} \delta^i_m \delta^l_j = g^{-1}_{im} \left(\frac{\lambda_a}{2i} \right)_{mk} g_{kj} = \left(\frac{\lambda_a}{2i} \right)_{ij}$$

□

Compare to Eq.9.3.12 we see immediately that the commutant of ω_{ij} and the base vectors D_μ of the horizontal part $H(P)$ of the tangent space $T(P)$ gives zero; the commutant of ω_{ij} and the base vectors $\frac{\partial}{\partial g_{ij}}$ of the vertical part $V(P)$ of $T(P)$ gives the constant matrix $\left(\frac{\lambda_a}{2i} \right) ij$.

4. The Principal Bundle's Connections and the Normal Transformation: Consider the intersecting region $U \cap U'$ of two neighborhoods U and U'. Suppose there is a transformation Φ between the local fibre coordinates g and g' of U and U', respectively, as follows:

$$g' = \Phi g , \qquad g'^{-1} = g^{-1} \Phi^{-1} \qquad (\Phi = \Phi_{UU'}) \tag{9.3.27}$$

To make the definitions of the connection 1-form ω in the two coordinate systems of the intersecting region $U \cap U'$ consistent, A of Eq.9.3.23 must have the following transformation in the intersecting region:

$$A' = \Phi \, A \, \Phi^{-1} + \Phi \, d\Phi^{-1} \tag{9.3.28}$$

Proof: $\omega = g^{-1} A \, g + g^{-1} dg = g'^{-1} \Phi \, A \, \Phi^{-1} g' + g'^{-1} \Phi \, d\left(\Phi^{-1} g^{-1}\right)$

$$= g'^{-1}\Phi\, A\, \Phi^{-1}\, g' + g'^{-1}\left(\Phi\, d\Phi^{-1}\right)g' + g'^{-1}dg'$$
$$= g'^{-1}\left(\Phi\, A\, \Phi^{-1} + \Phi\, d\Phi^{-1}\right)g' + g'^{-1}dg'$$
$$\omega' = g'^{-1}A'g' + g'^{-1}dg' .$$

Compare ω and ω' we see that if $\omega = \omega'$ in the intersecting region $U \cap U'$, then Eq.9.3.28 must be true. □

Definition 9.3.2: If the 1-form $A = A_\mu^a(x)\frac{\lambda_a}{2i}dx^\mu$ on the base manifold M of the principal bundle has a transformation of the form Eq.9.3.28 in the intersecting region $U \cap U'$, then $A_\mu^a(x)$ is called a *normal field.*

5. The Principal Bundle's Connection and the Curvature 2-Form: According to the definitions in Eq.3.3.6 and Eq.3.3.2, the curvature 2-form on a manifold is

$$\Omega_b^a = d\omega_b^a + \omega_c^a \wedge \omega_b^c .$$

where $\omega_b^a = \theta^d \Gamma_{db}^a$ is a connection 1-form.

Now let us take the principal bundle's connection 1-form in Eq.9.3.24

$$\omega = g^{-1}Ag + g^{-1}dg .$$

Then similar to Eq.3.3.6, the curvature of the principal bundle can be written as the following (Notice that A is a 1-form):

$$\Omega = d\omega + \omega \wedge \omega$$
$$= d\left(g^{-1}Ag + g^{-1}dg\right) + \left(g^{-1}Ag + g^{-1}dg\right) \wedge \left(g^{-1}Ag + g^{-1}dg\right)$$
$$= -g^{-1}dg\, g^{-1} \wedge Ag + g^{-1}dA\, g - g^{-1}A \wedge dg - g^{-1}(dg)g^{-1} \wedge dg$$
$$+ g^{-1}Ag \wedge g^{-1}Ag + g^{-1}dg \wedge g^{-1}dg + g^{-1}Ag \wedge g^{-1}dg + g^{-1}dg \wedge g^{-1}Ag$$

The first and the eighth terms are canceled out, and so are the third and the seventh terms, and the fourth and the sixth terms. We thus obtain

$$\Omega = g^{-1}(dA + A \wedge A)g + g^{-1}Fg \tag{9.3.29}$$

where $F = (dA + A \wedge A)$ is the curvature 2-form on the base manifold M.

Use Eq.9.3.23, we can rewrite F into the following 2-form:

$$F = d\left(A_\mu^a(x)\frac{\lambda_a}{2i}dx^\mu\right) + \left(A_\mu^a(x)\frac{\lambda_a}{2i}dx^\mu\right) \wedge \left(A_\gamma^b(x)\frac{\lambda_b}{2i}dx^\gamma\right)$$
$$= \frac{\partial A_\mu^a(x)}{\partial x^\gamma}\frac{\lambda_a}{2i}dx^\gamma \wedge dx^\mu + A_\mu^a(x)A_\gamma^b(x)\left(\frac{\lambda_a}{2i}\cdot\frac{\lambda_b}{2i}\right)dx^\mu \wedge dx^\gamma$$
$$= \frac{1}{2}\left(\frac{\partial A_\gamma^a(x)}{\partial x^\mu} - \frac{\partial A_\mu^a(x)}{\partial x^\gamma}\right)\frac{\lambda_a}{2i}dx^\mu \wedge dx^\gamma + \frac{1}{2}A_\mu^a(x)\,A_\gamma^b(x)\left[\frac{\lambda_a}{2i},\frac{\lambda_b}{2i}\right]_- dx^\mu \wedge dx^\gamma$$
$$= \frac{1}{2}\left(\frac{\partial A_\gamma^a(x)}{\partial x^\mu} - \frac{\partial A_\mu^a(x)}{\partial x^\gamma} + f_{abc}A_\mu^b(x)\,A_\gamma^c(x)\right)\frac{\lambda_a}{2i}dx^\mu \wedge dx^\gamma \tag{9.3.30}$$

Notice that there is no dg_{ij} (~ the vertical component) but only dx^μ (~ the horizontal component) in this curvature 2-form.

Start from $F = (dA + A \wedge A)$ we can also obtain

$$dF = dA \wedge A - A \wedge dA\,,\quad A \wedge F = A \wedge dA + A \wedge A \wedge A\,,\quad F \wedge A = dA \wedge A + A \wedge A \wedge A$$

$$\therefore\quad dF + A \wedge F - F \wedge A = 0 \tag{9.3.31}$$

where F is the curvature 2-form, and A is the connection 1-form. Therefore Eq.9.3.31 is consistent with the curvature integrability condition given in Eq.3.3.13 of Chapter 3.

§ 9.4 Connections on Adjoint Vector Bundles

Take a substantial field $\psi(x)$ which is equivalent to a section (see Section 8.3 **c**) of the adjoint vector bundle $E(M,F,G,\Pi,P)$ on the space-time manifold M.

The substantial field $\psi(x)$ is a function of the space-time coordinate x. However, it relates to the G group in a global sense. It has been shown in Eq.8.3.14 that the points in $E = P \times F / G$ are distinguished by the equivalent classes, not by some group element of G.

The substantial field $\psi(x)$ takes values in the vector space F represented by the G group. That is, if F is k-dimensional, then $\psi(x)$ has k components.

The base manifolds of the principal bundle P and the adjoint vector bundle E are both M, on which the curve $x(\tau)$ has its horizontal promotion $c(\tau)$ on the principal bundle:

$$c(\tau) = \left(x(\tau)\,,\, g(\tau)\right) \qquad g \in G \qquad \text{(see Eq.9.2.5).}$$

The horizontal promotion of the curve $x(\tau)$ of M on the direct product manifold $P \times F$ of the adjoint bundle E is $z(\tau)$:

$$z(\tau) = \left(x(\tau)\,,\, g(\tau)\,,\, \psi(\tau)\right) \qquad g \in G\,,\ \psi \in F \tag{9.4.1}$$

Compare to Eq.8.3.13, and write the representation matrix of $g(\tau)$ in F as $\bar{g}(\tau)$:

$$\begin{array}{llcl} G: & P \times F & \rightarrow & P \times F \\ g^{-1}(\tau): & \left(g(\tau), \psi(\tau)\right) & \rightarrow & \left(g(\tau)g^{-1}(\tau), \bar{g}(\tau)\psi(\tau)\right) = \left(I, \bar{g}(\tau)\psi(\tau)\right) \end{array}$$

Then $z(\tau)$ in Eq.9.4.1 can be represented by the equivalent class:

$$\left(x(\tau), g(\tau), \psi(\tau)\right) \xleftrightarrow{\text{Equivalent Classes}} \left(x(\tau), I, \bar{g}(\tau)\psi(\tau)\right) \tag{9.4.2}$$

where I is the identity element of G.

From Eq.9.4.2 we know that it is enough to represent the horizontal promotion of $x(\tau)$ on the adjoint bundle $\cong P \times F / G$ by $x(\tau)$ and $\bar{g}(\tau)\psi(\tau)$. Hence Eq.9.4.1 can be simplified as

$$z(\tau) = \left(x(\tau)\,,\, \bar{g}(\tau)\psi(\tau)\right) \tag{9.4.3}$$

Keeping in mind the fact that the group element $\bar{g}(\tau)$ of G does not appear in the group representation matrix of $P \times F / G$, we define, with t preserved,

$$\psi(x(\tau)) = \bar{g}(\tau)\psi(\tau). \tag{9.4.4}$$

Then Eq.9.4.3 can be written in the following form which is irrelevant to $\bar{g}(\tau)$:

$$z(\tau) = (x(\tau), \psi(x(\tau))) \tag{9.4.5}$$

where $z(\tau)$ is the horizontal promotion of $x(\tau)$ on the adjoint bundle $E = P \times F / G$.

In order to determine the relations between ψ and τ, we observe how ψ varies with τ. When $\tau = 0$, $\psi(\tau)|_{\tau=0} = \psi(x(\tau))|_{\tau=0}$. Then plug into Eq.9.4.4 we obtain

$$\bar{g}(0) = 1, \quad \bar{g}^{-1}(0) = 1 \tag{9.4.6}$$

The way $\psi(\tau)$ varies with τ can be derived from Eq.9.4.4:

$$\begin{aligned} D_\tau \psi(\tau)|_{\tau=0} &= \frac{d}{d\tau}\left(\bar{g}^{-1}(\tau)\psi(x(\tau))\right)\Big|_{\tau=0} \\ &= \bar{g}^{-1}(\tau)\frac{dx^\mu}{d\tau}\frac{\partial}{\partial x^\mu}\psi(x(\tau))\Big|_{\tau=0} + \left(\frac{d}{d\tau}\bar{g}^{-1}(\tau)\right)\psi(x(\tau))\Big|_{\tau=0} \end{aligned} \tag{9.4.7}$$

According to the fact that

$$\frac{d}{d\tau}\left(\bar{g}(\tau)\bar{g}^{-1}(\tau)\right) = 0 \to \left(\frac{d}{d\tau}\bar{g}(\tau)\right)\bar{g}^{-1}(\tau) + \bar{g}(\tau)\left(\frac{d}{d\tau}\bar{g}^{-1}(\tau)\right) = 0,$$

we have

$$\left(\frac{d}{d\tau}\bar{g}^{-1}(\tau)\right) = -\bar{g}^{-1}(\tau)\left(\frac{d}{d\tau}\bar{g}(\tau)\right)\bar{g}^{-1}(\tau) \tag{9.4.8}$$

Then use Eq.9.3.1, write g_{ik} as $\bar{g}$, and $A_\mu^a \dfrac{(\lambda_a)_{ij}}{2i}$ as $\bar{A}_\mu$, we can thus write Eq.9.3.1 as

$$\frac{d\bar{g}(\tau)}{d\tau} + \bar{A}_\mu(x)\frac{dx^\mu}{d\tau}\bar{g}(\tau) = 0 \qquad (x = x(\tau))$$

Plug into Eq.9.4.8 we obtain

$$\begin{aligned} \left(\frac{d}{d\tau}\bar{g}^{-1}(\tau)\right) &= -\bar{g}^{-1}(\tau)\left(-\bar{A}_\mu(x)\frac{dx^\mu}{d\tau}\bar{g}(\tau)\right)\bar{g}^{-1}(\tau) \qquad (x = x(\tau)) \\ &= \bar{g}^{-1}(\tau)\bar{A}_\mu(x)\frac{dx^\mu}{d\tau} \end{aligned} \tag{9.4.9}$$

Therefore plug it into Eq.9.4.7 we have

$$D_\tau \psi(\tau)|_{\tau=0} = \bar{g}^{-1}(\tau)\frac{dx^\mu}{d\tau}\left(\frac{\partial}{\partial x^\mu}\psi(x(\tau)) + \bar{A}_\mu(x)\psi(x(\tau))\right)\Big|_{\tau=0}$$

Because of Eq.9.4.6, $\bar{g}^{-1}(\tau = 0) = 1$. Hence $D_\tau \psi(\tau)|_{\tau=0}$ should be

$$D_\tau \psi(\tau)|_{\tau=0} = \frac{dx^\mu}{d\tau}\left(\frac{\partial}{\partial x^\mu} + \bar{A}_\mu(x)\right)\psi(x) = \frac{dx^\mu}{d\tau}D_\mu \psi(x) \tag{9.4.10}$$

where $x = x(\tau)\big|_{\tau=0}$. Therefore

$$D_\mu = \frac{\partial}{\partial x^\mu} + \overline{A}_\mu(x) \qquad (x = x(\tau = 0)) \tag{9.4.11}$$

where D_μ is the covariant differential operator which acts on the section $\psi(x)$ of the adjoint vector bundle E.

According to what we have previously mentioned, $\overline{A}_\mu$ is just $A_\mu^a \frac{(\lambda_a)_{ij}}{2i}$. Now compare

$$\left(\frac{\partial}{\partial x^\mu} + \overline{A}_\mu(x)\right) = \left(\frac{\partial}{\partial x^\mu} + A_\mu^a \frac{(\lambda_a)_{ij}}{2i}\right) \tag{9.4.12}$$

with the covariant differential operator in the gauge field of physics

$$\left(\partial_\mu - igA_\mu^a T_{ij}^a\right). \tag{9.4.13}$$

Obviously $(\lambda_a)_{ij}$ corresponds to T_{ij}^a, and the form of Eq.9.4.12 is exactly the same as that of Eq.9.4.13. Therefore the connection on the adjoint vector bundle is just $\overline{A}_\mu(x) = A_\mu^a(x) \frac{(\lambda_a)_{ij}}{2i}$.

Chapter 10. Characteristic Classes of Fibre Bundles

§ 10.1 *Invariant Polynomials and Characteristic Classes*

Definition 10.1.1: Let α be a $k\times k$ complex matrix, and $P(\alpha)$ be a polynomial formed by the matrix elements of α. If for any $g\in GL(k,c)$ (where $GL(k,c)$ represents a general nonsingular (that is, the inverse matrix exists) complex linear group with k rows and k columns), $P(\alpha)$ has the following invariant property

$$P(\alpha)=P\left(g^{-1}\alpha\, g\right) \tag{10.1.1}$$

Then $P(\alpha)$ is called an *invariant polynomial*, or a *characteristic polynomial.*

Suppose that $P(\alpha_1,\alpha_2,\cdots,\alpha_r)$ is a homogeneous invariant polynomial of degree r. With the invariance of $P(\alpha_1,\alpha_2,\cdots,\alpha_r)$ under small perturbations $g\cong 1+\eta g'$ and $g^{-1}\cong 1-\eta g'$, we should have, under the linear approximation for small η, that

$$\sum_{1\le i\le r} P\left(\alpha_1,\cdots,g'\alpha_i-\alpha_i g',\cdots,\alpha_r\right)=0 \tag{10.1.2}$$

However, take g' in Eq.10.1.2 as some generator of a Lie group, $g'=\lambda_a$; then multiply Eq.10.1.2 to the left by $A_\mu^a dx^\mu$ and translate it to the right to the position of $g'=\lambda_a$; assume that each α_i is a differential form of order d_i, we obtain

$$\begin{aligned}0&=\sum_{1\le i\le r}A_\mu^a dx^\mu P\left(\alpha_1,\cdots,g'\alpha_i-\alpha_i g',\cdots,\alpha_r\right)\\&=\sum_{1\le i\le r}(-1)^{d_1+d_2+\cdots+d_{i-1}}P\left(\alpha_1,\cdots,\alpha_{i-1},A_\mu^a\lambda_a dx^\mu\wedge\alpha_i\wedge\alpha_{i+1},\cdots,\alpha_r\right)\\&\quad-\sum_{1\le i\le r}(-1)^{d_1+d_2+\cdots+d_{i-1}+d_i}P\left(\alpha_1,\cdots,\alpha_{i-1},\alpha_i\wedge A_\mu^a\lambda_a dx^\mu\wedge\alpha_{i+1},\cdots,\alpha_r\right)\end{aligned} \tag{10.1.3}$$

Denote the 1-form $A_\mu^a\lambda_a dx^\mu$ in Eq.10.1.3 by θ, we have

$$\begin{aligned}0&=\sum_{1\le i\le r}(-1)^{d_1+d_2+\cdots+d_{i-1}}P\left(\alpha_1,\cdots,\alpha_{i-1}\wedge\theta\wedge\alpha_i\wedge\alpha_{i+1},\cdots,\alpha_r\right)\\&\quad-\sum_{1\le i\le r}(-1)^{d_1+d_2+\cdots+d_{i-1}+d_i}P\left(\alpha_1,\cdots,\alpha_{i-1},\alpha_i\wedge\theta\wedge\alpha_{i+1},\cdots,\alpha_r\right)\end{aligned} \tag{10.1.4}$$

Thus we see that Eq.10.1.4 is always true as long as θ being a 1-form and $P(\alpha)=P\left(g^{-1}\alpha\, g\right)$ being an invariant polynomial.

Now we differentiate $P(\alpha)=P(\alpha_1,\alpha_2,\cdots,\alpha_r)$ and obtain

$$dP\left(\alpha_1,\cdots,\alpha_r\right)=\sum_{1\le i\le r}(-1)^{d_1+\cdots+d_{i-1}}P\left(\alpha_1,\cdots,d\alpha_i,\cdots,\alpha_r\right) \tag{10.1.5}$$

Use Eq.10.1.4 and take θ as the connection 1-form ω, that is, $\omega_j^i=A_\mu^a\dfrac{\left(\lambda_a\right)_j^i}{2i}dx^\mu$. Then we have

$$0=\sum_{1\le i\le r}(-1)^{d_1+d_2+\cdots+d_{i-1}}P(\alpha_1,\cdots,\alpha_{i-1},\omega\wedge\alpha_i,\alpha_{i+1},\cdots,\alpha_r)$$
$$-\sum_{1\le i\le r}(-1)^{d_1+d_2+\cdots+d_i}P(\alpha_1,\cdots,\alpha_{i-1},\alpha_i\wedge\omega,\alpha_{i+1},\cdots,\alpha_r) \quad (10.1.6)$$

Add to Eq.10.1.5 we obtain

$$dP(\alpha_1,\cdots,\alpha_r)=\sum_{1\le i\le r}(-1)^{d_1+\cdots+d_{i-1}}\{P(\alpha_1,\cdots,d\alpha_i,\alpha_{i+1},\cdots\alpha_r)$$
$$+P(\alpha_1,\cdots,\omega\wedge\alpha_i,\alpha_{i+1},\cdots\alpha_r)$$
$$-P(\alpha_1,\cdots,\alpha_i\wedge\omega,\alpha_{i+1},\cdots\alpha_r)\}$$
$$=\sum_{1\le i\le r}(-1)^{d_1+\cdots+d_{i-1}}P(\alpha_1,\cdots,D\alpha_i,\cdots\alpha_r) \quad (10.1.7)$$

Take every α_i in Eq.10.1.7 as the curvature 2-form, that is, $\alpha_i=\Omega$. Then it is given by Eq.10.1.7 that

$$dP(\Omega,\cdots,\Omega)=\sum_{1\le i\le r}(-1)^{d_1+d_2+\cdots+d_{i-1}}P(\Omega,\cdots,d\Omega+\omega\wedge\Omega-\Omega\wedge\omega,\cdots,\Omega) \quad (10.1.8)$$

Hence by Bianchi's identity (see Eq.3.3.13 and Eq.9.3.31)

$$d\Omega+\omega\wedge\Omega-\Omega\wedge\omega=0 \quad (10.1.9)$$

we know that $P(\Omega,\cdots,\Omega)$ must be a closed form, that is,

$$dP(\Omega,\cdots,\Omega)=0 \quad (10.1.10)$$

which is the first essential feature of $P(\Omega,\cdots,\Omega)$.

Then assume that ω and ω' are distinct connections on the bundle, of which the corresponding curvatures are Ω and Ω'. Now take an interpolation ω_t between ω and ω' (where ω and ω' are both 1-forms):

$$\omega_t=\omega+t\eta \qquad \eta=\omega'-\omega \qquad 0\le t\le 1 \quad (10.1.11)$$

and define the covariant exterior differential operator D as follows:

Definition 10.1.2: Let V_b^a be a differential form of order p, where a and b are the sup- and subscripts respectively. Then the action of the covariant exterior differential operator D to V_b^a is

$$DV_b^a=dV_b^a+\omega_c^a\wedge V_b^c-(-1)^P V_c^a\wedge\omega_b^c \quad (10.1.12)_1$$

For example, take $V_b^a=\Omega_b^a$, the order of the curvature 2-form Ω_b^a is $p=2$, and

$$D\Omega_b^a=d\Omega_b^a+\omega_c^a\wedge\Omega_b^c-(-1)^2\Omega_c^a\wedge\omega_b^c$$
$$=d\Omega_b^a+\omega_c^a\wedge\Omega_b^c-\Omega_c^a\wedge\omega_b^c=0 \quad (10.1.12)_2$$

which is consistent with Bianchi's identity in Eq.3.3.13 and Eq.9.3.31[1].

With the interpolation ω_t we can define the interpolated curvature Ω_t:

$$\Omega_t=d\omega_t+\omega_t\wedge\omega_t=(d\omega+td\eta)+(\omega+t\eta)\wedge(\omega+t\eta)$$

---

[1] Here $\omega\wedge\omega$ and $\eta\wedge\eta$ are abbreviations of $\omega_c^a\wedge\omega_b^c$ and $\eta_c^a\wedge\eta_b^c$ respectively. Compare to definition of Γ_a^b in Eq.3.1.8, the abbreviations $\omega\wedge\omega$ and $\eta\wedge\eta$ are both nonzero.

$$= d\omega + \omega \wedge \omega + t(d\eta + \eta \wedge \omega + \omega \wedge \eta) + t^2 \eta \wedge \eta \tag{10.1.13}$$

where

$$\Omega = \Omega_{t=0} = d\omega + \omega \wedge \omega \tag{10.1.14}$$

Then according to Eq.10.1.11 which corresponds to η of order $P = 1$, we have

$$D\eta = d\eta + \omega \wedge \eta + \eta \wedge \omega \tag{10.1.15}$$

Therefore Eq.10.1.13 can be written as

$$\Omega_t = \Omega + tD\eta + t^2 \eta \wedge \eta \tag{10.1.16}$$

where η is a 1-form, Ω and $D\eta$ are 2-forms. Hence

$$\frac{d}{dt} P(\underbrace{\Omega_t \cdots \Omega_t}_{r \text{ entries}}) = P(D\eta, \underbrace{\Omega_t \cdots \Omega_t}_{(r-1) \text{ entries}}) + P(\Omega_t, D\eta, \cdots \Omega_t) + \cdots + P(\underbrace{\Omega_t \cdots \Omega_t}_{(r-1) \text{ entries}}, D\eta)$$

$$+ 2t \left[P(\eta \wedge \eta, \underbrace{\Omega_t \cdots \Omega_t}_{(r-1) \text{ entries}}) + P(\Omega_t, \eta \wedge \eta, \cdots \Omega_t) + \cdots + P(\underbrace{\Omega_t \cdots \Omega_t}_{(r-1) \text{ entries}}, \eta \wedge \eta) \right] \tag{10.1.17}$$

From Eq.10.1.16, since $D\Omega = 0$, then

$$D\Omega_t = tD^2\eta + t^2 (D\eta \wedge \eta - \eta \wedge D\eta) \tag{10.1.18}$$

Then from Eq.10.1.15 we obtain

$$\begin{aligned} D^2\eta &= D(d\eta + \omega \wedge \eta + \eta \wedge \omega) \\ &= dd\eta + D\omega \wedge \eta - \omega \wedge D\eta + D\eta \wedge \omega - \eta \wedge D\omega \\ &= d\omega \wedge \eta - \omega \wedge d\eta + d\eta \wedge \omega - \eta \wedge d\omega \end{aligned}$$

Use $d\omega = \Omega - \omega \wedge \omega$ given by Eq.10.1.14, we have

$$\begin{aligned} D^2\eta &= (\Omega - \omega \wedge \omega) \wedge \eta - \omega \wedge d\eta + d\eta \wedge \omega - \eta \wedge (\Omega - \omega \wedge \omega) \\ &= (\Omega - \omega \wedge \omega) \wedge \eta - \eta \wedge (\Omega - \omega \wedge \omega) \qquad (\text{Use } \alpha_p \wedge \beta_q = (-1)^{pq} \beta_q \wedge \alpha_p) \end{aligned}$$

$$\therefore \quad D^2\eta = \Omega \wedge \eta - \eta \wedge \Omega \tag{10.1.19}$$

Plug Eq.10.1.19 and Eq.10.1.16 into Eq.10.1.18, we have

$$\begin{aligned} D\Omega_t &= t(\Omega \wedge \eta - \eta \wedge \Omega) + t^2 (D\eta \wedge \eta - \eta \wedge D\eta) \\ &= t((\Omega + tD\eta) \wedge \eta - \eta \wedge (\Omega + tD\eta)) \\ &= t(\Omega_t \wedge \eta - \eta \wedge \Omega_t) \end{aligned} \tag{10.1.20}$$

Then similar to Eq.10.1.7, take $\alpha_1 = \eta$ (a 1-form), and $\alpha_2 = \alpha_3 = \cdots = \alpha_r = \Omega_t$ (a 2-form), we obtain

$$dP(\eta, \underbrace{\Omega_t \cdots \Omega_t}_{(r-1) \text{ entries}}) = P(D\eta, \underbrace{\Omega_t \cdots \Omega_t}_{(r-1) \text{ entries}}) - P(\eta, D\Omega_t, \cdots, \Omega_t) - \cdots - P(\eta, \Omega_t, \cdots, D\Omega_t) \tag{10.1.21}$$

Using Eq.10.1.4 with $\theta = \alpha_1 = \eta$ and $\alpha_2 = \cdots = \alpha_r = \Omega_t$ gives

$$0 = P(\underbrace{\eta\wedge\eta}_{\theta\wedge\alpha_1} \wedge \underbrace{\Omega_t\cdots\Omega_t}_{(r-1)\text{ entries}}) - P(\underbrace{\eta}_{\alpha_1}\wedge\underbrace{\eta\wedge\Omega_t}_{\theta\wedge\alpha_2}\cdots\Omega_t) - \cdots - P(\underbrace{\eta}_{\alpha_1}\wedge\Omega_t\cdots\wedge\underbrace{\eta\wedge\Omega_t}_{\theta\wedge\alpha_r})$$

$$-(-)P(\underbrace{\eta\wedge\eta}_{\alpha_1\wedge\theta}\wedge\underbrace{\Omega_t}_{\alpha_2}\cdots\underbrace{\Omega_t}_{\alpha_r}) + P(\underbrace{\eta}_{\alpha_1}\wedge\underbrace{\Omega_t\wedge\eta}_{\alpha_2\wedge\theta}\wedge\cdots\underbrace{\Omega_t}_{\alpha_r}) + \cdots + P(\underbrace{\eta}_{\alpha_1}\wedge\underbrace{\Omega_t}_{\alpha_2}\cdots\wedge\underbrace{\Omega_t\wedge\eta}_{\alpha_r\wedge\theta})$$

$$= 2P(\eta\wedge\eta\wedge\Omega_t\cdots\Omega_t)$$

$$+ P(\eta\wedge(\Omega_t\wedge\eta-\eta\wedge\Omega_t)\wedge\Omega_t\cdots\Omega_t) + \cdots + P(\eta\wedge\Omega_t\cdots\Omega_t\wedge(\Omega_t\wedge\eta-\eta\wedge\Omega_t)) \tag{10.1.22}$$

Plugging Eq.10.1.20 into Eq.10.1.22 gives

$$2P(\eta\wedge\eta\wedge\underbrace{\Omega_t\cdots\Omega_t}_{(r-1)\text{ entries}}) + P(\eta\wedge\frac{D\Omega_t}{t},\Omega_t\cdots\Omega_t) + \cdots + P(\eta\wedge\Omega_t\cdots\Omega_t,\frac{D\Omega_t}{t}) = 0 \tag{10.1.23}$$

Compare Eq.10.1.21 and Eq.10.1.23 we obtain

$$dP(\eta,\underbrace{\Omega_t\cdots\Omega_t}_{(r-1)\text{ entries}}) = P(D\eta,\underbrace{\Omega_t\cdots\Omega_t}_{(r-1)\text{ entries}}) + 2tP(\eta\wedge\eta,\underbrace{\Omega_t\cdots\Omega_t}_{(r-1)\text{ entries}}) \tag{10.1.24}$$

Similarly,

$$dP(\Omega_t\wedge\eta\wedge\Omega_t\cdots\Omega_t) = P(\Omega_t\wedge D\eta,\Omega_t\cdots\Omega_t) + 2tP(\Omega_t\wedge\eta\wedge\eta\cdots\Omega_t) \tag*{(10.1.24)$_a$}$$

$$\cdots \qquad \cdots \qquad \cdots$$

$$dP(\underbrace{\Omega_t\cdots\Omega_t}_{(r-1)\text{ entries}}\wedge\eta) = P(\Omega_t\wedge\Omega_t\cdots\Omega_t\wedge D\eta) + 2tP(\Omega_t\wedge\Omega_t\cdots\Omega_t\wedge\eta\wedge\eta) \tag*{(10.1.24)$_b$}$$

Definition 10.1.3: $$q(\eta,\Omega_t) = P(\eta,\Omega_t\cdots\Omega_t) + \cdots + P(\Omega_t,\cdots\Omega_t,\eta) \tag{10.1.25}$$

Hence take Eq.10.1.24 + Eq.10.1.24$_a$ + $\cdots$ + Eq.10.1.24$_b$, we obtain

$$dq(\eta,\Omega_t) = q(D\eta,\Omega_t) + 2t\,q(\eta\wedge\eta,\Omega_t) \tag{10.1.26}$$

Then differentiate Eq.10.1.16 with respect to t, we obtain

$$\frac{d}{dt}\Omega_t = D\eta + 2t\eta\wedge\eta \tag{10.1.27}$$

Plug in $\frac{d}{dt}P(\underbrace{\Omega_t\cdots\Omega_t}_{r\text{ entries}})$, we have

$$\frac{d}{dt}P(\underbrace{\Omega_t\cdots\Omega_t}_{r\text{ entries}}) = P(D\eta\wedge\underbrace{\Omega_t\cdots\Omega_t}_{(r-1)\text{ entries}}) + \cdots + P(\underbrace{\Omega_t\cdots\Omega_t}_{(r-1)\text{ entries}}\wedge D\eta)$$

$$+ 2tP(\eta\wedge\eta\wedge\underbrace{\Omega_t\cdots\Omega_t}_{(r-1)\text{ entries}}) + \cdots + 2tP(\underbrace{\Omega_t\cdots\Omega_t}_{(r-1)\text{ entries}}\wedge\eta\wedge\eta) \tag{10.1.28}$$

Compare Eq.10.1.28 to the right hand side of Eq.10.1.25, $\frac{d}{dt}P(\underbrace{\Omega_t\cdots\Omega_t}_{r\text{ entries}})$ can also be written as

$$\frac{d}{dt}P(\underbrace{\Omega_t\wedge\cdots\wedge\Omega_t}_{r\text{ entries}})=q(D\eta,\Omega_t)+2tq(\eta\wedge\eta,\Omega_t) \qquad (10.1.29)$$

Then by Eq.10.1.26 we obtain

$$\frac{d}{dt}P(\underbrace{\Omega_t\wedge\cdots\wedge\Omega_t}_{r\text{ entries}})=q(D\eta,\Omega_t)+2tq(\eta\wedge\eta,\Omega_t)=dq(\eta,\Omega_t) \qquad (10.1.30)$$

Let $\Omega_{t=1}=\Omega'$ and $\Omega_{t=0}=\Omega$. By Eq.10.1.30 and use η in Eq.10.1.10,

$$P(\underbrace{\Omega',\Omega',\cdots,\Omega'}_{r\text{ entries}})-P(\underbrace{\Omega,\Omega,\cdots,\Omega}_{r\text{ entries}})=d\int_0^1 q(\omega'-\omega,\Omega_t)dt\,, \qquad (10.1.31)$$

hence proves that the characteristic polynomials written by different connections only differ by an exact form

$$d\int_0^1 q(\omega'-\omega,\Omega_t)dt\,. \qquad (10.1.32)$$

This is the second essential feature of $P(\Omega,\Omega,\cdots,\Omega)$.

Eq.10.1.10 and Eq.10.1.32 also give the following two important conclusions:

1. By Eq.10.1.10, $dP(\Omega,\Omega,\cdots,\Omega)=0$. Therefore $P(\Omega,\Omega,\cdots,\Omega)$ *is a closed form.*
2. By Eq.10.1.32 and Stokes' Theorem, and assume that M is a compact manifold without boundary, $\partial M=0$. Then

$$\int_M\left(P(\Omega',\Omega',\cdots,\Omega')-P(\Omega,\Omega,\cdots,\Omega)\right)=\int_{\partial M}\int_0^1 q(\omega'-\omega,\Omega_t)dt=0 \qquad (10.1.33)_1$$

$$\therefore\quad \int_M\left(P(\Omega',\Omega',\cdots,\Omega')\right)=\int_M\left(P(\Omega,\Omega,\cdots,\Omega)\right) \qquad (10.1.33)_2$$

It is indicated that the integral of a characteristic polynomial on a compact base manifold M without boundary — the characteristic number — will not be varied by switching the connection ω with some other ω'. Therefore Eq.10.1.33$_2$ is a topologically invariant integral.

§ 10.2 Chern Classes of Complex Vector Bundles

Suppose that on the base manifold M there is a k-dimensional complex vector space E, of which the structure group is the general k-dimensional complex linear group $GL(k,c)$. Then by introducing the connection, we can write the invariant polynomial which satisfies Eq.10.1.1 in the last section as follows[1]:

---

1 Notice that $Det\ g\left(I+\frac{i}{2\pi}\Omega\right)g^{-1}=Det\ g\cdot Det\left(I+\frac{i}{2\pi}\Omega\right)\cdot Det\ g^{-1}=Det\left(I+\frac{i}{2\pi}\Omega\right)$. Therefore (10.2.1) does

$$C(E) = Det\left(I + \frac{i}{2\pi}\Omega\right) = C_0 + C_1(E) + C_2(E) + \cdots \tag{10.2.1}$$

where $C(E)$ is called the *complete Chern class*, and C_0 , $C_1(E)$, $C_2(E)$,$\cdots$ are given by

$$\left.\begin{aligned} &C_0 = 1 \\ &C_1(E) = \frac{i}{2\pi} Tr\,\Omega \\ &C_2(E) = \frac{1}{8\pi^2}\left(Tr(\Omega\wedge\Omega) - Tr\,\Omega\wedge Tr\,\Omega\right) \\ &\vdots \end{aligned}\right\} \tag{10.2.2}$$

respectively. In general,

$$C_k(E) = \left(\frac{i}{2\pi}\right)^k S_k(\Omega), \tag{10.2.3}$$

$$S_k(\Omega) = \frac{1}{k!}\sum_{(i)(j)} \delta^{j_1\cdots j_k}_{i_1\cdots i_k}\Omega^{i_1}_{j_1}\wedge\Omega^{i_2}_{j_2}\wedge\cdots\wedge\Omega^{i_k}_{j_k}. \tag{10.2.4}$$

For example, we have

$$\begin{aligned} k = 1,\quad & S_1(\Omega) = \sum_{(i)}\Omega^i_i = Tr\,\Omega \\ k = 2,\quad & S_2(E) = \frac{1}{2}\left(\delta^{j_1j_2}_{i_1i_2}\Omega^{i_1}_{j_1}\wedge\Omega^{i_2}_{j_2} - \delta^{j_1j_2}_{i_2i_1}\Omega^{i_1}_{j_1}\wedge\Omega^{i_2}_{j_2}\right) \\ & = \frac{1}{2}\left(\Omega^{i_1}_{i_1}\wedge\Omega^{i_2}_{i_2} - \Omega^{i_1}_{i_2}\wedge\Omega^{i_2}_{i_1}\right) \\ & = -\frac{1}{2}\left(Tr(\Omega\wedge\Omega) - Tr\,\Omega\wedge Tr\,\Omega\right) \\ \cdots\ \cdots \end{aligned} \tag{10.2.5}$$

where k in $S_k(\Omega)$ is a finite integer. Since that the curvature Ω is a 2-form, then if the dimension of M is n, we can have $S_k(\Omega)$ only if $2k \le n$. Therefore

$$C_j = 0, \text{ when } 2j > n = \dim M \tag{10.2.6}$$

From Eq.10.2.1 and Eq.10.2.2 we see that any invariant polynomial $P(\alpha)$ consists of Chern forms C_0, $C_1(E)$, $C_2(E)$, $\cdots$ (by (10.2.6), there are only finitely many terms in $P(\alpha)$). Hence the set of C_0, $C_1(E)$, $C_2(E)$, $\cdots$ is a ring, denoted by $I(G)$, of a characteristic class (including the addition and multiplication of $C_i(E)$ and $C_j(E)$).

Chern Character: Another example of the invariant polynomial of the group $GL(k,c)$ is the Chern character, which is defined as follows:

give an invariant polynomial.

$$\begin{aligned}\mathrm{ch}(E) &= Tr\left(\exp\frac{i}{2\pi}\Omega\right)\\ &= k+\frac{i}{2\pi}Tr\,\Omega+\frac{i^2}{2!(2\pi)^2}Tr(\Omega\wedge\Omega)+\cdots\\ &= k+C_1(E)+\frac{1}{2!}(C_1\wedge C_1-2C_2)(E)+\cdots\end{aligned} \tag{10.2.7}$$

The Chern characters of the direct sum and direct product of two vector bundles on the same base manifold M have the following properties:

$$\begin{aligned}\mathrm{ch}(E\oplus F) &= \mathrm{ch}(E)\oplus\mathrm{ch}(F)\\ \mathrm{ch}(E\otimes F) &= \mathrm{ch}(E)\cdot\mathrm{ch}(F)\end{aligned} \tag{10.2.8}$$

Proof: (i) Since $\Omega_{E\oplus F}=\begin{pmatrix}\Omega_E & 0\\ 0 & \Omega_F\end{pmatrix}$ and $\Omega_{E\oplus F}\wedge\Omega_{E\oplus F}=\begin{pmatrix}\Omega_E\wedge\Omega_E & 0\\ 0 & \Omega_F\wedge\Omega_F\end{pmatrix}$,

$$\therefore\quad Tr\,\Omega_{E\oplus F}=Tr\,\Omega_E+Tr\,\Omega_F, \tag{10.2.9$_1$}$$

$$\text{and}\quad Tr\,\Omega_{E\oplus F}\wedge\Omega_{E\oplus F}=Tr\,\Omega_E\wedge\Omega_E+Tr\,\Omega_F\wedge\Omega_F. \tag{10.2.9$_2$}$$

Similar to Eq.10.2.7, we obtain

$$\begin{aligned}\mathrm{ch}(E\oplus F) &= Tr\left(\exp\frac{i}{2\pi}\Omega_{E\oplus F}\right)\\ &= (k_E+k_F)+\frac{i}{2\pi}(Tr\,\Omega_E+Tr\,\Omega_F)+\frac{i^2}{2!(2\pi)^2}(Tr\,\Omega_E\wedge\Omega_E+Tr\,\Omega_F\wedge\Omega_F)\\ &\quad+\cdots\cdots\\ &= \mathrm{ch}(E)+\mathrm{ch}(F).\end{aligned}$$

Therefore the first equation of Eq.10.2.8 holds.

$$\text{(ii)}\quad \Omega_{E\otimes F}=\begin{pmatrix}\Omega_E & 0\\ 0 & I_F\end{pmatrix}\otimes\begin{pmatrix}I_E & 0\\ 0 & \Omega_F\end{pmatrix} \tag{10.2.10}$$

$$\begin{aligned}\mathrm{ch}(E\otimes F) &= Tr\left(\exp\frac{i}{2\pi}\Omega_{E\otimes F}\right)\\ &= Tr\left(\exp\frac{i}{2\pi}[(\Omega_E\otimes I_E)+(I_F\otimes\Omega_F)]\right)\\ &= Tr\left(\exp\frac{i}{2\pi}(\Omega_E\otimes I_E)\right)\cdot Tr\left(\exp\frac{i}{2\pi}(I_F\otimes\Omega_F)\right)\\ &= Tr\left(\exp\frac{i}{2\pi}(\Omega_E)\right)\cdot Tr\left(\exp\frac{i}{2\pi}(\Omega_F)\right)=\mathrm{ch}(E)\cdot\mathrm{ch}(F).\end{aligned}$$

Hence the second equation of Eq.10.2.8 holds. □

Decomposition of Complex Vector Bundles: If the complex vector bundle E can be decomposed into a direct sum of complex line bundles, then we have

$$E = \oplus\sum_{j=1}^{k} L_j, \quad C(L_j) = 1 + x_j, \quad x_j = \frac{i}{2\pi}\Omega_j = C_1(L_j).$$

Notice that for each complex line bundle L_j, the dimension is $k = 1$, and thus Ω_j has only one row and one column:

$$C_2(L_j) = C_3(L_j) = \cdots = 0.$$

Recall Eq.10.2.1 and generalize it as follows:

$$C(E \oplus F) = Det\left(I + \frac{i}{2\pi}\Omega_{E\oplus F}\right) = Det\begin{pmatrix} I_E + \frac{i}{2\pi}\Omega_E & 0 \\ 0 & I_F + \frac{i}{2\pi}\Omega_F \end{pmatrix}$$

$$= Det\left(I_E + \frac{i}{2\pi}\Omega_E\right) \wedge Det\left(I_F + \frac{i}{2\pi}\Omega_F\right)$$

$$= C(E) \wedge C(F) \tag{10.2.11}$$

Similarly we obtain

$$C(E) = C\left(\oplus\sum_{j=1}^{k} L_j\right) = C(L_1) \wedge C(L_2) \wedge \cdots \wedge C(L_k)$$

$$= (1 + x_1) \wedge (1 + x_2) \wedge \cdots \wedge (1 + x_k) \tag{10.2.12}$$

Not all the complex vector bundles can be decomposed into direct sums of complex line bundles, but in our case, we may use the decomposition theorem.

Decomposition Theorem: When analyzing or proving some equivalent relations for the characteristic classes, we can represent the complex vector bundles with rank *k* by the direct sums of the line bundles $L_1, L_2, \cdots, L_k$ [4] [5].

For example, suppose that a *k*-dimensional complex vector bundle *E* can be decomposed into the complex line bundles $L_1, L_2, \cdots, L_k$. That is,

$$E = L_1 \oplus L_2 \oplus \cdots L_k \tag{10.2.13}$$

Then according to Eq.10.2.11, the complete Chern class is

$$C(E) = C(L_1) \wedge C(L_2) \wedge \cdots \wedge C(L_k).$$

It has been given previously that the complex line bundles are $C(L_j) = 1 + x_j$, where $x_j = C_1(L_j)$; and $C_2(L_j) = C_3(L_j) = 0$.

$$\therefore \quad C(E) = \prod_{j=1}^{k}\left(1 + C_1(L_j)\right) = \prod_{j=1}^{k}(1 + x_j)$$

$$= 1 + C_1(E) + C_2(E) + \cdots + C_k(E) \tag{10.2.14}$$

where

$$C_1(E) = \sum_{j=1}^{k} x_j, \quad C_2(E) = \sum_{i<j}^{k} x_i \wedge x_j, \quad \cdots, \quad C_k(E) = x_1 \wedge x_2 \wedge \cdots \wedge x_k \tag{10.2.15}$$

§ 10.3 Pontrjagin Classes of Real Vector Bundles

The fibres of the transition matrix and the adjoint principal bundle (i.e. the frame bundle) of a *k*-dimensional vector bundle belong to the general *k*-dimensional real linear group $GL(k,R)$. If a Riemannian metric is defined on the fibre space of a real vector bundle, then the transition matrix can be reduced to a $O(k)$ group, and its adjoint principal bundle is a $O(k)$ principal bundle (i.e. an orthogonal frame bundle).

According to Eq.10.1.31, the characteristic polynomials of two real vector bundles must only differ by an exact form, if their structure groups are the $O(k)$ group and the $GL(k,R)$ group, respectively. Hence the characteristic polynomials of the $GL(k,R)$ bundle and the $O(k)$ bundle have the same integral, which is topologically invariant, on the compact base manifold M without boundary (see Eq.10.1.33).

Now let us take a look at Pontrjagin Classes of the $O(k)$ bundle. Take a Riemannian metric on a real vector bundle. From Eq.4.2.15 we have

$$\Omega_{ij} = -\Omega_{ji} \to g^{ik}\Omega_{ij} = \Omega^k_j = -g^{ik}\Omega_{ji} = -\Omega^k_j$$

$$\therefore \quad Tr\,\Omega = \Omega^i_i = -\Omega^i_i = -\Omega^j_k g_{ij} g^{ik} = -\Omega^j_k \delta^k_j = -\Omega^j_j = 0$$

$$\therefore \quad Tr\,\Omega = -Tr\,\Omega = 0 \tag{10.3.1}$$

Then since $\Omega^k_j = -\Omega^k_j$, also notice that Ω is a 2-form, and that Ω and Ω are commutable, we have

$$\begin{aligned} Tr\,\Omega\wedge\Omega\wedge\Omega &= \Omega^i_j \wedge \Omega^j_k \wedge \Omega^k_i = \left(-\Omega^i_j\right)\wedge\left(-\Omega^j_k\right)\wedge\left(-\Omega^k_i\right) \\ &= -\Omega^i_j \wedge \Omega^j_k \wedge \Omega^k_i = -\Omega^{ji} \wedge \Omega_{kj} \wedge \Omega^k_i \\ &= -\Omega^j_i \wedge \Omega_{kj} \wedge \Omega^{ik} = -\Omega^j_i \wedge \Omega^k_j \wedge \Omega^i_k \\ &= -\Omega^j_i \wedge \Omega^i_k \wedge \Omega^k_j = -Tr\,\Omega\wedge\Omega\wedge\Omega \end{aligned}$$

$$\therefore \quad Tr\,\Omega\wedge\Omega\wedge\Omega = -Tr\,\Omega\wedge\Omega\wedge\Omega = 0 \tag{10.3.2}$$

Deducing from analogy we see that $Tr\,\Omega\wedge\cdots\wedge\Omega \neq 0$ only if the number of Ω's is even.

The complete Pontrjagin Class is defined as the following invariant polynomial:

$$P(E) = Det\left(I - \frac{1}{2\pi}\Omega\right) = I + P_1(E) + P_2(E) + \cdots \tag{10.3.3}$$

Then $P(E)$ has the following property:

$$P(E \oplus F) = P(E)\cdot P(F) \tag{10.3.4}$$

Proof:

$$Det\left(I_{E\oplus F} - \frac{1}{2\pi}\Omega_{E\oplus F}\right) = Det\left(I_E - \frac{1}{2\pi}\Omega_E\right)\cdot Det\left(I_F - \frac{1}{2\pi}\Omega_F\right)$$

$$=\begin{vmatrix} I_E-\frac{1}{2\pi}\Omega_E & 0 \\ 0 & I_F-\frac{1}{2\pi}\Omega_F \end{vmatrix}=\left|I_E-\frac{1}{2\pi}\Omega_E\right|\cdot\left|I_F-\frac{1}{2\pi}\Omega_F\right| \tag{10.3.5.}$$

Therefore Eq.10.3.4 holds. □

Take an expansion of Eq.10.3.4 we obtain

$$P_1(E)=\frac{1}{2!(2\pi)^2}\,\delta^{i_1i_2}_{j_1j_2}\,\Omega_{i_1j_1}\wedge\Omega_{i_2j_2} \tag{10.3.6}$$

$$\cdots \quad \cdots \quad \cdots$$

$$P_k(E)=\frac{1}{(2k)!(2\pi)^{2k}}\quad\delta^{i_1\cdots i_{2k}}_{j_1\cdots j_{2k}}\quad\Omega_{i_1j_1}\wedge\cdots\wedge\Omega_{i_{2k}j_{2k}} \tag{10.3.7}$$

$$\cdots \quad \cdots \quad \cdots$$

The reason of $(2k)!$ being in the denominator of the right hand side of Eq.10.3.7 is that there are totally $(2k)!$ equivalent permutations and combinations of $j_1,\cdots,j_{2k}$ when $i_1,\cdots,i_{2k}$ are fixed in $\delta^{i_1\cdots i_{2k}}_{j_1\cdots j_{2k}}$. If we do not divide Eq.10.3.7 by $(2k)!$, then the left hand side should be $(2k)!\,P_k(E)$.

Other invariant polynomials of the real vector bundle E are omitted here.

§ 10.4 Euler Classes of Real Oriented Even Dimensional Vector Bundles

According to the definitions of the real orthogonal groups $O(k)$ and $SO(k)$, we take the basic representation to be of dimensional $k\times k$ and then have

$$O(k)=\left\{A\,\middle|\,A=k\times k\text{ matrix},\ A^TA=E\right\},$$

$$\text{and}\quad SO(k)=\left\{A\,\middle|\,A\in O(k),\ \det A=1>0\right\}.$$

The structure function of any real *k*-dimensional oriented vector bundle can usually be reduced to a member of the $SO(k)$ group, with its basic representation being a real $k\times k$ anti-symmetric matrix. In the case of an odd dimensional real anti-symmetric matrix, the corresponding determinant must be zero. For example,

$$\begin{vmatrix} 0 & a_{12} & a_{13} \\ -a_{12} & 0 & a_{23} \\ -a_{13} & -a_{23} & 0 \end{vmatrix}=-a_{12}a_{23}a_{13}+a_{13}a_{12}a_{23}=0 \tag{10.4.1}$$

Hence if we take the $SO(k)$ group and require that $Det\left(a_{ij}\right)\neq 0$, then k of $SO(k)$ must be even. That is, it is only when $k=2r$ that we may define a real oriented vector bundle, i.e., a real oriented even dimensional vector bundle.

Conclusions: Let $\left(a_{ij}\right)$ be a real anti-symmetric matrix.

$$Det\left(a_{ij}\right)=\begin{vmatrix} 0 & a_{12} & a_{13} & \cdots & a_{1k} \\ a_{21} & 0 & a_{23} & \cdots & a_{2k} \\ a_{31} & a_{32} & 0 & \cdots & a_{3k} \\ \cdots & \cdots & \cdots & \cdots & \cdots \\ a_{k1} & a_{k2} & a_{k3} & \cdots & 0 \end{vmatrix}=\begin{vmatrix} 0 & -a_{21} & -a_{31} & \cdots & -a_{k1} \\ -a_{12} & 0 & -a_{32} & \cdots & -a_{k2} \\ -a_{13} & -a_{23} & 0 & \cdots & -a_{k3} \\ \cdots & \cdots & \cdots & \cdots & \cdots \\ -a_{1k} & -a_{2k} & -a_{3k} & \cdots & 0 \end{vmatrix}$$

$$=(-1)^k\,Det\left(a_{ij}\right)$$

$$Det\left(a_{ij}\right)=-Det\left(a_{ij}\right)=0 \quad \text{if } k \text{ is odd.} \tag{10.4.2}$$

$$Det\left(a_{ij}\right)=\begin{vmatrix} 0 & a_{12} & a_{13} & \cdots & a_{1k} \\ -a_{12} & 0 & a_{23} & \cdots & a_{2k} \\ -a_{13} & -a_{23} & 0 & \cdots & a_{3k} \\ \cdots & \cdots & \cdots & \cdots & \cdots \\ -a_{1k} & -a_{2k} & -a_{3k} & \cdots & 0 \end{vmatrix}$$

$$=\frac{1}{\left(2^r r!\right)^2}\left(\varepsilon_{i_1\cdots i_{2r}}\, a_{i_1 i_2}\cdots a_{i_{2r-1}i_{2r}}\right)^2=\left(e(a)\right)^2 \quad \text{if } k \text{ is even } (k=2r). \tag{10.4.3}$$

The proof is omitted.

Take $a_{ij}=\dfrac{\Omega^{ij}}{2\pi}$, we can define the Euler classes as follows, which happens to be the square root of the right hand side of Eq.10.4.3:

In the case of k-dimensional Riemann manifold where $k=2r$:

$$e\left(\frac{\Omega}{2\pi}\right)=\frac{1}{2^r r!}\frac{1}{(2\pi)^r}\left(\varepsilon_{i_1,i_2\cdots i_{2r}}\Omega^{i_1i_2}\wedge\Omega^{i_3i_4}\wedge\cdots\wedge\Omega^{i_{2r-1}i_{2r}}\right) \tag{10.4.4}$$

In the case of 2-dimensional Riemann manifold where $r=1$ and $k=2$:

$$e\left(\frac{\Omega}{2\pi}\right)=\frac{1}{4\pi}\left(\varepsilon_{12}\Omega^{12}+\varepsilon_{21}\Omega^{21}\right)=\frac{1}{2\pi}\Omega^{12} \tag{10.4.5}$$

In the case of 4-dimensional Riemann manifold where $r=2$ and $k=4$:

$$e\left(\frac{\Omega}{2\pi}\right)=\frac{1}{32\pi^2}\varepsilon_{abcd}\Omega^{ab}\wedge\Omega^{cd} \tag{10.4.6}$$

… … …

Discussion 1. *The Comparisons of Euler and Pontrjagin Classes (k=2r)*

Beginning from Eq.10.3.3, we notice that $\Omega_{ij}=-\Omega_{ji}\ \rightarrow\ \Omega^{ij}=-\Omega^{ji}$. Hence

$$P(E) = Det\left(I - \frac{1}{2\pi}\Omega\right) = \begin{vmatrix} 1 & -\frac{\Omega^{12}}{2\pi} & -\frac{\Omega^{13}}{2\pi} & \cdots & -\frac{\Omega^{1k}}{2\pi} \\ \frac{\Omega^{12}}{2\pi} & 1 & -\frac{\Omega^{23}}{2\pi} & \cdots & -\frac{\Omega^{2k}}{2\pi} \\ \frac{\Omega^{13}}{2\pi} & \frac{\Omega^{23}}{2\pi} & 1 & \cdots & -\frac{\Omega^{3k}}{2\pi} \\ \cdots & \cdots & \cdots & \cdots & \cdots \\ \frac{\Omega^{1k}}{2\pi} & \frac{\Omega^{2k}}{2\pi} & \frac{\Omega^{3k}}{2\pi} & \cdots & 1 \end{vmatrix}$$

$$= 1 + P_1(E) + P_2(E) + \cdots \tag{10.4.7}$$

By Eq.10.4.3, we have

$$P_k(E) = \begin{vmatrix} 0 & -\frac{\Omega^{12}}{2\pi} & -\frac{\Omega^{13}}{2\pi} & \cdots & -\frac{\Omega^{1k}}{2\pi} \\ \frac{\Omega^{12}}{2\pi} & 0 & -\frac{\Omega^{23}}{2\pi} & \cdots & -\frac{\Omega^{2k}}{2\pi} \\ \frac{\Omega^{13}}{2\pi} & \frac{\Omega^{23}}{2\pi} & 0 & \cdots & -\frac{\Omega^{3k}}{2\pi} \\ \cdots & \cdots & \cdots & \cdots & \cdots \\ \frac{\Omega^{1k}}{2\pi} & \frac{\Omega^{2k}}{2\pi} & \frac{\Omega^{3k}}{2\pi} & \cdots & 0 \end{vmatrix}$$

$$= Det\left(\frac{\Omega^{ij}}{2\pi}\right) = \left(e\left(\frac{\Omega}{2\pi}\right)\right)^2 \tag{10.4.8}$$

a. Compare Eq.10.4.8 with Eq.10.4.4, we see that $P_1(E)$ is a $4r$-form when the dimension of the Riemann manifold is $k = 2r$. However, since the manifold is only $k = 2r$ dimensional, thus $P_k(E) = \left(e\left(\frac{\Omega}{2\pi}\right)\right)^2 = 0$ for any integer r (since there must be some repeated indices such that $dx^i \wedge dx^i = -dx^i \wedge dx^i = 0$).

Now let us observe Eq.10.4.4. Euler class $e\left(\frac{\Omega}{2\pi}\right)$ happens to be a $k = 2r$ form when the Riemann manifold is $k = 2r$ dimensional. Thus $e\left(\frac{\Omega}{2\pi}\right) \neq 0$, and it is a $2r$-form of a $k = 2r$ dimensional manifold.

b. Pontrjagin classes are invariant polynomials with respect to the changes of groups with either the Riemannian connection or the general linear connection (both cases give the same cohomology class).

Euler classes, however, is invariant only with respect to the changes of $SO(k)$ with the Riemannian connection, but not with the general $GL(k,R)$.

Discussion 2. *The Comparisons of Euler and Chern Classes*

From the investigation of the complex manifolds in Section 5.1, we know that a complex manifold must be even dimensional and real oriented. In fact a complex vector bundle has this very property similar to that of a complex manifold. That is, a complex vector bundle is even dimensional and real oriented as well.

From Section 10.3 we can also see that the realization of an r-dimensional complex vector bundle is a real vector bundle E of dimension $2r = k$, if the complex structure can be omitted.

Example 1.

$$\begin{aligned} \underline{\varphi_1}: \quad & C^1 \longrightarrow C^1 \qquad (z = x_1 + ix_2) \\ & z \longrightarrow z' = (\alpha + i\beta)z = (\alpha x_1 - \beta x_2) + i(\alpha x_2 + \beta x_1) \end{aligned}$$

Express in the real space, we have

$$\underline{\varphi_{R_2}}: \quad \begin{pmatrix} x_1 \\ x_2 \end{pmatrix} \longrightarrow \begin{pmatrix} x_1' \\ x_2' \end{pmatrix} = \begin{pmatrix} \alpha & -\beta \\ \beta & \alpha \end{pmatrix} \begin{pmatrix} x_1 \\ x_2 \end{pmatrix} \tag{10.4.9}$$

where $\begin{pmatrix} \alpha & -\beta \\ \beta & \alpha \end{pmatrix}$ is called the realization of ($\alpha + i\beta$), and is denoted by

$$(\alpha + i\beta)_R = \begin{pmatrix} \alpha & -\beta \\ \beta & \alpha \end{pmatrix} \tag{10.4.10}$$

Example 2.

$$\underline{\varphi_2}: \qquad C^2 \longrightarrow C^2 \qquad (z_1 = x_1 + ix_2,\ z_2 = x_3 + ix_4)$$

$$\begin{pmatrix} z_1 \\ z_2 \end{pmatrix} \longrightarrow \begin{pmatrix} z_1' \\ z_2' \end{pmatrix} = \begin{pmatrix} \alpha_1 + i\beta_1 & \alpha_2 + i\beta_2 \\ \alpha_3 + i\beta_3 & \alpha_4 + i\beta_4 \end{pmatrix} \begin{pmatrix} z_1 \\ z_2 \end{pmatrix} = \begin{pmatrix} (\alpha_1 + i\beta_1)z_1 + (\alpha_2 + i\beta_2)z_2 \\ (\alpha_3 + i\beta_3)z_1 + (\alpha_4 + i\beta_4)z_2 \end{pmatrix}$$

That is,

$$\begin{pmatrix} z_1 \\ z_2 \end{pmatrix} = \begin{pmatrix} x_1 + ix_2 \\ x_3 + ix_4 \end{pmatrix} \longrightarrow \begin{pmatrix} x_1' + ix_2' \\ x_3' + ix_4' \end{pmatrix} = \begin{pmatrix} (\alpha_1 + i\beta_1)(x_1 + ix_2) + (\alpha_2 + i\beta_2)(x_3 + ix_4) \\ (\alpha_3 + i\beta_3)(x_1 + ix_2) + (\alpha_4 + i\beta_4)(x_3 + ix_4) \end{pmatrix}$$

Express in the real space, we have

$$\underline{\varphi_{R_4}}: \quad \begin{pmatrix} x_1 \\ x_2 \\ x_3 \\ x_4 \end{pmatrix} \longrightarrow \begin{pmatrix} x_1' \\ x_2' \\ x_3' \\ x_4' \end{pmatrix} = \begin{pmatrix} \alpha_1 & -\beta_1 & \alpha_2 & -\beta_2 \\ \beta_1 & \alpha_1 & \beta_2 & \alpha_2 \\ \alpha_3 & -\beta_3 & \alpha_4 & -\beta_4 \\ \beta_3 & \alpha_3 & \beta_4 & \alpha_4 \end{pmatrix} \begin{pmatrix} x_1 \\ x_2 \\ x_3 \\ x_4 \end{pmatrix} \tag{10.4.11}$$

where $\begin{pmatrix} \alpha_1 & -\beta_1 & \alpha_2 & -\beta_2 \\ \beta_1 & \alpha_1 & \beta_2 & \alpha_2 \\ \alpha_3 & -\beta_3 & \alpha_4 & -\beta_4 \\ \beta_3 & \alpha_3 & \beta_4 & \beta_4 \end{pmatrix}$ is called the realization of $\begin{pmatrix} \alpha_1 + i\beta_1 & \alpha_2 + i\beta_2 \\ \alpha_3 + i\beta_3 & \alpha_4 + i\beta_4 \end{pmatrix}$, and is denoted by

$$\begin{pmatrix} \alpha_1 + i\beta_1 & \alpha_2 + i\beta_2 \\ \alpha_3 + i\beta_3 & \alpha_4 + i\beta_4 \end{pmatrix}_R = \begin{pmatrix} \alpha_1 & -\beta_1 & \alpha_2 & -\beta_2 \\ \beta_1 & \alpha_1 & \beta_2 & \alpha_2 \\ \alpha_3 & -\beta_3 & \alpha_4 & -\beta_4 \\ \beta_3 & \alpha_3 & \beta_4 & \alpha_4 \end{pmatrix} \tag{10.4.12}$$

With the same idea of realizations in Eq.10.4.10 and Eq.10.4.12, we can write the complex vector bundle E as the real vector bundle E_R. Compare Eq.10.4.12 to Eq.10.4.8, since $\Omega^{ab} = -\Omega^{ba}$ is anti-symmetric, we have $\Omega^{aa} = \Omega^{bb} = \cdots = 0$. Therefore the invariant polynomials of the realization E_R can still be written in the form of Eq.10.4.8, that is,

$$P_k(E_R) = \left(e\left(\frac{\Omega}{2\pi}\right)\right)^2, \tag{10.4.13}$$

or

$$e\left(\frac{\Omega}{2\pi}\right) = \left(P_k(E_R)\right)^{\frac{1}{2}} = C_k(E) \tag{10.4.14}$$

where $C_k(E) = x_1 \wedge \cdots \wedge x_k$ is the highest Chern class of the realization vector bundle E_R (see Eq.10.2.15).

Thus we know that the Euler class $e\left(\frac{\Omega}{2\pi}\right)$ of the realization vector bundle E_R of the complex vector bundle E is equal to the highest Chern class $C_k(E)$ of the realization vector bundle E_R.

§ 10.5 Stifel-Whitney Classes of Real Vector Bundles

Being different than the previously mentioned three classes, the Stifel-Whitney classes are not expressed by the polynomials of the curvatures Ω^{ij}. Let M be a base manifold, and ξ be an n-dimensional real vector bundle on M. Then the complete Stifel-Whitney class, denoted by $W(\xi)$, can be defined as follows:

$$W(\xi) = 1 + W_1(\xi) + W_2(\xi) + \cdots + W_n(\xi) \tag{10.5.1}$$

where

$$\left.\begin{aligned} &W_i(\xi) \in H^i(M, Z_2) \quad (i = 1, 2, \cdots, n) \\ &W_i(\xi) = 0 \qquad\qquad\quad (i > \dim M = n) \end{aligned}\right\} \tag{10.5.2}$$

In the first line of Eq.10.5.2, Z_2 represents the Z_2 group. In fact the Z_2 group equals the $O(1)$ group, since what be described by the $O(k)$ group is a k-dimensional sphere

$$a_1^2 + a_2^2 + \cdots + a_k^2 = 1 \tag{10.5.3}$$

and hence what be described by the $O(1)$ group is a 1-dimensional "sphere"

$$a_1^2 = 1 \longrightarrow a_1 = \pm 1 \tag{10.5.4}$$

Therefore we can see that the structure group of a Stifel-Whitney class is

$$Z_2 \text{ group} = O(1) \text{ group} \tag{10.5.5}$$

Regarding the applications of the Stifel-Whitney classes, here we give two interesting examples (the proofs are omitted).

Example 1: M is an oriented manifold if and only if

$$W_1(T(M)) = 0 \tag{10.5.6}$$

where $T(M)$ is the tangent bundle of M.

Example 2. M is a spin manifold (i.e. the spinor field exists on the manifold M) if and only if the first two Stifel-Whitney classes are zero. That is,

$$W_2(T(M)) = W_1(T(M)) = 0 \tag{10.5.7}$$

§ 10.6 Chern-Simons Form — an Example of Secondary Classes

Starting from Eq.10.1.8 and Eq.10.1.9, we have obtained Eq.10.1.10 which proves that the invariant polynomial $P(\Omega, \cdots, \Omega)$ must be a closed form.

Let ω' and ω be two distinct connections, of which the corresponding curvatures are Ω' and Ω, respectively. Then according to Eq.10.1.31, $P(\Omega', \cdots, \Omega') - P(\Omega, \cdots, \Omega)$ can be locally expressed by an exact form. That is,

$$\underbrace{P(\Omega', \cdots, \Omega')}_{r\text{ entries}} - \underbrace{P(\Omega, \cdots, \Omega)}_{r\text{ entries}} = \underbrace{dQ(\omega', \omega)}_{\text{An exact form}} \tag{10.6.1}$$

Then compare Eq.10.1.31 to Eq.10.6.1, we have

$$Q(\omega', \omega) = \int_0^1 q(\omega' - \omega, \Omega_t)\, dt \tag{10.6.2}$$

where the definition of $q(\eta, \Omega_t)$ is given by Eq.10.1.25.

Hence by Stokes' formula we obtain

$$\int_M dQ(\omega', \omega) = \int_{\partial M} Q(\omega', \omega) \tag{10.6.3}$$

The above integral is zero when the base manifold M is compact without boundary $(\partial M = 0)$[1]. In this case $Q(\omega', \omega)$ itself is a characteristic class with some independent meanings, and is called the *Chern-Simons class*, or simply the Chern

[1] However, it is not necessarily zero if $\partial M \neq 0$.

-Simons.

Ω is a 2-form and $P(\underbrace{\Omega_t,\cdots,\Omega_t}_{r\text{ entries}})$ is a $2r$-form. Thus $P(\Omega,\cdots,\Omega)$ is *the class of even forms.*

ω is a 1-form and $q(\omega'-\omega,\underbrace{\Omega_t,\cdots,\Omega_t}_{(r-1)\text{ entries}})$ is a $2r-1$ form. Thus $Q(\omega',\omega)$ is *the class of odd forms.*

Here we notice the definition of ω_t in Eq.10.1.11

$$\omega_t = \omega = +t\eta\,, \qquad \eta = \omega' - \omega\,, \qquad 0 \le t \le 1 \tag{10.6.4}$$

and the definitions of Ω_t in Eq.10.1.13 and Eq.10.1.16

$$\Omega_t = d\omega_t + \omega_t \wedge \omega_t = \Omega + tD\eta + t^2\eta\wedge\eta \tag{10.6.5}$$

With the definitions of ω_t and Ω_t as the prerequisite, we can take

$$\omega' = A \text{ (a 1-form)},\ \ \omega = 0,\ \ \text{and}\ \ \eta = \omega' - \omega = A\,. \tag{10.6.6}$$

Then we have

$$\omega_t = t\eta = At.\ \ (\omega = 0) \tag{10.6.7}$$

$$\Omega_t = d\omega_t + \omega_t \wedge \omega_t = t\,dA + t\,A \wedge t\,A \tag{10.6.8}$$

Therefore the Chern-Simons of a $(2r-1)$ form can be defined as (use Eq.10.6.2):

$$Q(A,0) = \int_0^1 q(A,\underbrace{\Omega_t\cdots\Omega_t}_{(r-1)\text{ entries}})dt = \omega_{2r-1}(A) \tag{10.6.9}$$

For example, take $r = 2$ and define $F = dA + A \wedge A$, we have

$$\begin{aligned}\omega_3(A) &= \int_0^1 q(A,\underbrace{\Omega_t}_{1\text{ entry}})dt = \int_0^1 Tr(A\wedge\Omega_t + \Omega_t \wedge A)dt \\ &= \int_0^1 Tr(A\wedge t\,dA + A\wedge t\,A\wedge t\,A + t\,dA\wedge A + t\,A\wedge t\,A\wedge A)dt \\ &= \int_0^1 Tr(2t\,A\wedge dA + 2t^2 A\wedge A\wedge A)dt \\ &= Tr\left(A\wedge dA + \frac{2}{3}A\wedge A\wedge A\right) = Tr\left(F\wedge A - \frac{1}{3}A\wedge A\wedge A\right)\end{aligned} \tag{10.6.10}$$

Remarks:

1. According to Definition 10.1.3 (see Eq.10.1.25), we have
$$q(A,\Omega_t) = P(A,\Omega_t) + P(A,\Omega_t)\,.$$
where the meaning of P is given in Eq.10.1.1.
2. From Eq.10.2.1, Eq.10.2.2 and Eq.10.2.5 we know that an invariant polynomial must take Tr.
3. A is a 1-form, and dA is a 2-form. Hence $A \wedge dA = dA \wedge A$.
4. There are infinitely many Chern-Simons' since r can be any positive integers.

Part III. Index Theorems and 4-Manifolds

Chapter 11. Index Theorems for Manifolds without Boundary

§ 11.1 Analytic Index and Elliptic Differential Operator

Let M be an n-dimensional compact smooth manifold without boundary. Let E and F be two complex vector bundles on M of dimensions m and l, respectively. Take

$$\Gamma(E)=\{x;\ \xi^1(x),\cdots,\xi^m(x)\}$$
$$\Gamma(F)=\{x;\ \eta^1(x),\cdots,\eta^l(x)\} \qquad (11.1.1)$$

to represent the sets of the sections on the complex vector bundles E and F, respectively.

The operator D which maps the section $\Gamma(E)$ into the section $\Gamma(F)$ is called the *differential operator*, of which the action in the neighborhood of each point x of M is:

$$D(\xi^1(x),\cdots,\xi^m(x))=(\eta^1(x),\cdots,\eta^l(x)) \qquad (11.1.2)$$

$$\eta^i(x)=\sum_j a^i_{j,\mu_1,\cdots,\mu_n}(x)\frac{\partial^{\mu_1+\cdots+\mu_n}}{\partial x_1^{\mu_1}\cdots\partial x_n^{\mu_n}}\xi^j(x) \qquad (11.1.3)$$

where $i=1,2,\cdots,l$ and $j=1,2,\cdots,m$. Eq.11.1.3 may also be written as

$$\left.\begin{aligned}&\eta^i(x)=D^i_j\xi^j(x),\\ &D^i_j=\left(a^i_j\right)_{\mu_1\cdots\mu_n}(x)\frac{\partial^{\mu_1+\cdots+\mu_n}}{\partial x_1^{\mu_1}\cdots\partial x_n^{\mu_n}}\end{aligned}\right\} \qquad (11.1.4)$$

where D^i_j is an $l\times m$ matrix.

Denote the Fourier transformation of D^i_j's differential term of the highest order (i.e. the leading term of the differential operator with $\sum\limits_{\mu_1+\mu_2+\cdots+\mu_n=\max}$) by (using Eq.11.1.4)

$$\widetilde{D}^i_j(x,k)=\sum_{\mu_1+\mu_2+\cdots+\mu_n=\max}\left(a^i_j\right)_{\mu_1\mu_2\cdots\mu_n}(x)\,k_1^{\mu_1}\cdots k_n^{\mu_n} \qquad (11.1.5)$$

where $k=(k_1,k_2,\cdots,k_n)$, and both sides are $l\times m$ matrices.

Definition 11.1.1: D is called an *elliptic differential operator* if $m=l$ and $Det\,\widetilde{D}(x,k)\neq 0$ $(k\in R^n-\{0\})$. Otherwise D is not elliptic.

Example 1. Let $M=R^3$. We consider the following three differential operators:

grad : Take $\Gamma(E)$ to be a 1-dimensional vector (i.e. a scalar with the index $i=1$), and $\Gamma(F)$ a 3-dimensional vector (with $j=1,2,3$). From Eq.11.1.5 we obtain

$$\widetilde{D}(x,k) \sim \begin{pmatrix} k_1 \\ k_2 \\ k_3 \end{pmatrix} \begin{matrix} j=1 \\ j=2 \\ j=3 \end{matrix} \quad i=1$$

where $m-1$, $l=3$, and $l \neq m$.
Therefore **grad** is not an elliptic operator.

curl : Take $\Gamma(E)$ and $\Gamma(F)$ to be 3-dimensional vectors (with the indices $i=1,2,3$ and $j=1,2,3$ respectively). From Eq.11.1.5 we obtain

$$\widetilde{D}(x,k) \sim \begin{pmatrix} 0 & -k_3 & k_2 \\ k_3 & 0 & -k_1 \\ -k_2 & k_1 & 0 \end{pmatrix} \begin{matrix} j=1 \\ j=2 \\ j=3 \end{matrix} \quad i=1 \;\; i=2 \;\; i=3$$

where $Det\ \widetilde{D}(x,k)=0\ \ (k \neq 0)$.
Therefore **curl** is not an elliptic operator.

div : Take $\Gamma(E)$ to be a 3-dimensional vector (with $i=1,2,3$), and $\Gamma(F)$ a 1-dimensional vector (with $j=1$). From Eq.11.1.5 we obtain

$$\widetilde{D}(x,k) = \begin{pmatrix} k_1, & k_2, & k_3 \end{pmatrix} \quad i=1 \;\; i=2 \;\; i=3$$

where $m=3$, $l=1$, and $l \neq m$.
Therefore **div** is not an elliptic operator.

Example 2. The second ordered partial differential mapping between the real line bundles of a 2-dimensional base manifold has three typical types:

(i) **Hyperbolic** type : $\dfrac{\partial^2 u}{\partial x_1^2} - \dfrac{\partial^2 u}{\partial x_2^2} + A\dfrac{\partial u}{\partial x_1} + B\dfrac{\partial u}{\partial x_2} + Cu + G$.

(ii) **Parabolic** type : $\dfrac{\partial^2 u}{\partial x_1 \partial x_2} + A\dfrac{\partial u}{\partial x_1} + B\dfrac{\partial u}{\partial x_2} + Cu + G$.

(iii) **Elliptic** type : $\dfrac{\partial^2 u}{\partial x_1^2} + \dfrac{\partial u^2}{\partial x_2^2} + A\dfrac{\partial u}{\partial x_1} + B\dfrac{\partial u}{\partial x_2} + Cu + G$.

The essential features of the above three types are basically illustrated by the differential terms of order 2, and hence can be represented by the second ordered terms of k in $\widetilde{D}(x,k)$. Compare to Definition 11.1.1, we have

(i) $\widetilde{D}(x,k) = k_1^2 - k_2^2$ when $k_1 = k_2 \neq 0$ and $\widetilde{D} = 0$. Thus in this case the operator is not elliptic.

(ii) $\widetilde{D}(x,k) = k_1 k_2$ when $k_1(\text{or } k_2) = 0$ and $\widetilde{D} = 0$. Thus in this case the operator is not elliptic.

(iii) $\widetilde{D}(x,k) = k_1^2 + k_2^2$ when $k_1, k_2 \neq 0$ and $\widetilde{D} \neq 0$. Therefore this operator is indeed elliptic.

Thus we see that the second ordered partial differential operator of the *elliptic type* is an elliptic operator (see Definition 11.1.1), and it is where the later one has obtained its name from. Let us also notice that the elliptic operator in (iii) is the *Laplacian.*

Example 3. Mappings between surface bundles (i.e. 2-dimensional vector bundles) on 2-manifolds.

Here we give an example in the form of Eq.11.1.3:

$$\begin{aligned}\eta^1(x_1,x_2) &= \frac{\partial\xi^1}{\partial x_1} - \frac{\partial\xi^2}{\partial x_2} \\ \eta^2(x_1,x_2) &= \frac{\partial\xi^1}{\partial x_2} + \frac{\partial\xi^2}{\partial x_1}\end{aligned} \Rightarrow \begin{pmatrix}\eta^1(x_1,x_2)\\ \eta^2(x_1,x_2)\end{pmatrix} \sim \begin{pmatrix}\frac{\partial}{\partial x_1} & -\frac{\partial}{\partial x_2}\\ \frac{\partial}{\partial x_2} & \frac{\partial}{\partial x_1}\end{pmatrix}\begin{pmatrix}\xi^1(x_1,x_2)\\ \xi^2(x_1,x_2)\end{pmatrix}$$

Similar to Eq.11.1.5 we obtain

$$\tilde{D}(x,k) \sim \begin{pmatrix} k_1 & -k_2 \\ k_2 & k_1 \end{pmatrix} \begin{matrix} j=1 \\ j=2 \end{matrix} \qquad i=1 \;\; i=2$$

where $m=2$, $l=2$, and thus $l=m$.

$\therefore \; Det\ \tilde{D}(x,k) \sim k_1^2 + k_2^2 > 0$

Therefore according to Definition 11.1.1, $\tilde{D}(x,k)$ of this example is an elliptic operator which is called the Cauchy-Riemann operator.

Example 4. Mappings between real line bundles on 4-manifolds (i.e. the 4-dimensional space-time):

$$\eta(x_0,x_1,x_2,x_3) = \left(\frac{\partial^2}{\partial x_0^2} - \frac{\partial^2}{\partial x_1^2} - \frac{\partial^2}{\partial x_2^2} - \frac{\partial^2}{\partial x_3^2}\right)\xi(x_0,x_1,x_2,x_3)$$

Here $Det\ \tilde{D}(x,k) = k_0^2 - k_1^2 - k_2^2 - k_3^2$. It is possible for $Det\ \tilde{D}(x,k)$ to be zero when $k \neq 0$. Therefore $\tilde{D}(x,k)$ is not an elliptic operator, but is called the d'Alember operator.

Now we introduce the analytic index of the operator D.

Definition 11.1.2: The kernel of the operator, Ker D, is defined to be the space of the zero-modular solutions. That is, let $\Gamma(E)$ be the entire space on which D is operating, then

$$\text{Ker } D = \{\ \xi \in \Gamma(E) \mid D\xi = 0\ \}. \qquad (11.1.6)$$

Definition 11.1.3: The image of the operator D, Imag D, is defined as

$$\text{Imag } D = \{\ \eta \in \Gamma(F) \mid \exists\xi \in \Gamma(E),\ \eta = D\xi\ \}. \qquad (11.1.7)$$

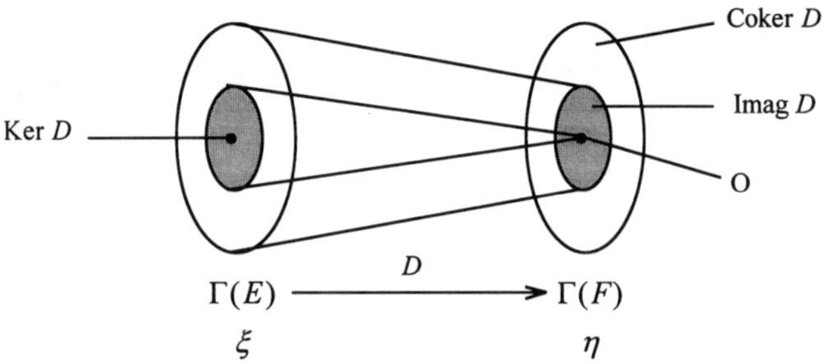

Figure 11.1.1

Definition 11.1.2 and Definition 11.1.3 can both be illustrated by Figure 11.1.1.

The subspaces of $\Gamma(F)$ include not only Imag D, but also Coker D which is the complement of Imag D in $\Gamma(F)$.

When D is the elliptic operator given by Definition 11.1.1 on a compact manifold, it can be proved that *both* Ker D *and* Coker D *are finite dimensional, and are called Fredholm operators*. (The proof is omitted.)

For Fredholm operators, "Ker D is finite dimensional" means that the solution space of the homogeneous equation $D\xi = 0$ is finite dimensional (i.e. there are finitely many linearly independent solutions); "Coker D is finite dimensional" means that a solution of the non-homogeneous equation $D\xi = \eta$ can be determined when η satisfies finitely many conditions.

Definition 11.1.4: The analytic index of the operator D is defined as

$$\text{Index } D = \dim \text{Ker } D - \dim \text{Coker } D \tag{11.1.8}$$

where "dim" represents "dimension". From Figure 11.1.1 we see that

$$\dim \Gamma(E) = \dim \text{Ker } D + \dim \text{Imag } D, \tag{11.1.9}$$

and $$\dim \Gamma(F) = \dim \text{Coker } D + \dim \text{Imag } D. \tag{11.1.10}$$

Hence Eq.11.1.8 can also be written as

$$\text{Index } D = \dim \Gamma(E) - \dim \Gamma(F) \tag{11.1.11}$$

This is a key formula. It seems that Index D on the left hand side of the equation should depend on the choice of D since it is the analytic index of the operator D. However, the right hand side of the equation is independent of D, which shows that Index D is a topological invariant, and is independent of the choice of metric as well.

Corresponding to Figure 11.1.1, we may define the dual operator D^+ of D. See the following figure:

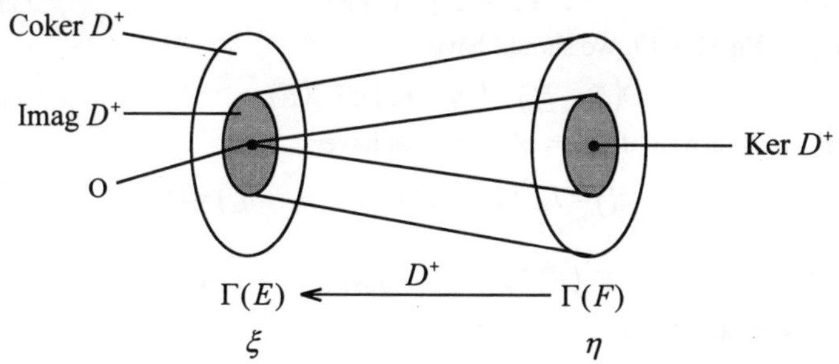

Figure 11.1.2

When a metric is defined in a fibre space, so is the inner product $(D\xi, \eta)$.

Definition 11.1.5: The dual operator D^+ of D has the following property:

$$(D\xi, \eta) = (\xi, D^+\eta) \tag{11.1.12}$$

where $D^+: \quad \Gamma(F) \to \Gamma(E) \; : \; \eta \to D^+\eta \in \Gamma(E)$.

D and D^+ form a pair of dual elliptic operators Δ_E and Δ_F.

Definition 11.1.6:

$$\Delta_E = D^+ D\colon\ \Gamma(E) \to \Gamma(E)$$
$$\Delta_F = DD^+\colon\ \Gamma(F) \to \Gamma(F) \tag{11.1.13}$$

In $\Gamma(E)$ and $\Gamma(F)$, we can define $\Gamma_\lambda(E)$, $\Gamma_\lambda(F)$, $\Gamma_0(E)$ and $\Gamma_0(F)$ as follows:

Definition 11.1.7:

$$\Gamma_\lambda(E) = \{\xi \in \Gamma(E),\, \Delta_E \xi = \lambda\xi\} \quad (\lambda \neq 0);\quad \Gamma_0(E) = \mathrm{Ker}D$$
$$\Gamma_F(F) = \{\eta \in \Gamma(F),\, \Delta_F \eta = \lambda\eta\} \quad (\lambda \neq 0);\quad \Gamma_0(F) = \mathrm{Ker}D^+ \tag{11.1.14}$$

Here we prove that the spaces of the $\lambda\ (\neq 0)$ eigensections, $\Gamma_\lambda(E)$ and $\Gamma_\lambda(F)$, of the operators Δ_E and Δ_F, respectively, are isomorphic to each other. Hence

$$\dim \Gamma_\lambda(E) = \dim \Gamma_\lambda(F), \quad \text{for } \lambda \neq 0 \tag{11.1.15}$$

Proof: According to Definition 11.1.7, we may take $\xi_\lambda \in \Gamma_\lambda(E)$, for $\lambda \neq 0$. Thus

$$\Delta_E \xi_\lambda = D^+ D\xi_\lambda = \lambda\xi_\lambda \tag{11.1.16}$$
$$\therefore\quad \Delta_F(D\xi_\lambda) = DD^+ D\xi_\lambda = D\Delta_E \xi_\lambda = \lambda(D\xi_\lambda)$$

Hence we have

$$D\xi_\lambda = \eta\lambda \in \Gamma_\lambda(F) \tag{11.1.17}$$

From Eq.11.1.16 we know that $D\xi_\lambda = \eta_\lambda \neq 0$ when $\lambda \neq 0$, otherwise it is implied by Eq.11.1.16 that $\lambda = 0$, which contradicts our assumption of $\lambda \neq 0$.

Then suppose that $\xi_\lambda \neq \xi'_\lambda$. According to Eq.11.1.14 we obtain

$$\Delta_E(\xi_\lambda - \xi'_\lambda) = \lambda(\xi_\lambda - \xi'_\lambda) \neq 0, \quad \text{for } \lambda \neq 0 \tag{11.1.18}$$
$$\therefore\quad (\xi_\lambda - \xi'_\lambda) \in \Gamma_\lambda(E) \tag{11.1.18'}$$

According to Eq.11.1.17, we should have

$$D(\xi_\lambda - \xi'_\lambda) = (\eta_\lambda - \eta'_\lambda) \in \Gamma_\lambda(F) \tag{11.1.19}$$

where $\eta_\lambda \neq \eta'_\lambda$. Otherwise if $\eta_\lambda = \eta'_\lambda$, we must have

$$\Delta_E(\xi_\lambda - \xi'_\lambda) = D^+ D(\xi_\lambda - \xi'_\lambda) = D^+(\eta_\lambda - \eta'_\lambda) = 0$$

which contradicts Eq.11.1.18.

Thus we know that $D\colon\ \Gamma_\lambda(E) \to \Gamma_\lambda(F)$ must be a 1-1 mapping, that is, $\xi_\lambda \neq \xi'_\lambda$ $\to$ $\eta_\lambda \neq \eta'_\lambda$. Therefore $\dim \Gamma_\lambda(E) = \dim \Gamma_\lambda(F)$ for $\lambda \neq 0$, which proves Eq.11.1.15. □

We see from Figure 11.1.1 and Figure 11.1.2 that

$$\left.\begin{aligned} \dim \mathrm{Imag}\, D &= \sum_\lambda \dim \Gamma_\lambda(E) \\ \dim \mathrm{Imag}\, D^+ &= \sum_\lambda \dim \Gamma_\lambda(F) \end{aligned}\right\} \tag{11.1.20}$$

Compare Eq.11.1.15 to Eq.11.1.20, we have

$$\dim \text{Imag } D = \dim \text{Imag } D^{+} \tag{11.1.21}$$

Similar to the way we obtained Eq.11.1.9, we can write the corresponding $\dim \Gamma(F)$ as

$$\dim \Gamma(F) = \dim \text{Ker } D^{+} + \dim \text{Imag } D^{+} \tag{11.1.22}$$

Compare Eq.11.1.10 to Eq.11.1.22, we obtain

$$\dim \text{Ker } D^{+} + \dim \text{Imag } D^{+} = \dim \text{Co}\ker D + \dim \text{Imag } D$$

By Eq.11.1.21, $\dim \text{Imag } D^{+}$ on the left hand side of the equation and $\dim \text{Imag } D$ on the right hand side of the equation are canceled out. Hence we have

$$\dim \text{Ker } D^{+} = \dim \text{Co}\ker D \tag{11.1.23}$$

Then plug Eq.11.1.9 and Eq.11.1.10 into Eq.11.1.11, we obtain

$$\text{Index } D = \dim \text{Ker } D - \dim \text{Coker } D.$$

Use Eq.11.1.23, it follows that

$$\text{Index } D = \dim \text{Ker } D - \dim \text{Ker } D^{+} \tag{11.1.24}$$

which is the definition of the analytic index of the differential operator D, when a metric is defined on the fibre bundle. In case where the operator is self-dual (i.e. $D = D^{+}$), Index $D = 0$.

§ 11.2 Elliptic Complexes and Atiyah-Singer's Index Theorem

Let $\{E_p\}$ be a finite sequence of the vector bundles E_p $(p = 1,2,\cdots)$ of *M*. Suppose that

$$D_p : C^{\infty}(E_p) \longrightarrow C^{\infty}(E_{p+1}) \quad \text{for } p = 1,2,3,\cdots \tag{11.2.1}$$

is a finite sequence of the differential operators D_p corresponding to E_p. $C^{\infty}(E_p)$ in Eq.11.2.1 represents the *space of the smooth p-forms*.

Definition 11.2.1: The sequence of the pair E_p and D_p is called a *complex* if $D_p D_{p-1} = 0$ (that is, $\text{Ker } D_p \supset \text{Imag } D_{p-1}$. See Figure 11.2.1.).

According to Definition 11.1.5, if an inner product is defined, then we may define the dual operator D_p^{+} of D_p as follows:

$$D_p^{+} : C^{\infty}(E_{p+1}) \longrightarrow C^{\infty}(E_p) \quad \text{for } p = 1,2,3,\cdots \tag{11.2.2}$$

and

$$(D_p \xi_p, \eta_{p+1}) \longrightarrow (\xi_{p+1}, \eta_{p+1}) \longleftarrow (\xi_{p+1}, D_{p+1}^{+} \eta_{p+2}) \tag{11.2.3}$$

Definition 11.2.2: The Laplacian Δ_p of the sequence D_p is taken as

$$\Delta_p = D_p^{+} D_p + D_{p-1} D_{p-1}^{+} : \ \Gamma(E_p) \to \Gamma(E_p) \tag{11.2.4}$$

where Δ_p's form a sequence as well.

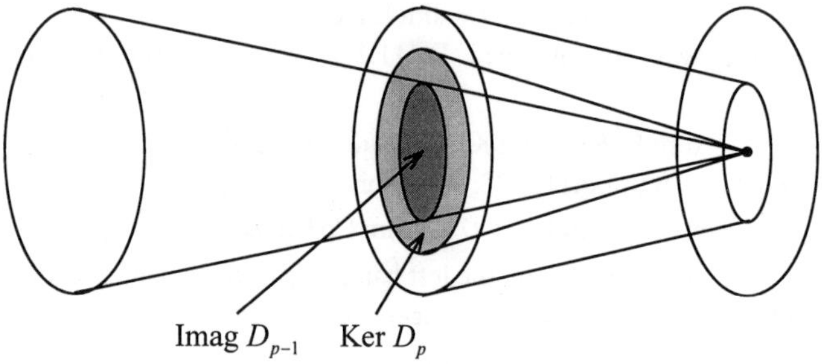

$$\longrightarrow \Gamma(E_{p-1}) \xrightarrow{\quad D_{p-1}\quad} \Gamma(E_p) \xrightarrow{\quad D_p\quad} \Gamma(E_{p+1}) \longrightarrow$$

$$\longleftarrow \Gamma(E_{p-1}) \xleftarrow{\quad D_{p-1}^{+}\quad} \Gamma(E_p) \xleftarrow{\quad D_p^{+}\quad} \Gamma(E_{p+1}) \longleftarrow$$

Figure 11.2.1

Definition 11.2.3: If Δ_p is an elliptic differential operator on $C^{\infty}(E_p)$ (see Definition 11.1.1), then the sequence indicated by Figure 11.2.1 is called an *elliptic complex*, and is denoted by $\{E,D\}=(\{E_p\},\{D_p\})$.

Cohomology of Elliptic Complexes:

The Hodge Decomposition Theorem (see Eq.3.4.21) can be generalized to the elliptic complexes. Let $f_p \in C^{\infty}(E_p)$. Then f_p can be uniquely decomposed as follows (The proof is omitted.):

$$f_p = D_{p-1}f_{p-1} + D_p^{+}f_{p+1} + h_p \tag{11.2.5}$$

where h_p is called the *harmonic section*, satisfying

$$\Delta_p h_p = 0. \tag{11.2.6}$$

The cohomology group of the elliptic complex $\{E,D\}$ can be defined in a similar way of defining the de Rham cohomology group (see Eq.7.4.7):

$$H^p(E,D) = \text{Ker } D_p/\text{Imag } D_{p-1} \tag{11.2.7}$$

Compare Eq.11.2.7 to Eq.7.4.7 we can see that

(i) Ker D_p is similar to $Z_{dR}^p(M,R)$ of $H_{dR}^p(M,R)$, i.e. Ker d_p;

(ii) Imag D_{p-1} is similar to $B_{dR}^p(M,R)$ of $H_{dR}^p(M,R)$, i.e. Imag d_{p-1}.

It can be proved that the cohomology $H^p(E,D)$ in Eq.11.2.7 has only one harmonic form (The proof is omitted.). Furthermore, according to Eq.11.2.6, if $H^p(E,D)$ is a harmonic form, then

$$\Delta_p H^p(E,D) = 0. \tag{11.2.8}$$

Therefore we have the following isomorphic relation:

$$H^p(E,D) \approx \text{Ker}\,\Delta_p \tag{11.2.9}$$

satisfying $\text{Ker}\,H^p(E,D) \approx \text{Ker}\left(\text{Ker}\,\Delta_p\right) = 0$, which is similar to $\text{Ker}\left(\text{Ker}\,D_p\right) = 0$ in Figure 11.2.1.

Recall what we have mentioned in Section 11.1 that $\text{Ker}\,D$ and $\text{Coker}\,D$ are both finite dimensional Fredholm operators if D is an elliptic operator on a compact manifold M. Use Eq.11.2.4, $\text{Ker}\,\Delta_p$ can be written as

$$\begin{aligned}\text{Ker}\,\Delta_p &= \text{Ker}\left(D_p^+ D_p + D_{p-1} D_{p-1}^+\right)\\ &= D_p^+\left(\text{Ker}\,D_p\right) + \left(\text{Ker}\,D_{p-1}\right) D_{p-1}^+\end{aligned} \tag{11.2.10}$$

where $\text{Ker}\,D_p$ and $\text{Ker}\,D_{p-1}$ are both finite dimensional. Therefore $\text{Ker}\,\Delta_p$ is finite dimensional as well.

Definition 11.2.4: The index of an elliptic complex $\{E,D\}$ is (according to Eq.11.2.9)

$$\text{Index}\,(E,D) = \sum_p (-1)^p \dim H^p(E,D) = \sum_p (-1)^p \dim \text{Ker}\,\Delta_p \tag{11.2.11}$$

$$(\dim H^p(E,D) = \dim \Delta_p)$$

Contraction of Complexes An elliptic complex can be contracted into two vector bundles F_0 and F_1, which are defined as follows:

$$F_0 = \bigoplus_p E_{2p} \qquad F_1 = \bigoplus_p E_{2p+1} \qquad \text{where } p \text{ is taken as positive integers.} \tag{11.2.12}$$

Then define the operator A and its dual operator A^+ as follows:

$$A = \bigoplus_q \left(D_{2q} + D_{2q-1}^+\right) \qquad A^+ = \bigoplus_q \left(D_{2q}^+ + D_{2q-1}\right) \qquad \text{where } q \text{ is taken as integers.} \tag{11.2.13}$$

Then we have

$$\begin{aligned} A\ &:\ C^\infty(F_0) \Rightarrow C^\infty(F_1)\\ A^+&:\ C^\infty(F_1) \Rightarrow C^\infty(F_0)\end{aligned} \tag{1.2.14}$$

Proof: Eq.11.2.14 can be proved according to Figure 11.2.1 as follows:

$$\begin{aligned} AF_0 &= \bigoplus_q \left(D_{2q} + D_{2q-1}^+\right) \cdot \bigoplus_p E_{2p} = \bigoplus_p \left(D_{2q} E_{2p} + D_{2q-1}^+ E_{2p}\right)\\ &= \bigoplus_p E_{2p+1} + \bigoplus_p E_{2p-1} = F_1 \end{aligned} \tag{11.2.15}$$

$$A^+ F_1 = \bigoplus_q \left(D_{2q}^+ + D_{2q-1}\right) \cdot \bigoplus_p E_{2p+1} = \bigoplus_p \left(D_{2p}^+ E_{2p+1} + D_{2p+1} E_{2p+1}\right)$$

$$= \bigoplus_p E_{2p} + \bigoplus_p E_{2p+2} = F_0 \tag{11.2.16}$$

□

Besides, A and A^+ have the following property as well:

$$\square_0 = A^+ A = \bigoplus_p \Delta_{2p}$$

$$\square_1 = A A^+ = \bigoplus_p \Delta_{2p+1} \tag{11.2.17}$$

Proof: According to Eq.11.2.13, Eq.11.2.15 and Eq.11.2.16, together with the definition of Δ_p in Eq.11.2.4, we have

$$\square_0 = A^+ A = \bigoplus_p \left(D_{2p}^+ + D_{2p-1} \right) \cdot \bigoplus_q \left(D_{2q} + D_{2q-1}^+ \right)$$

$$= \bigoplus_p \left(D_{2p}^+ D_{2p} + D_{2p-1} D_{2p-1}^+ \right) = \bigoplus_p \Delta_{2p} \tag{11.2.18}$$

There were two other products in $\square_0$ of Eq.11.2.18, namely $D_{2p-1}D_{2q}$ and $D_{2p}^+ D_{2q-1}^+$. They are eliminated for the following reasons:

First, if we assume $D_{2p-1}D_{2q}E_{2q} = D_{2p-1}E_{2q+1} \neq 0$, then E_{2q+1} is in the domain of D_{2p-1} only if it equals E_{2p-1}. Hence we have

$$D_{2p-1}D_{2q}E_{2q} = D_{2p-1}E_{2p-1} = D_{2p-1}D_{2p-2}E_{2p-2}.$$

However, according to Definition 11.2.1, for an elliptic complex we should have

$$D_{2p-1}D_{2p-2} = 0.$$

Therefore in Eq.11.2.18 we must have

$$D_{2p-1}D_{2q} = 0. \tag{11.2.19}$$

Secondly, if we assume $D_{2p}^+ D_{2q-1}^+ E_{2q} = D_{2p}^+ E_{2q-1} \neq 0$, then E_{2q-1} is in the domain of D_{2p}^+ only if it equals E_{2p+1}. Thus we must replace $2q-1$ and $2q$ by $2p+1$ and $2p+2$, respectively, to obtain the result which makes sense as follows:

$$D_{2p}^+ D_{2q-1}^+ E_{2q} \Rightarrow D_{2p}^+ D_{2p+1}^+ E_{2p+2} \tag{11.2.20}$$

Notice that in Eq.11.2.3, "$\rightarrow$" and "$\leftarrow$" have the same meaning as the equivalent sign "$\approx$". Hence for an elliptic complex, similarly we have

$$\left(D_{p+1} D_p \xi_p ,\ \eta_{p+2} \right) \approx \left(D_p \xi_p ,\ D_{p+2}^+ \eta_{p+2} \right) \approx \left(\xi_p ,\ D_{p+1}^+ D_{p+2}^+ \eta_{p+2} \right) \tag{11.2.21}$$

Thus we see that if $D_{p+1}D_p = 0$, then from Eq.11.2.21 we must have

$$D_{p+1}^+ D_{p+2}^+ = 0 \longrightarrow D_{2q-2}^+ D_{2q-1}^+ = 0.$$

Therefore from Eq.11.2.20 we know that in Eq.11.2.18,

$$D_{2p}^+ D_{2q-1}^+ = 0 \tag{11.2.22}$$

and that completes the proof for the first formula of Eq.11.2.17, that is, $\square_0 = A^+ A = \bigoplus_p \Delta_{2p}$.

Similar to Eq.11.2.18, we can prove

$$\square_1 = A\,A^+ = \bigoplus_p \left(D_{2p} + D_{2p-1}^+\right) \bigoplus_q \left(D_{2q}^+ + D_{2q-1}\right)$$
$$= \bigoplus_p \left(D_{2p} D_{2p}^+ + D_{2p-1}^+ D_{2p-1}\right) = \bigoplus_p \Delta_{2p+1}, \qquad (11.2.23)$$

and hence accomplish the proof for the second formula of Eq.11.2.17, that is, $\square_1 = AA^+ = \bigoplus_p \Delta_{2p+1}$. □

Compare $\{E, D\}$ of Figure 11.2.1 to $\{F, A\}$ of Eq.11.2.15 and Eq.11.2.16, we see that they have the following similarities:

$D : E_p \longrightarrow E_{p+1}$	$A : F_0 \longrightarrow F_1$
$D^+ : E_{p+1} \longrightarrow E_p$	$A^+ : F_1 \longrightarrow F_0$

Recall Eq.11.1.24, since D is defined on E_p, thus Eq.11.1.24 should be written more precisely as

$$\text{Index}\,(E, D) = \dim \text{Ker}\, D - \dim \text{Ker}\, D^+. \qquad (11.2.24)$$

Then according to the above similarities between $\{E, D\}$ and $\{F, A\}$, Index (F, A) should be given by

$$\text{Index}\,(F, A) = \dim \text{Ker}\, A - \dim \text{Ker}\, A^+. \qquad (11.2.25)$$

Although this is merely a conjecture for now, we shall prove later that indeed it is true.

First we are going to prove the following two propositions:

(i) $\dim \text{Ker}\, A = \dim \text{Ker}\, \square_0$ $\qquad (11.2.26)_1$

(ii) $\dim \text{Ker}\, A^+ = \dim \text{Ker}\, \square_1$ $\qquad (11.2.26)_2$

Proof (i): Suppose ω satisfies $\square_0 \omega = 0$. Then according to Eq.11.2.18, follow the rule of Eq.11.2.21 we obtain

$$0 = \left(\square_0 \omega, \omega\right) = \left(A^+ A\omega, \omega\right) = \left(A\omega, A\omega\right) \qquad (11.2.27)_1$$

By definition of inner product, we know $A\omega = 0$. That is, $\square_0 \omega = 0$ gives $A\omega = 0$. Hence Ker $\square_0$ = Ker A, proves Eq.11.2.26$_1$. □

Proof (ii): Suppose ω satisfies $\square_1 \omega = 0$. Then according to Eq.11.2.23, follow the rule of Eq.11.2.21 we obtain

$$0 = \left(\square_1 \omega, \omega\right) = \left(AA^+ \omega, \omega\right) = \left(A^+\omega, A^+\omega\right) \qquad (11.2.27)_2$$

Again, by definition of inner product, we know $A^+\omega = 0$. That is, $\square_1 \omega = 0$ gives $A^+\omega = 0$. Therefore Ker $\square_1$ = Ker A^+ and Eq.11.2.26$_2$ holds. □

Then plug Eq.11.2.26$_1$ and Eq.11.2.26$_2$ into Eq.11.2.25, we obtain

$$\text{Index}\,(F, A) = \dim \text{Ker}\, \square_0 - \dim \text{Ker}\, \square_1 \qquad (11.2.28)$$

Use Eq.11.2.17, we have

$$\text{Index}\,(F, A) = \dim \text{Ker} \bigoplus_p \Delta_{2p} - \dim \text{Ker} \bigoplus_p \Delta_{2p+1}$$

$$\begin{aligned}\text{Index}\,(F,A) &= \dim\text{Ker}\,\bigoplus_p \Delta_{2p} - \dim\,\text{Ker}\,\bigoplus_p \Delta_{2p+1} \\ &= \sum_q (-1)^q \dim\,\text{Ker}\,\Delta_q \end{aligned} \tag{11.2.29}$$

Compare to Eq.11.2.11 we see

$$\text{Index}\,(E,D) = \sum_p (-1)^p \dim\,\text{Ker}\,\Delta_p = \text{Index}\,(F,A) \tag{11.2.30}$$

By the A-S Index Theorem [7], $\text{Index}\,(E,D) = \text{Index}\,(F,A)$ should be (The proof is omitted.):

$$\text{Index}\,(E,D) = \text{Index}\,(F,A) = (-1)^{\frac{n}{2}(n+1)} \int_M \frac{\text{ch}F_0 - \text{ch}F_1}{e(T(M))} td(T(M)\otimes C)$$

Use Eq.11.2.12 $$= (-1)^{\frac{n}{2}(n+1)} \int_M \frac{\text{ch}\bigoplus_p E_{2p} - \text{ch}\bigoplus_p E_{2p+1}}{e(T(M))} td(T(M)\otimes C)$$

$$= (-1)^{\frac{n}{2}(n+1)} \int_M \frac{\sum_p (-1)^p \text{ch}(E_p)}{e(T(M))} td(T(M)\otimes C) \tag{11.2.31}$$

Here we explain the meanings of some of the symbols in Eq.11.2.31:

(i) $\text{ch}E$ and $\text{ch}F$ are the Chern characters of the vector bundles E and F, respectively (see the definition in Eq.10.2.7). $\text{ch}F_0$, $\text{ch}F_1$ and $\text{ch}E_p$ are the Chern characters of the vector bundles F_0, F_1 and E_p, respectively.

(ii) $td(T(M)\otimes C)$ is the Todd class of the tangent bundle $T(M)\times C$ of the base manifold. The Todd class of a vector bundle E is defined by

$$\begin{aligned} td\,(E) &= \prod_{j=1}^{k} \frac{x_j}{1-e^{-x_j}} \qquad (\text{where } x_j = \frac{i}{2\pi}\Omega_j) \\ &= 1 + \frac{1}{2}C_1(E) + \frac{1}{12}\left(C_1^2(E) + C_2(E)\right) + \frac{1}{24}C_1(E)C_2(E) + \cdots \end{aligned} \tag{11.2.32}$$

where the definition of $C_1(E), C_2(E), \cdots, C_k(E)$ is given by Eq.10.2.15.

The Todd class also has the following property:

$$td(E\oplus F) = td(E)\,td(F). \tag{11.2.33}$$

(iii) $e(T(M))$ is the Euler class of the tangent bundle $T(M)$ of the base manifold.

(iv) $\int_M$ is an integral over the entire base manifold M. $n = \dim M$ if M is a real n-manifold; $n = 2l$ if M is a complex l-manifold.

The transformations of complex vectors in an l-dimensional complex space can be expressed as the transformations of real vectors in an $n = 2l$-dimensional real space. Hence a complex mapping of dimension l can be transferred into a real mapping of dimension $n = 2l$.

Example 1. φ_1: $C^1 \longrightarrow C^1 \qquad (z = x_1 + ix_2)$

By the complex mapping : $z \longrightarrow z' = (\alpha + i\beta)z = (\alpha x_1 - \beta x_2) + i(\alpha x_2 + \beta x_1)$. That is, $x_1 + ix_2 \longrightarrow x_1' + ix_2' = (\alpha x_1 - \beta x_2) + i(\alpha x_2 - \beta x_1)$.

The corresponding real mapping is

$$\begin{pmatrix} x_1 \\ x_2 \end{pmatrix} \longrightarrow \begin{pmatrix} x_1' \\ x_2' \end{pmatrix} = \begin{pmatrix} \alpha & -\beta \\ \beta & \alpha \end{pmatrix} \begin{pmatrix} x_1 \\ x_2 \end{pmatrix}$$

where $\begin{pmatrix} \alpha & -\beta \\ \beta & \alpha \end{pmatrix}$ is called the realization of the complex mapping $(\alpha + i\beta)$, and is denoted by

$$\begin{pmatrix} \alpha & -\beta \\ \beta & \alpha \end{pmatrix} = (\alpha + i\beta)_R . \tag{11.2.34}$$

Example 2. φ_2: $C^2 \longrightarrow C^2 \qquad (z_1 = x_1 + ix_2$ and $z_2 = x_3 + ix_4)$

By the complex mapping :

$$\begin{pmatrix} z_1 \\ z_2 \end{pmatrix} \longrightarrow \begin{pmatrix} z_1' \\ z_2' \end{pmatrix} = \begin{pmatrix} \alpha_1 + i\beta_1 & \alpha_2 + i\beta_2 \\ \alpha_3 + i\beta_3 & \alpha_4 + i\beta_4 \end{pmatrix} \begin{pmatrix} z_1 \\ z_2 \end{pmatrix}$$

$$\therefore \quad \begin{pmatrix} x_1 + ix_2 \\ x_3 + ix_4 \end{pmatrix} \longrightarrow \begin{pmatrix} x_1' + ix_2' \\ x_3' + ix_4' \end{pmatrix} = \begin{pmatrix} (\alpha_1 + i\beta_1)(x_1 + ix_2) + (\alpha_2 + i\beta_2)(x_3 + ix_4) \\ (\alpha_3 + i\beta_3)(x_1 + ix_2) + (\alpha_4 + i\beta_4)(x_3 + ix_4) \end{pmatrix}$$

The corresponding real mapping is

$$\begin{pmatrix} x_1 \\ x_2 \\ x_3 \\ x_4 \end{pmatrix} \longrightarrow \begin{pmatrix} x_1' \\ x_2' \\ x_3' \\ x_4' \end{pmatrix} = \begin{pmatrix} \alpha_1 & -\beta_1 & \alpha_2 & -\beta_2 \\ \beta_1 & \alpha_1 & \beta_2 & \alpha_2 \\ \alpha_3 & -\beta_3 & \alpha_4 & -\beta_4 \\ \beta_3 & \alpha_3 & \beta_4 & \alpha_4 \end{pmatrix} \begin{pmatrix} x_1 \\ x_2 \\ x_3 \\ x_4 \end{pmatrix}$$

where $\begin{pmatrix} \alpha_1 & -\beta_1 & \alpha_2 & -\beta_2 \\ \beta_1 & \alpha_1 & \beta_2 & \alpha_2 \\ \alpha_3 & -\beta_3 & \alpha_4 & -\beta_4 \\ \beta_3 & \alpha_3 & \beta_4 & \alpha_4 \end{pmatrix}$ is called the realization of the complex mapping $\begin{pmatrix} \alpha_1 + i\beta_1 & \alpha_2 + i\beta_2 \\ \alpha_3 + i\beta_3 & \alpha_4 + i\beta_4 \end{pmatrix}$, and is denoted by

$$\begin{pmatrix} \alpha_1 & -\beta_1 & \alpha_2 & -\beta_2 \\ \beta_1 & \alpha_1 & \beta_2 & \alpha_2 \\ \alpha_3 & -\beta_3 & \alpha_4 & -\beta_4 \\ \beta_3 & \alpha_3 & \beta_4 & \alpha_4 \end{pmatrix} = \begin{pmatrix} \alpha_1 + i\beta_1 & \alpha_2 + i\beta_2 \\ \alpha_3 + i\beta_3 & \alpha_4 + i\beta_4 \end{pmatrix}_R . \tag{11.2.35}$$

The above equations Eq.11.2.34 and Eq.11.2.35 represent the realizations for the cases of $l=1$ and $l=2$, respectively. For other positive integers l, there are formulae similar to Eq.11.2.34 and Eq.11.2.35 which indicate that complex l-dimensional mappings can be realized to real $n=2l$-dimensional mappings.

§ 11.3 de Rham Complex and Gauss-Bonnet Theorem

Recall Eq.11.2.34 and Eq.11.2.35, we find out that

$$B_2=\frac{1}{\sqrt{2}}\begin{pmatrix}1 & 1\\ -i & i\end{pmatrix},\quad B_2^{-1}=\frac{1}{\sqrt{2}}\begin{pmatrix}1 & i\\ 1 & -i\end{pmatrix},\quad B_2B_2^{-1}=\begin{pmatrix}1 & 0\\ 0 & 1\end{pmatrix};$$

$$B_4=\frac{1}{\sqrt{2}}\begin{pmatrix}1 & 0 & 1 & 0\\ -i & 0 & i & 0\\ 0 & 1 & 0 & 1\\ 0 & -i & 0 & i\end{pmatrix},\quad B_4^{-1}=\frac{1}{\sqrt{2}}\begin{pmatrix}1 & i & 0 & 0\\ 0 & 0 & 1 & i\\ 1 & -i & 0 & 0\\ 0 & 0 & 1 & -i\end{pmatrix},\quad B_4B_4^{-1}=\begin{pmatrix}1 & 0 & 0 & 0\\ 0 & 1 & 0 & 0\\ 0 & 0 & 1 & 0\\ 0 & 0 & 0 & 1\end{pmatrix}.$$

(11.3.1)

By induction, we can give the general forms for $2k\times 2k$ matrices B_{2k} and B_{2k}^{-1}:

$$B_{2k}=\frac{1}{\sqrt{2}}\left.\begin{pmatrix}\begin{matrix}1 & & & \\ -i & & & \bigcirc\\ & 1 & & \\ & -i & & \\ & & 1 & \\ & & -i & \\ & & & \ddots \\ \bigcirc & & & 1\\ & & & -i\end{matrix} & \vdots & \begin{matrix}1 & & & \\ i & & & \bigcirc\\ & 1 & & \\ & i & & \\ & & 1 & \\ & & i & \\ & & & \ddots \\ \bigcirc & & & 1\\ & & & i\end{matrix}\end{pmatrix}\right\}2k\text{ rows},$$

$$\underbrace{\qquad\qquad}_{k\text{ columns}}\qquad\underbrace{\qquad\qquad}_{k\text{ columns}}$$

$$B_{2k}^{-1}=\frac{1}{\sqrt{2}}\begin{pmatrix}\left.\begin{matrix}1 & i & & & & & & & \\ & & 1 & i & & & & \bigcirc & \\ & \bigcirc & & & 1 & i & & & \\ & & & & & & \ddots & & \\ & & & & & & & 1 & i\end{matrix}\right\}k\text{ rows}\\ \cdots\cdots\cdots\cdots\cdots\cdots\cdots\cdots\\ \left.\begin{matrix}1 & -i & & & & & & & \\ & & 1 & -i & & & & \bigcirc & \\ & \bigcirc & & & 1 & -i & & & \\ & & & & & & \ddots & & \\ & & & & & & & 1 & -i\end{matrix}\right\}k\text{ rows}\end{pmatrix}$$

(11.3.2)

$$B_{2k}B_{2k}^{-1}=\underbrace{\begin{pmatrix}1 & & & & \\ & 1 & & \bigcirc & \\ & & \ddots & & \\ & \bigcirc & & 1 & \\ & & & & 1\end{pmatrix}}_{2k\text{ columns}}\Bigg\}\,2k\text{ rows} \tag{11.3.3}$$

According to Eq.11.2.34 and Eq.11.3.1,

$$B_2^{-1}(\alpha+i\beta)_R B_2=\frac{1}{2}\begin{pmatrix}1 & i\\ 1 & -i\end{pmatrix}\begin{pmatrix}\alpha & -\beta\\ \beta & \alpha\end{pmatrix}\begin{pmatrix}1 & 1\\ -i & i\end{pmatrix}=\begin{pmatrix}\alpha+i\beta & 0\\ 0 & \alpha-i\beta\end{pmatrix} \tag{11.3.4}$$

Then according to Eq.11.2.35 and Eq.11.3.1,

$$B_4^{-1}\begin{pmatrix}\alpha_1+i\beta_1 & \alpha_2+i\beta_2\\ \alpha_3+i\beta_3 & \alpha_4+i\beta_4\end{pmatrix}B_4=\frac{1}{2}\begin{pmatrix}1 & i & 0 & 0\\ 0 & 0 & 1 & i\\ 1 & -i & 0 & 0\\ 0 & 0 & 1 & -i\end{pmatrix}\begin{pmatrix}\alpha_1 & -\beta_1 & \alpha_2 & -\beta_2\\ \beta_1 & \alpha_1 & \beta_2 & \alpha_2\\ \alpha_3 & -\beta_3 & \alpha_4 & -\beta_4\\ \beta_3 & \alpha_3 & \beta_4 & \alpha_4\end{pmatrix}\begin{pmatrix}1 & 0 & 1 & 0\\ -i & 0 & i & 0\\ 0 & 1 & 0 & 1\\ 0 & -i & 0 & i\end{pmatrix}$$

$$=\begin{pmatrix}\alpha_1+i\beta_1 & \alpha_2+i\beta_2 & 0 & 0\\ \alpha_3+i\beta_3 & \alpha_4+i\beta_4 & 0 & 0\\ 0 & 0 & \alpha_1-i\beta_1 & \alpha_2-i\beta_2\\ 0 & 0 & \alpha_3-i\beta_3 & \alpha_4-i\beta_4\end{pmatrix} \tag{11.3.5}$$

Generalize Eq.11.3.4 and Eq.11.3.5 to $2k\times 2k$ matrices, we have

$$B_{2k}^{-1}\begin{pmatrix}\alpha_1+i\beta_1 & \cdots & \alpha_k+i\beta_k\\ \vdots & & \vdots\\ \alpha_{k+1}+i\beta_{k+1} & \cdots & \alpha_{2k}+i\beta_{2k}\end{pmatrix}B_{2k}=\begin{pmatrix}\begin{matrix}\alpha_1+i\beta_1 & \cdots & \alpha_k+i\beta_k\\ \vdots & & \vdots\\ \alpha_{k+1}+i\beta_{k+1} & \cdots & \alpha_{2k}+i\beta_{2k}\end{matrix} & 0\\ 0 & \begin{matrix}\alpha_1-i\beta_1 & \cdots & \alpha_k-i\beta_k\\ \vdots & & \vdots\\ \alpha_{k+1}-i\beta_{k+1} & \cdots & \alpha_{2k}-i\beta_{2k}\end{matrix}\end{pmatrix} \tag{11.3.6}$$

Therefore we may draw our conclusions as follows:

Let M be a real oriented compact smooth $n=2l$-manifold, and $T(M)$ be the cotangent bundle (of dimension $n=2l$) of M. After complexification and a transformation by B_{2l}, x^1, x^2, $\cdots$, x^l, y^1, y^2, $\cdots$, y^l and their transformation matrix are replaced by z^1, $\cdots$, z^l, $\bar{z}^1$, $\cdots$, $\bar{z}^l$ and their transformation matrix, respectively. Thus we obtain $T(M)\otimes C=E_l\oplus\overline{E}_l$ where E_l and $\overline{E}_l$ are conjugate complex cotangent bundles of dimension l.

Define $\Lambda^k T(M) \otimes C$ to be the set of smooth complex-valued de Rham k-forms on $M \otimes C$. In the related subsequence $\{d_k\}$, each d_k is taken as

$$d_k = d = (\partial + \bar{\partial}) \tag{11.3.7}$$

According to Eq.5.2.5 and Eq.5.2.6,

$$df = \left(dx^k \frac{\partial}{\partial x^k} + dy^k \frac{\partial}{\partial y^k} \right) f = \left(dz^k \frac{\partial}{\partial z^k} + d\bar{z}^k \frac{\partial}{\partial \bar{z}^k} \right) f = (\partial + \bar{\partial}) f \tag{11.3.8}$$

Hence d_k and Λ^k can form a sequence which is similar to that of Figure 11.2.1:

$$\cdots \xrightarrow{d_{k-1}} \Lambda^k \xrightarrow{d_k} \Lambda^{k+1} \xrightarrow{d_{k+1}} \cdots \tag{11.3.9}$$

Since ∂ and $\bar{\partial}$ have the following properties:

$$\partial \cdot \partial = 0, \quad \bar{\partial} \cdot \bar{\partial} = 0, \text{ and } \partial \cdot \bar{\partial} = -\bar{\partial} \cdot \partial.$$

Thus we may obtain

$$dd = (\partial + \bar{\partial})(\partial + \bar{\partial}) = \partial \cdot \partial + \bar{\partial} \cdot \bar{\partial} + \partial \cdot \bar{\partial} + \bar{\partial} \cdot \partial = 0 \tag{11.3.10}$$

satisfying $d_k d_{k-1} = 0$. Therefore (Λ, d) is a complex.

Then similar to Figure 11.2.1, we may demonstrate the sequence of Eq.11.3.9 by the following figure:

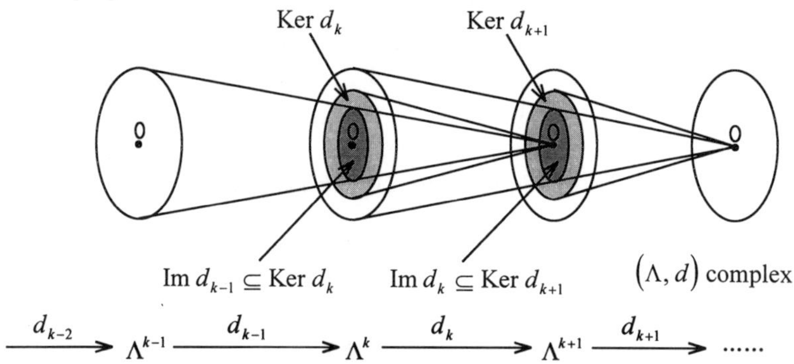

Figure 11.3.1

Definition 11.3.1: If $d_k d_{k-1} = 0$ (where k is an integer) in the sequence of the vector bundles Λ^k, i.e. Im $d_{k-1} \subseteq$ Ker d_k, then the complex in Figure 11.3.1 is called an *elliptic complex*. Im $d_{k-1} \subseteq$ Ker d_k is the necessary and sufficient condition for a complex to be elliptic. The complex of Figure 11.3.1 is called the (Λ, d) complex, of which the operator matrix has only one element d (i.e. d is a 1×1 matrix). According to Eq.11.1.5, $\tilde{d} \sim k \neq 0$ when $k \neq 0$, satisfying the condition for elliptic differential operators. Therefore the (Λ, d) complex is indeed an elliptic complex.

Now we sum up Λ^k for even numbers k and odd numbers k respectively, as follows:

$$\left.\begin{aligned}\Lambda^{\text{even}} &= \Lambda^0 \oplus \Lambda^2 \oplus \Lambda^4 \oplus \cdots \\ \Lambda^{\text{odd}} &= \Lambda^1 \oplus \Lambda^3 \oplus \Lambda^5 \oplus \cdots\end{aligned}\right\} \tag{11.3.11}$$

where Λ^{even} and Λ^{odd} form de Rham complex. Let $d+\delta$ (see Eq.3.4.10 for the definition of δ) be the differential operator of de Rham complex. Then

$$\begin{aligned}(d+\delta):\ & C^\infty\left(\Lambda^{\text{even}}\right) \to C^\infty\left(\Lambda^{\text{odd}}\right) \\ (d+\delta):\ & C^\infty\left(\Lambda^{\text{odd}}\right) \to C^\infty\left(\Lambda^{\text{even}}\right)\end{aligned} \tag{11.3.12}$$

where C^∞ means that the differential manifold Λ is smooth.

According to Eq.11.2.11, the index of de Rham complex should be

$$\text{Index}\,(\Lambda, d) = \sum_p (-1)^p \dim_C H^p_{dR}(T(M) \otimes C) \tag{11.3.13}$$

Then from Eq.11.2.7, we have

$$\begin{aligned}\dim_C H^p_{dR}(T(M) \otimes C) &= \dim_C \frac{\text{Ker } d_p}{\text{Imag } d_{p-1}} \\ &= \dim_R H^p_{dR}(T(M), R) = b_p\end{aligned} \tag{11.3.14}$$

(Here the dimension of the complex manifold = the dimension of the real manifold obtained by taking away the imaginary part of the coordinates of the complex manifold.)

Plug Eq.11.3.14 into Eq.11.3.13, we obtain

$$\begin{aligned}\text{Index}\,(\Lambda, d) &= \sum_p (-1)^p \dim_R H^p_{dR}(T(M), R) \\ &= \sum_p (-1)^p b_p = \chi(M)\end{aligned} \tag{11.3.15}$$

where $b_p = \dim H^p(M, R) = \dim H_p(M, R)$ is the pth ordered Betti number of the manifold M. The "alternating sum" of the Betti numbers is called the Euler's characteristic $\chi(M)$ of the manifold M.

Index (Λ, d) on the left hand side of Eq.11.3.15 can be obtained by the Index Theorem. According to Eq.11.2.31, Index (Λ, d) can be given as follows:

$$\text{Index}\,(\Lambda, d) = (-1)^{\frac{n}{2}(n+1)} \sum_p (-1)^p \int_{\substack{M \\ \dim n = 2l}} \frac{\text{ch}\left(\Lambda^p T(M) \otimes C\right) \cdot td(T(M) \otimes C)}{e(T(M))} \tag{11.3.16}$$

To compute the right hand side of Eq.11.3.16, we refer to the result of Eq.11.3.6, that is,

$$T(M) \otimes C \cong E_l \oplus \overline{E}_l\,, \tag{11.3.17}$$

and use Eq.10.2.8, we have

$$\mathrm{ch}(T(M)\otimes C)=\mathrm{ch}(E_l)\oplus\mathrm{ch}(\overline{E}_l) \tag{11.3.18}$$

where E_l and $\overline{E}_l$ are conjugate complex cotangent bundles of dimension l.

Then by Decomposition Theorem in Eq.10.2.13, E_l and $\overline{E}_l$ can be decomposed as

$$E_l=\oplus\sum_{j=1}^{l}L_j \quad \text{and} \quad \overline{E}_l=\oplus\sum_{j=1}^{l}\overline{L}_j\,. \tag{11.3.19}$$

In fact E_l and $\overline{E}_l$ are not only conjugate, but also dual. The reasons are stated as follows:

If Hermite's metric is taken, then the *unitary group* $U(n)$ can always be taken as the structure group of a manifold.

$$U(n)=\left\{\ A \mid A^{+}A=I\,,\ A\in GL(n)\ \right\} \tag{11.3.20}$$

Since the matrices of the unitary group $U(n)$ have the property $A^{+}=A^{-1}$, hence

$$\left.\begin{array}{c} A^{-1}=A^{+}=\left(A^{*}\right)^{T},\ A^{+}A=\left(A^{*}\right)^{T}A=I \\ A^{*}=\left(A^{+}\right)^{T}=\left(A^{-1}\right)^{T} \end{array}\right\} \tag{11.3.21}$$

Use the above features in Eq.11.3.21, we have that if E_l has a transformation matrix C, that is,

$$E_l'=C\,E_l, \tag{11.3.22}$$

then the transformation of its conjugate must be (according to Eq.11.3.21)

$$\overline{E}_l'=\overline{C}\,\overline{E}_l=\left(C^{+}\right)^{T}\overline{E}_l=\left(C^{-1}\right)^{T}\overline{E}_l=\overline{E}_l\,C^{-1}. \tag{11.3.23}$$

Thus by Eq.11.3.22 and Eq.11.3.23 we have

$$\overline{E}_l'\cdot E_l'=\left(\overline{E}_l'C^{-1}\right)\cdot\left(C\,E_l\right)=\overline{E}_l\cdot E_l\,. \tag{11.3.24}$$

Definition 11.3.1: If the vector bundles E_l and $\overline{E}_l$ follow the rules of Eq.11.3.22 and Eq.11.3.23, respectively, then E_l and $\overline{E}_l$ are said to be *dual* to each other. L_j and $\overline{L}_j$ in Eq.11.3.19 must be dual as well.

By the rule of multiplication for Chern characters in Eq.10.2.8, we have

$$\mathrm{ch}\left(L_j\otimes\overline{L}_j\right)=\mathrm{ch}\left(L_j\right)\cdot\mathrm{ch}\left(\overline{L}_j\right) \tag{11.3.25}$$

For the 1-dimensional line bundles L_j and $\overline{L}_j$, of which the curvatures Ω_j's are all 1×1 real matrices, we may define

$$x_j=\frac{i}{2\pi}\Omega_j \quad \text{and} \quad \overline{x}_j=\frac{-i}{2\pi}\Omega_j\,. \tag{11.3.26}$$

By the definition of $\mathrm{ch}(E)$ in Eq.10.2.7, we obtain

$$\left.\begin{aligned} \operatorname{ch}\left(L_j\right) &= e^{\frac{i}{2\pi}\Omega_j} = e^{x_j} \\ \operatorname{ch}\left(\overline{L}_j\right) &= e^{-\frac{i}{2\pi}\Omega_j} = e^{\overline{x}_j} \end{aligned}\right\} \tag{11.3.27}$$

For the sake of convenience, we may replace $\overline{x}_1 = -x_1$, $\overline{x}_2 = -x_2$, $\cdots$, $\overline{x}_l = -x_l$ by x_{l+1} $(= -x_1)$, x_{l+2} $(= -x_2)$, $\cdots$, x_{2l} $(= -x_l)$. Then by the Decomposition Theorem which states that a complex vector bundle can be decomposed into a direct sum of complex line bundles (see Eq.10.2.13), we can also write the curvature matrix $\Omega_{T(M)\otimes C} = \Omega_{E_l \oplus \overline{E}_l}$ of $T(M) \otimes C = E_l \oplus \overline{E}_l$ after the complex line bundle decomposition (see Eq.11.3.6) in the form of the following (also according to Eq.11.3.26):

$$\frac{i}{2\pi}\Omega_{T(M)\otimes C} = \frac{i}{2\pi}\Omega_{E_l \oplus \overline{E}_l} = \begin{pmatrix} x_1 & & & & & & \\ & \ddots & & & & \mathbf{0} & \\ & & \ddots & & & & \\ & & & x_l & & & \\ & & & & x_{l+1} & & \\ & \mathbf{0} & & & & \ddots & \\ & & & & & & \ddots \\ & & & & & & & x_{2l} \end{pmatrix} \tag{11.3.28}$$

According to Eq.11.3.27, Eq.11.3.35, and Eq.10.2.8, the Chern characters (see Eq.10.2.7) can be given as follows:

$$\begin{aligned} &\operatorname{ch}\left(L_j\right) = e^{x_j}, \\ &\operatorname{ch}\left(L_j \otimes L_k\right) = \operatorname{ch}\left(L_j\right)\cdot\operatorname{ch}\left(L_k\right) = e^{x_j}\cdot e^{x_k}, \\ &\operatorname{ch}\left(L_j \otimes L_k \otimes L_l\right) = \operatorname{ch}\left(L_j\right)\cdot\operatorname{ch}\left(L_k\right)\cdot\operatorname{ch}\left(L_l\right) = e^{x_j}\cdot e^{x_k}\cdot e^{x_l} \\ &\cdots \quad \cdots \quad \cdots \end{aligned} \tag{11.3.29}$$

The integrand in Eq.11.3.16 can be computed using Eq.11.3.28 and Eq.11.3.29.

$$\sum_{p=0}^{p=n=2l}(-1)^p \operatorname{ch}\left(\Lambda^p T(M)\otimes C\right) = +\operatorname{ch}\left(\Lambda^0 T(M)\otimes C\right) - \operatorname{ch}\left(\Lambda^1 T(M)\otimes C\right)$$

$$+\operatorname{ch}\left(\Lambda^2 T(M)\otimes C\right) - \cdots + \operatorname{ch}\left(\Lambda^{2l} T(M)\otimes C\right) \tag{11.3.30}$$

where

$$p=0: \quad \operatorname{ch}\left(\Lambda^0 T(M)\otimes C\right) = \operatorname{ch}(0) = e^0 = 1$$

$$p=1: \quad \operatorname{ch}\left(\Lambda^1 T(M)\otimes C\right) = \operatorname{ch}\left(\sum_{j=1}^{n=2l} L_j\right) = \sum_{j=1}^{n=2l} \operatorname{ch} L_j = \sum_{j=1}^{n=2l} e^{x_j}$$

$$p=2: \quad \mathrm{ch}\left(\Lambda^2 T(M)\otimes C\right)=\mathrm{ch}\left(\sum_{j=1}^{n=2l}\sum_{k=1}^{n=2l} L_j\otimes L_k\right)$$

$$=\sum_{j=1}^{n=2l}\sum_{k=1}^{n=2l}\mathrm{ch}\left(L_j\otimes L_k\right)=\sum_{j=1}^{n=2l}\sum_{k=1}^{n=2l} e^{x_j}e^{x_k}$$

$$\cdots \quad \cdots \quad \cdots$$

$$p=n=2l: \quad \mathrm{ch}\left(\Lambda^{2l} T(M)\otimes C\right)=\mathrm{ch}\left(L_1\otimes L_2\otimes\cdots\otimes L_{2l}\right)$$

$$=e^{x_1}e^{x_2}\cdots e^{x_l}e^{x_{l+1}}e^{x_{l+2}}\cdots e^{x_{2l}} \qquad (11.3.31)$$

Plug Eq.11.3.31 into Eq.11.3.30, we obtain

$$\sum_{p=0}^{p=n=2l}(-1)^p\,\mathrm{ch}\left(\Lambda^p T(M)\otimes C\right)$$

$$=1-\sum_{j=1}^{n}e^{x_j}+\sum_{j=1}^{n}\sum_{k=1}^{n}e^{x_j}e^{x_k}-\cdots+e^{x_1}e^{x_2}\cdots e^{x_l}e^{x_{l+1}}e^{x_{l+2}}\cdots e^{x_{2l}}$$

$$=\prod_{i=1}^{n=2l}\left(1-e^{x_i}\right)=\prod_{i=1}^{l}\left(1-e^{x_i}\right)\prod_{j=1}^{l}\left(1-e^{-x_i}\right)\left(T(M)\otimes C\right) \qquad (11.3.32)$$

Here we remark on the Todd class $td\left(T(M)\otimes C\right)$ in Eq.11.3.16.

Definition 11.3.2: Todd class is defined by

$$td(E)=\prod_{i=1}^{k}\frac{x_i}{1-e^{-x_i}}=1+\frac{1}{2}C_1(E)+\frac{1}{12}\left(C_1^2+C_2\right)(E)+\cdots \qquad (11.3.33)$$

where $C_1(E), C_2(E), \cdots, C_k(E)$ are defined in Eq.10.2.2, Eq.10.2.3, and Eq.10.2.4.

For the direct sum $E\oplus F$, Todd class has the following property of multiplication:

$$td(E\oplus F)=td(E)\cdot td(F) \qquad (11.3.34)$$

Hence

$$td\left(T(M)\otimes C\right)=td\left(E_l\oplus\overline{E}_l\right)=td(E)\cdot td\left(\overline{E}\right)$$

$$=\prod_{i=1}^{l}\frac{x_i}{1-e^{-x_i}}\cdot(-1)^l\prod_{j=1}^{l}\frac{x_j}{1-e^{x_j}}\left(T(M)\otimes C\right) \qquad (11.3.35)$$

$e\left(T(M)\right)$ in the integrand of Eq.11.3.16 is the Euler class of the real M (of dimension $n=2l$).

Use Eq.10.4.14, we obtain

$$e\left(T(M)\right)=e\left(\frac{\Omega}{2\pi}\right)=C_l(E)=x_1\wedge\cdots\wedge x_l=\prod_{i=1}^{l}x_i\left(T(M)\otimes C\right) \qquad (11.3.36)$$

Plug Eq.11.3.32, Eq.11.3.35, and Eq.11.3.36 into the integrand on the right hand side of Eq.11.3.16, we obtain

$$\sum_p\frac{(-1)^p\,\mathrm{ch}\left(\Lambda^p T(M)\otimes C\right)td\left(T(M)\otimes C\right)}{e\left(T(M)\right)}$$

$$
\begin{aligned}
&= \prod_{i=1}^{l}\left(1-e^{x_i}\right)\prod_{j=1}^{l}\left(1-e^{-x_j}\right)(T(M)\otimes C) \\
&\quad \times \prod_{i=1}^{l}\frac{x_i}{1-e^{x_i}}(-1)^l\prod_{j=1}^{l}\frac{x_j}{\left(1-e^{-x_j}\right)}(T(M)\otimes C) \\
&\quad \times \frac{1}{\prod_{i=1}^{l} x_i}(T(M)\otimes C) \\
&= (-1)^l\prod_{i=1}^{l} x_i(T(M)\otimes C) = (-1)^l e(T(M))
\end{aligned}
\tag{11.3.37}
$$

Thus from Eq.11.3.15 and Eq.11.3.16 we obtain

$$
\begin{aligned}
\chi(M) &= \text{Index } (\Lambda, d) \\
&= (-1)^{\frac{n}{2}(n+1)}\sum_p(-1)^p\int_{\substack{M\\ n=2l}}\frac{\text{ch}\left(\Lambda^p T(M)\otimes C\right)\cdot td(T(M)\otimes C)}{e(T(M))}
\end{aligned}
$$

Then plug Eq.11.3.37 into the right hand side of the above equation, we have

$$
\begin{aligned}
\chi(M) &= \text{Index } (\Lambda, d) \\
&= (-1)^{\frac{2l}{2}(2l+1)}\cdot(-1)^l\int_{\substack{M\\ n=2l}} e(T(M)) = \int_{\substack{M\\ n=2l}} e(T(M))
\end{aligned}
\tag{11.3.38}
$$

which is the Gauss-Bonnet Theorem.

Note: If M is odd dimensional (i.e. $n \neq 2l$), then $e(T(M)) = 0$ and $\chi(M) = 0$, since $n = 2l$ is the prerequisite for Eq.11.3.31 and Eq.11.3.32 to hold!

§ 11.4 Index Theorems of Other Typical Elliptic Complexes

Here we list three typical elliptic complexes and their Index Theorems.

Example 1. Index Theorem of Dolbeault complex on a compact manifold without boundary.

Example 2. Index Theorem of Hirzebruch complex on a compact manifold without boundary.

Example 3. Index Theorem of Dirac complex on a compact manifold without boundary.

A reference book on more details of these three common elliptic complexes and their Index Theorems is given in [6].

Chapter 12. Essential Features of 4-Manifolds

§ 12.1 *Nontrivial Instanton Solutions(*F=F) and Bianchi Identity on* S^4

A general gauge field 2-form can be written as[1]:

$$F^a = \frac{1}{2} F^a_{\mu\nu} dx^\mu \wedge dx^\nu . \tag{12.1.1}$$

The dual of F^a (on the 4-manifold S^4) is

$$* F^a = \frac{1}{2} * F^a_{\lambda\rho} dx^\lambda \wedge dx^\rho \tag{12.1.2}$$

where

$$* F^a_{\lambda\rho} = \frac{1}{2} F^a_{\mu\nu} \varepsilon^{\mu\nu}{}_{\lambda\rho} = \frac{1}{2} \varepsilon_{\mu\nu\lambda\rho} F^{a\mu\nu} \tag{12.1.3}$$

(See the definition of the Hodge∗ operator in Eq.3.4.3.)

If the gauge field F is not coupled with other substantial fields, then the *action* of F is

$$S = -\frac{1}{4} \int d^4x \, F^a_{\mu\nu} F^{a\mu\nu} = -\frac{1}{4} \int Tr(F \wedge *F) . \tag{12.1.4}$$

The second part of the equation is proved according to Eq.12.1.2 and Eq.12.1.3 as follows:

$$\begin{aligned} Tr(F \wedge *F) &= \frac{1}{2} F^a_{\mu\nu} dx^\mu \wedge dx^\nu \wedge \frac{1}{2} F^{a\sigma\tau} \varepsilon_{\sigma\tau\lambda\rho} \frac{1}{2} dx^\lambda \wedge dx^\rho \\ &= \frac{1}{8} F^a_{\mu\nu} F^{a\sigma\tau} \varepsilon_{\sigma\tau\lambda\rho} dx^\mu \wedge dx^\nu \wedge dx^\lambda \wedge dx^\rho \\ &= \frac{1}{8} F^a_{\mu\nu} F^{a\sigma\tau} \varepsilon_{\sigma\tau\lambda\rho} \varepsilon^{\mu\nu\lambda\rho} dx^4 \\ &= \frac{1}{8} F^a_{\mu\nu} F^{a\sigma\tau} \delta^\mu_\sigma \delta^\nu_\tau \cdot 2 \cdot 2 \cdot 2 \, dx^4 = F^a_{\mu\nu} F^{a\mu\nu} d^4x \end{aligned} \tag{12.1.5}$$

(For $\varepsilon_{\sigma\tau\lambda\rho}\varepsilon^{\mu\nu\lambda\rho}$, there are 2 kinds of permutations of $\lambda\rho$, 2 kinds of permutations $\mu\nu = \sigma\tau$ and $\mu\nu = \tau\sigma$, and 2 kinds of combinations when $\mu\nu$ and $\lambda\rho$ are switched. Hence comes the factor $2 \cdot 2 \cdot 2 = 8$, and Eq.12.1.5 holds.)

Take the gauge potential to be $A^a_\mu(x)$ (a vector field of which a is the index for gauge groups). Take the tensor of field intensity to be $F^a_{\mu\nu}(x)$ (a tensor field of which a is the index for gauge groups). Then we may define

$$F^a_{\mu\nu} = \left(\partial_\mu A^a_\nu - \partial_\nu A^a_\mu + f^a_{bc} A^b_\mu A^c_\nu \right) \tag{12.1.6}$$

where f^a_{bc} is the structure constant of the gauge group, and has another form

---

[1] Here μ and ν are indices for time-space; a, b, and c are indices for groups.

$$f_{abc} \equiv f_{ab}^{n} g_{nc}. \tag{12.1.7}$$

By a linear transformation, g_{nc} can be diagonalized as follows:

$$g_{nc} = \delta_{nc} \tag{12.1.8}$$

Hence from Eq.12.1.7, we obtain

$$f_{abc} = f_{ab}^{c}. \tag{12.1.9}$$

Plug it into Eq.12.1.4, we can write the action S as

$$\begin{aligned} S &= -\frac{1}{4}\int d^4x\, F_{\mu\nu}^{a} F^{a\mu\nu} \\ &= -\frac{1}{4}\int d^4x \left(\partial_\mu A_\nu^a - \partial_\nu A_\mu^a + f_{abc} A_\mu^b A_\nu^c\right) F^{a\mu\nu} \end{aligned} \tag{12.1.10}$$

By the variational theory of stationary waves, $\delta S = 0$ when $A_\mu^i \to A_\mu^i + \delta A_\mu^i$. We obtain

$$0 = \delta S = -\frac{1}{4}\int d^4x \left[\partial_\mu \delta A_\nu^a - \partial_\nu \delta A_\mu^a + f_{abc}\left(\delta A_\mu^b\right) A_\nu^c + f_{abc} A_\mu^b \left(\delta A_\nu^c\right)\right] F^{a\mu\nu} \tag{12.1.11}$$

Since $F^{i\mu\nu} = -F^{i\nu\mu}$,

$$\delta A_\mu^i \cdot \partial_\nu F^{i\mu\nu} = \delta A_\nu^i \cdot \partial_\mu F^{i\nu\mu} = -\delta A_\nu^i \cdot \partial_\mu F^{i\mu\nu}.$$

Plug it into Eq.12.1.11. Notice that the terms such as δA_ν^a, δA_μ^a, etc. are all constants. Thus we have $\partial_\mu \delta A_\nu^a F^{a\mu\nu} = \delta A_\nu^a \partial_\mu F^{a\mu\nu}$, etc.. Hence S takes the extremum of the variation of A. That is,

$$\begin{aligned} 0 = \delta S &= -\frac{1}{4}\int d^4x \left[\delta A_\nu^a \cdot \partial_\mu F^{a\mu\nu} - \delta A_\mu^a \cdot \partial_\nu F^{a\mu\nu}\right. \\ &\qquad \left. + f_{abc}\left(\delta A_\mu^b\right) A_\nu^c F^{a\mu\nu} + f_{acb} A_\nu^c \left(\delta A_\mu^b\right) F^{a\nu\mu}\right] \\ &= -\frac{1}{4}\int d^4x \left[2\left(\delta A_\nu^a\right)\partial_\mu F^{a\mu\nu} + 2 f_{abc}\left(\delta A_\mu^b\right) A_\nu^c F^{a\mu\nu}\right] \\ &= -\frac{1}{2}\int d^4x \left[\partial_\mu F^{a\mu\nu} + f_{abc} A_\mu^b F^{c\mu\nu}\right]\left(\delta A_\nu^a\right) \end{aligned} \tag{12.1.12}$$

where δA_ν^a are arbitrary constants. Hence we obtain the equation of motion for the field F as follows:

$$\partial_\mu F^{a\mu\nu} + g\, f_{abc} A_\mu^b F^{c\mu\nu} = 0 \tag{12.1.13}$$

Let $F^{\mu\nu} = F^{a\mu\nu} T^a$, $A_\mu = A_\mu^b T^b$, and $\left[T^b, T^c\right] = i f_{bca} T^a$. Then the action of D_μ can be defined as (using Eq.12.1.13):

$$D_\mu F^{\mu\nu} = \partial_\mu F^{\mu\nu} - i\, g\left[T^b, T^c\right] A_\mu^b F^{c\mu\nu} = \partial_\mu F^{\mu\nu} - i\, g\left[A_\mu, F^{\mu\nu}\right] = 0 \tag{12.1.14}$$

Eq.12.1.14 can also be simplified as the Bianchi identity:

$$DF = dF - i\, g\left[A, F\right] = 0. \tag{12.1.14$_1$}$$

Hence we have

$$D = d - i\,g[A, \ \ ] \tag{12.1.15}$$

Using Eq.12.1.15, we obtain

$$\begin{aligned}
&D_\mu F^{\lambda\rho} + D_\lambda F^{\rho\mu} + D_\rho F^{\mu\lambda} \\
&= \partial_\mu F^{\lambda\rho} - i[A_\mu, F^{\lambda\rho}] + \partial_\lambda F^{\rho\mu} - i[A_\lambda, F^{\rho\mu}] + \partial_\rho F^{\mu\lambda} - i[A_\rho, F^{\mu\lambda}] \\
&= \partial_\mu F^{c\lambda\rho} T^c - i[A_\mu^a T^a, F^{b\lambda\rho} T^b] \\
&\quad + \partial_\lambda F^{b\rho\mu} T^b - i[A_\lambda^c T^c, F^{a\rho\mu} T^a] \\
&\quad + \partial_\rho F^{a\mu\lambda} T^a - i[A_\rho^b T^b, F^{c\mu\lambda} T^c] \\
&= \partial_\mu F^{c\lambda\rho} T^c + f_{abc} A_\mu^a F^{b\lambda\rho} T^c \\
&\quad + \partial_\lambda F^{b\rho\mu} T^b + f_{cab} A_\lambda^c F^{a\rho\mu} T^b \\
&\quad + \partial_\rho F^{a\mu\lambda} T^a + f_{bca} A_\rho^b F^{c\mu\lambda} T^a
\end{aligned} \tag{12.1.16}$$

Comparing Eq.12.1.16 with Eq.12.1.13, we obtain immediately

$$D_\mu F^{\lambda\rho} + D_\lambda F^{\rho\mu} + D_\rho F^{\mu\lambda} = 0. \tag{12.1.17}$$

Eq.12.1.17 is called *the tensor form of the Bianchi identity.*

Consider the solution of the motion equation of the field F, satisfying

$$F = \lambda * F \tag{12.1.18}$$

(According to Eq.3.4.3, in a 4-manifold, $*F$ is a 2-form if F is a 2-form.)

Since the Bianchi identity holds in Eq.12.1.14,

$$D_\mu F^{\mu\nu} = \partial_\mu F^{\mu\nu} - i\,g[A_\mu, F^{\mu\nu}] = 0.$$

Then by Eq.12.1.18, we must have

$$D_\mu F^{\mu\nu} = D_\mu(\lambda * F)^{\mu\nu} = \lambda\, D_\mu (*F)^{\mu\nu} = 0.$$

Hence we know that

$$D_\mu (*F)^{\mu\nu} = 0. \tag{12.1.19}$$

Apply the $*$ operation again using Eq.12.1.18, we obtain

$$F = \lambda * F = \lambda \ * \lambda * F = \lambda^2 ** F \tag{12.1.20}$$

According to Eq.4.1.24, we take the time-space dimension $n = 4$ (even). With F being a 2-form, i.e. $r = 2$, we have

$$** F = \text{sign}\,(g)\, F \quad \text{where} \quad \begin{array}{ll} \text{sign}\,(g) = +1 & \text{for } E^4 \text{ metric} \\ \text{sign}\,(g) = -1 & \text{for } M^4 \text{ metric} \end{array} \tag{12.1.21}$$

(E^4 and M^4 represent respectively the 4-dimensional Euclidean space and Minkowski space.)

Plug Eq.12.1.21 into Eq.12.1.20, we obtain

$$F = \begin{cases} +\lambda^2 F & (\text{for } E^4 \text{ metric}) \\ -\lambda^2 F & (\text{for } M^4 \text{ metric}) \end{cases} \tag{12.1.22}$$

Thus we see that for E^4 metric, we require $\lambda = \pm 1$, that is,

$$*F = \pm F; \tag{12.1.23$_1$}$$

and for M^4 metric, we require $\lambda = \pm i$, that is

$$*F = \pm iF. \tag{12.1.23$_2$}$$

where F and $*F$ both have the group indices (also see F^a and $*F^a$ of Eq.12.1.1 and Eq.12.1.2, respectively, where a is the group index), and are the differential forms defined on some G group.

Notice that for M^4 metric, it is required that $*F = \pm iF$. Hence the G group must not be a compact Lie group, but a noncompact Lie group, e.g. $SL(n,c)$.

On the other hand, for E^4 metric it is required that $*F = \pm F$. In this case there is no restriction on the G group, which may then be taken as any compact Lie group.

The following discussions of the instanton solutions are constrained to E^4 metric only. Let G be any compact Lie group. Under the condition of E^4 metric, we may have the following definition:

Definition 12.1.1:

$$\begin{aligned} *F^+ &= F^+ \quad (F^+ \text{ is a self-dual field.}) \\ *F^- &= -F^- \quad (F^- \text{ is an anti-self-dual field.}) \end{aligned}$$

With E^4 metric, by Eq.12.1.21 we have

$$**F = F \tag{12.1.24}$$

The eigenvalue of $**$ is $+1$, and the eigenvalues of $*$ are ± 1.

Definition 12.1.2:

$$\left.\begin{aligned} F^\pm &= \frac{1}{2}(F \pm *F), \\ \therefore\ F &= F^+ + F^-, \quad *F = F^+ - F^-. \end{aligned}\right\} \tag{12.1.25}$$

Thus we have (according to Eq.12.1.24 and Eq.12.1.25)

$$*F^\pm = \frac{1}{2}(*F \pm **F) = \frac{1}{2}(*F \pm F) = \pm\frac{1}{2}(F \pm *F) = \pm F^\pm \tag{12.1.26}$$

From Eq.12.1.4, we take the integral over the S^4 manifold, and use Eq.12.1.25:

$$\begin{aligned} -S &= +\frac{1}{4}\int_{S^4} d^4x\, F^a_{\mu\nu} F^{a\mu\nu} = \frac{1}{4}\int_{S^4} d^4x\, Tr\, F \wedge *F \\ &= \frac{1}{4}\int_{S^4} Tr\left(F^+ + F^-\right)\wedge\left(F^+ - F^-\right) \end{aligned}$$

$$=\frac{1}{4}\int_{S^4} Tr\left(F^+\wedge F^+ - F^-\wedge F^- - F^+\wedge F^- + F^-\wedge F^+\right)$$

$$=\frac{1}{4}\int_{S^4} Tr\left(F^+\wedge * F^+ + F^-\wedge * F^-\right)=\frac{1}{4}\left(\left\|F^+\right\|^2+\left\|F^-\right\|^2\right)>0 \qquad (12.1.27)$$

where $\|\ \|^2$ is defined by

$$\int_{S^4} Tr(F\wedge * F)=\|F\|^2\geq 0 \qquad (12.1.28)$$

On S^4, both F and $*F$ are 2-forms.

Definition 12.1.3: Let F be a curvature 2-form. Then the Pontrjagin kernel is defined as (using Eq.12.1.25):

$$k=\int_{S^4} P_1(E)=\frac{g^2}{8\pi^2}\int_{S^4} Tr\, F\wedge F=\frac{g^2}{8\pi^2}\int_{S^4} Tr\left(F^+ + F^-\right)\wedge\left(* F^+ - * F^-\right)$$

$$=\frac{g^2}{8\pi^2}\int_{S^4} Tr\left(F^+\wedge * F^+ - F^-\wedge * F^- - F^+\wedge * F^- + F^-\wedge * F^+\right)$$

With E^4 metric as the prerequisite, and according to Definition 12.1.1, we obtain

$F^+\wedge * F^- = F^+\wedge\left(-F^-\right)=-F^-\wedge F^+$ (where F^+ and F^- are 2-forms),

$$F^-\wedge * F^+ = F^-\wedge F^+ \longrightarrow F^+\wedge * F^- + F^-\wedge * F^+ = 0 .$$

Hence $k=\dfrac{g^2}{8\pi^2}\displaystyle\int_{S^4}\left(Tr\left(F^+\wedge * F^+\right)-Tr\left(F^-\wedge * F^-\right)\right).$

According to the definition given in Eq.12.1.28, the above equation can be written as

$$k=\frac{g^2}{8\pi^2}\left(\left\|F^+\right\|^2-\left\|F^-\right\|^2\right) \qquad (12.1.29)$$

Thus k is a topological invariant.

From Eq.12.1.27 we see that $(-S)$ must be bounded below, and the lower bound must not be zero. Otherwise we have $F^+ = F^- = 0$ which is meaningless. To obtain this lower bound, we consider the following inequality (using Eq.12.1.28):

$$\int_{S^4} Tr(F\pm * F)\wedge *(F\pm * F)=\left\| F\pm * F\right\|^2\geq 0 \qquad (12.1.30)$$

The left hand side of the equation is

$$\int_{S^4} Tr\left(F\wedge * F + * F\wedge F \pm F\wedge F \pm * F\wedge * F\right)$$

$$=\int_{S^4} Tr\Big[\left(F^+ + F^-\right)\wedge\left(F^+ - F^-\right)+\left(F^+ - F^-\right)\wedge\left(F^+ + F^-\right)$$

$$\pm\left(F^+ + F^-\right)\wedge\left(F^+ + F^-\right)\pm\left(F^+ - F^-\right)\wedge\left(F^+ - F^-\right)$$

$$= \int_{S^4} Tr\left(2F^+ \wedge F^+ - 2F^- \wedge F^- \pm 2F^+ \wedge F^+ \pm 2F^- \wedge F^-\right) \tag{12.1.31}$$

Use Definition 12.1.1, that is, $*F^+ = F^+$, and $*F^- = -F^-$, we plug Eq.12.1.31 into Eq.12.1.30 and obtain

$$\begin{aligned}\|F \pm * F\|^2 &= 2\int_{S^4} Tr\left(F^+ \wedge F^+ - F^- \wedge F^-\right) \pm Tr\left(F^+ \wedge F^+ + F^- \wedge F^-\right)\\ &= 2\int_{S^4} Tr\left(F^+ \wedge * F^+ + F^- \wedge * F^-\right) \pm Tr\left(F^+ \wedge * F^+ - F^- \wedge * F^-\right)\\ &= 2\left(\|F^+\|^2 + \|F^-\|^2\right) \pm 2\left(\|F^+\|^2 - \|F^-\|^2\right) \geq 0\end{aligned} \tag{12.1.32}$$

From Eq.12.1.27 we have

$$\|F^+\|^2 + \|F^-\|^2 = -4S > 0 \tag{12.1.33}$$

From Eq.12.1.29 we have

$$\|F^+\|^2 - \|F^-\|^2 = \frac{8\pi^2 k}{g^2} \tag{12.1.34}$$

where k in both Eq.12.1.29 and Eq.12.1.34 is a real number and can be either positive or negative.

Recall that in Eq.12.1.32, if we take the lower bound $\|F \pm * F\|^2 = 0$, then Eq.12.1.32 gives the following formula for F^+ and F^-:

$$\left(\|F^+\|^2 + \|F^-\|^2\right) \pm \left(\|F^+\|^2 - \|F^-\|^2\right) = 0 \tag{12.1.35}$$

From Eq.12.1.35 we can then find another two groups of solutions of F^+ and F^-:

(i) $F^- = 0$ and $F = F^+$. $\therefore$ $*F = *F^+ = F^+ = F$ ········ self-dual solution

(ii) $F^+ = 0$ and $F = F^-$. $\therefore$ $*F = *F^- = -F^- = -F$ ··· anti-self-dual solution

Therefore we can see that the self-dual solution of F

$$F = F^+ = * F \tag{12.1.36$_1$}$$

and the anti-self-dual solution of F

$$F = F^- = -* F \tag{12.1.36$_2$}$$

are both solutions of the equation of motion Eq.12.1.13 introduced by S. The *self-dual solution* $*F = F$ is the *instanton solution* which we are about to discuss.

Theorem 12.1.1: In the theory of pure Y-M gauge fields, the static soliton solution exists only if the space is 4-dimensional and the time is 1-dimensional (The proof is in Kerson Huang, *Quarks Leptons and Gauge Fields* (1982).).

The following are some examples of exterior differential forms of gauge fields, and related gauge groups and gauge transformations.

(i) **Gauge Group** g's transformation matrices are Hermitian matrices $T^a, T^b, \cdots$, and have the following properties:

$$\left[T^a, T^b\right] = i\, f_{abc}\, T^c \tag{12.1.37}$$

where θ^a is a group of small numbers. $\Theta = \theta^a T^a$, $U = e^{-i\Theta}$ $\cdots$ $\in \Lambda^0_g(M)$ are 0-forms.

(ii) **Gauge Potential**'s matrix representation is

$$A_\mu = A^a_\mu T^a \tag{12.1.38}$$

$A = A_\mu dx^\mu = A^a_\mu T^a dx^\mu$ $\cdots$ $\in \Lambda^1_g(M)$ are 1-forms.

(iii) **The Local Gauge Transformation of Gauge Potential** is

$$\begin{aligned}\delta A_\mu &= \delta\left(A^a_\mu T^a\right) = -\partial_\mu \theta^a T^a + g f_{abc}\theta^b A^c_\mu T^a \\ &= -\partial_\mu \Theta - ig[\theta^b T^b, A^c_\mu T^c] = -\partial_\mu \Theta - ig[\Theta, A_\mu]\end{aligned} \tag{12.1.39}$$

where θ^a is a group of small numbers.

Take $\delta(A) = \delta\left(A_\mu dx^\mu\right)$. Then from (iii) we obtain

(iv) $$\delta(A) = \delta\left(A_\mu dx^\mu\right) = -\partial_\mu \Theta dx^\mu - ig[\Theta, A_\mu dx^\mu] \tag{12.1.40}$$

Replace Θ by $-\Theta$. Then Eq.12.1.40 can be written as

$$\begin{aligned}\delta(A) &= \partial_\mu \Theta dx^\mu + ig[\Theta, A_\mu dx^\mu] \\ &= d\Theta + ig\theta^a A^b_\mu dx^\mu [T^a, T^b] \\ &= d\Theta - ig\theta^a A^b_\mu dx^\mu [T^b, T^a] \\ &= d\Theta - ig[A, \Theta] \qquad \text{(according to Eq.12.1.37)} \\ &= D_0 \Theta \qquad \cdots\cdots \in \Lambda^1_g(M) \quad \text{1-form}\end{aligned}$$

$$\therefore\ D_0 = d - ig[A,\ ] \tag{12.1.41}$$

(v) **The Local Gauge Transformation of Gauge Field.**

For the **gauge field** we have

$$\begin{aligned}F &= \frac{1}{2} F^a_{\mu\nu} T^a dx^\mu \wedge dx^\nu \qquad \cdots\cdots 2-\text{form} \\ &= \frac{1}{2}\left\{\left(\frac{\partial A^a_\nu}{\partial x^\mu} - \frac{\partial A^a_\mu}{\partial x^\nu}\right) T^a + g f_{abc} A^b_\mu A^c_\nu T^a\right\} dx^\mu \wedge dx^\nu \\ &= \frac{\partial A^a_\nu}{\partial x^\mu} T^a dx^\mu \wedge dx^\nu - \frac{i}{2} g[A^b_\mu T^b, A^c_\nu T^c] dx^\mu \wedge dx^\nu \\ &= d\left(A^a_\nu T^a dx^\nu\right) - \frac{i}{2} g[A^b_\mu T^b, A^a_\nu T^a] dx^\mu \wedge dx^\nu \\ &= dA - \frac{i}{2} g\, A_\mu dx^\mu \wedge A_\nu dx^\nu + \frac{i}{2} g\, A_\nu dx^\nu \wedge A_\mu dx^\mu \\ &= dA - \frac{i}{2} g\, A_\mu dx^\mu \wedge A_\nu dx^\nu \times 2 = dA - igA \wedge A\end{aligned}$$

$$\cdots\cdots \in \Lambda_g^2(M) \qquad 2\text{-form} \tag{12.1.42}$$

The gauge transformation of the gauge field F can be written according to Eq.12.1.42:

$$\delta F = d(\delta A) - \frac{i}{2}g[\delta A_\mu, A_\nu]dx^\mu \wedge dx^\nu - \frac{i}{2}g[A_\mu, \delta A_\nu]dx^\mu \wedge dx^\nu$$

$$= d(\delta A) + \frac{i}{2}g[A_\nu, \delta A_\mu]dx^\mu \wedge dx^\nu + \frac{i}{2}g[A_\nu, \delta A_\mu]dx^\mu \wedge dx^\nu$$

$$= d(\delta A) - ig[A_\mu, \delta A_\nu]dx^\mu \wedge dx^\nu$$

$$= d(\delta A) - ig[A, \delta A] = D_1(\delta A) \quad \cdots\cdots \in \Lambda_g^2(M) \qquad 2-\text{form}$$

$$\therefore \quad D_1 = d - ig[A, \ ] \tag{12.1.43}$$

Compare D_0 of Eq.12.1.41 to D_1 of Eq.12.1.43 and D of Eq.12.1.15, we see that they have the following similarity:

$$D_0 = D_1 = D = d - ig[A, \ ] \tag{12.1.44}$$

and at the mean time, some differences:

(i) D_0 applies to 0-forms only, e.g.

$$\Theta = \theta^a T^a \in \Lambda_g^0(M), \quad \text{and} \quad D_0\Theta \in \Lambda_g^1(M). \tag{12.1.45}$$

(ii) D_1 applies to 1-forms only, e.g.

$$A = A_\mu dx^\mu \in \Lambda_g^1(M), \quad \text{and} \quad D_1 A \in \Lambda_g^2(M). \tag{12.1.46}$$

(iii) The action of D gives the tensor form of the Bianchi's equation Eq.12.1.17 and the Bianchi identity Eq.12.1.14$_1$.

§ 12.2 Dimension of the Moduli Space of the Self-dual Connection $A\ (\in \Lambda_g^1)$

Take a connection A which is also called a gauge potential. The gauge field F can then be induced by A. If the induced gauge field F is self-dual (i.e. $F = *F$), then the connection A is said to be a self-dual gauge potential, and $F = *F$ is the *instanton solution.* The self-dual property of the instanton solution is topologically invariant. After a G group transformation, $A \to g(A)$. The gauge field F_g induced by the connection $g(A)$ is self-dual as well, that is, $F_g = *F_g$.

From the vector space of all the connections which produce self-dual gauge fields, the moduli space of the self-dual connection A can be obtained after the degree of freedom of the gauge group G is omitted. We are about to discuss the dimension of this moduli space.

Take $D_0 = D_1 = D = d - ig[A, \ ]$ (see Eq.12.1.44) to be the differential operator, where A is the gauge potential of the self-dual solution. Then we have the following

3-complexe

$$0 \longrightarrow \Lambda_g^0(M) \xrightarrow{D_0} \Lambda_g^1(M) \xrightarrow{P_- D_1} \Lambda_{g_-}^2(M) \longrightarrow 0 \qquad (12.2.1)$$

where M represents the manifold.

$\Lambda_g^0(M)$ contains all the possible Θ's (see Eq.12.1.45).

$\Lambda_g^1(M)$ contains all the 1-forms $D_0\Theta$. Since that the potential A taken from D_0 is self-dual, then by the definition of self-dual potential, we know that $D_0\Theta$ is self-dual as well.

$\Lambda_{g_-}^2(M)$ contains the 2-forms $P_- D_1 D_0 \Theta$. Since that the potencies taken from D_1 and D_0 are both the self-dual A (see $d - ig[A, \ ]$ in Eq.12.1.44), then $D_1 D_0 \Theta$ *is self-dual*. However, since that P_- is the *anti-self-dual projection*, thus

$$P_- D_1 D_0 \Theta = 0 \qquad (12.2.2)$$

Hence we can see that there is only one 0-element in $\Lambda_{g_-}^2(M)$.

The 3-complexe in Eq.12.2.1 can thus be illustrated by the following figure.

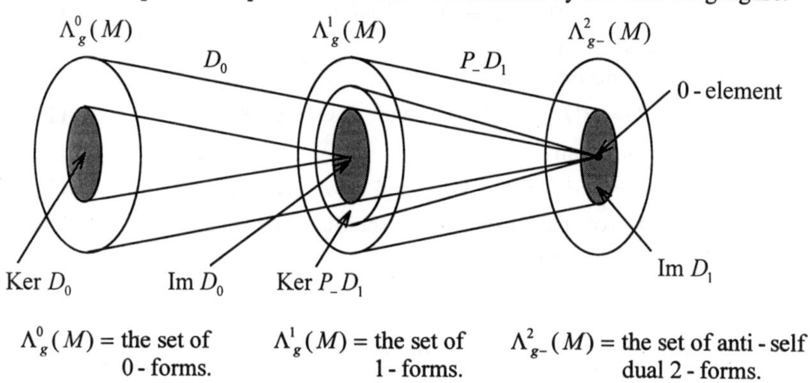

Figure 12.2.1

The Yang-Mills Complex The differential operators $\mathcal{D}_0$ and $\mathcal{D}_1$ are defined as follows:

$$\mathcal{D}_0 = D_0, \quad \text{and} \quad \mathcal{D}_1 = P_- D_1.$$

Similar to Eq.11.1.5, we take Fourier transformations and obtain the k representations $\widetilde{D}_0$ and $\widetilde{D}_1$ of D_0 and D_1, respectively. Since $\widetilde{D}_0$ and $\widetilde{D}_1$ are not zero when $k \neq 0$, the 3-complexe in Figure 12.2.1 is an *elliptic complex* (The definition of a complex requires $(P_- D_1) \cdot D_0 = 0$. See Definition 11.2.1 and Figure 11.2.1.), which is called the *YM elliptic complex*. The analytic index of this elliptic complex is defined by

$$\text{Index}\left(D_i, \Lambda_g^i(M)\right) = \sum_{i=0,1,2} \text{Index}\left(D_i, \Lambda_g^i(M)\right) = h^0 - h^1 + h^2 \qquad (12.2.3)$$

where Definition 11.2.4 and Eq.11.2.11 are used.

Compare Eq.12.2.3 to Eq.11.2.11, we obtain

$$\begin{aligned} h^0 &= (-1)^0 \dim H^0(E,D) = (-1)^0 \dim \operatorname{Ker} D_0 \\ -h^1 &= (-1)^1 \dim H^1(E,D) = (-1)^1 \dim \operatorname{Ker} D_1 \\ h^2 &= (-1)^2 \dim H^2(E,D) = (-1)^2 \dim \operatorname{Ker} D_2 \end{aligned} \tag{12.2.4}$$

Recall that h^k is the dimension of the cohomology group $H^k_A(M)$, which is consistent with Eq.12.2.4. That is,

$$h^k = \dim H^k(M). \qquad (12.2.4)_1$$

Hence we can prove

(i) For an irreducible gauge field $SU(2)$, $h^0 = 0$ (The details of the proof are omitted.).

(ii) When a self-dual instanton solution exists, we can obtain $h^2 = 0$ by Bochner-Weitzenbech's technique (The details of the proof are omitted.).

Therefore we obtain from Eq.12.2.3 that

$$\text{Index}\left(D_i, \Lambda^i_g(M)\right) = -h^1 \tag{12.2.5}$$

where h^1 is the degree of freedom of the independent instanton solution.

By the *A-S Index Theorem* Eq.11.2.31, we notice that D_i corresponds to D, and $\Lambda^i_g(M)$ corresponds to E. Thus Eq.12.2.5 can also be written as:

$$\begin{aligned} &\text{Index}\left(D_i, \Lambda^i_g(M)\right) \\ &= (-1)^{\frac{n}{2}(n+1)} \int_{M=S^4} \frac{\text{ch}\left(\bigoplus_{i=0}^{2}(-1)^i \Lambda^i_g(M)\right)}{e(T(M))} td(T(M)\otimes C) \\ &= (-1)^{\frac{n}{2}(n+1)} \int_{M=S^4} \frac{\text{ch}\left(\Lambda^0_g(M) - \Lambda^1_g(M) + \Lambda^2_{g-}(M)\right)}{e(T(M))} td(T(M)\otimes C) \end{aligned} \tag{12.2.6}$$

where S^4 is taken as the manifold M, and $\Lambda^0_g(M)$, $\Lambda^1_g(M)$, and $\Lambda^2_{g-}(M)$ are consistent with those of Figure 12.2.1.

The complex vector bundle on S^4 can be complexified and be written, using the Decomposition Theorem, as follows:

$$T(M)\otimes C = L_1 \oplus \overline{L}_1 \oplus L_2 \oplus \overline{L}_2 \tag{12.2.7}$$

Hence by the special property of the Todd class given in Eq.11.2.33, we obtain

$$\begin{aligned} td(T(M)\otimes C) &= td\left(L_1 \oplus \overline{L}_1 \oplus L_2 \oplus \overline{L}_2\right) \\ &= td(L_1)\cdot td(\overline{L}_1)\cdot td(L_2)\cdot td(\overline{L}_2) \end{aligned} \tag{12.2.8}$$

According to Eq.11.2.32, $\overline{x}_j = -x_j$ (i.e. x_j is a purely imaginary number). Furthermore, there are only 4 factors on the right hand side of the equation. Therefore from Eq.11.2.32 we have

$$td(T(M)\otimes C)=\left(\frac{x_1}{1-e^{-x_1}}\right)\left(\frac{-x_1}{1-e^{x_1}}\right)\left(\frac{x_2}{1-e^{-x_2}}\right)\left(\frac{-x_2}{1-e^{x_2}}\right) \tag{12.2.9}$$

Similar to Eq.10.4.3, we write $(e(a))^2$ as $(e(T(M)))^2$ and obtain

$$(e(T(M)))^2=\begin{vmatrix}0 & x_a & 0 & 0\\ -x_a & 0 & 0 & 0\\ 0 & 0 & 0 & x_b\\ 0 & 0 & -x_b & 0\end{vmatrix}=(x_a x_b)^2$$

$$\therefore \quad e(T(M))=x_a x_b \tag{12.2.10}$$

Using Eq.12.2.9 and Eq.12.2.10, let us take $x_1=x_a$, $x_2=x_b$, and obtain

$$\frac{td(T(M)\times C)}{e(T(M))}=\frac{1}{x_a x_b}\cdot\frac{x_a}{1-e^{-x_a}}\cdot\frac{-x_a}{1-e^{x_a}}\cdot\frac{x_b}{1-e^{-x_b}}\cdot\frac{-x_b}{1-e^{x_b}}$$

$$=x_a x_b\frac{1}{x_a-\frac{x_a^2}{2!}+\frac{x_a^3}{3!}-\cdots}\cdot\frac{1}{x_b-\frac{x_b^2}{2!}+\frac{x_b^3}{3!}-\cdots}\cdot\frac{1}{x_a+\frac{x_a^2}{2!}+\frac{x_a^3}{3!}+\cdots}\cdot\frac{1}{x_b+\frac{x_b^2}{2!}+\frac{x_b^3}{3!}+\cdots}$$

$$=\frac{1}{x_a x_b}\cdot\frac{1}{\left(1+\frac{x_a^2}{3!}+\frac{x_a^4}{5!}\cdots\right)^2-\left(\frac{x_a}{2!}+\frac{x_a^3}{4!}\cdots\right)^2\cdots}\cdot\frac{1}{\left(1+\frac{x_b^2}{3!}+\frac{x_b^4}{5!}\cdots\right)^2-\left(\frac{x_b}{2!}+\frac{x_b^3}{4!}\cdots\right)^2\cdots}$$

$$=\frac{1}{x_a x_b}\cdot\frac{1}{\left(1+\frac{2x_a^2}{3!}+\frac{2x_a^4}{5!}+\frac{x_a^4}{(3!)^2}\cdots\right)-\left(\frac{x_a^2}{(2!)^2}+\frac{2x_a^4}{2!4!}+\cdots\right)}\cdot$$

$$\cdot\frac{1}{\left(1+\frac{2x_b^2}{3!}+\frac{2x_b^4}{5!}+\frac{x_b^4}{(3!)^2}\cdots\right)-\left(\frac{x_b^2}{(2!)^2}+\frac{2x_b^4}{2!4!}+\cdots\right)}$$

(Here the principal contribution of x_a and x_b to the integral over the interval <<1 is remained, while their subordinate contribution to the integral over the interval which are not <<1 is omitted.)

$$=\frac{1}{x_a x_b}\cdot\frac{1}{\left(1+\frac{2x_a^2}{6}-\frac{x_a^2}{4}\cdots\right)}\cdot\frac{1}{\left(1+\frac{2x_b^2}{6}-\frac{x_b^2}{4}\cdots\right)}$$

$$=\frac{1}{x_a x_b}\left(1-\frac{x_a^2}{12}\right)\left(1-\frac{x_b^2}{12}\right)\cdots \tag{12.2.11}$$

So far we have found the specific form of the factor $\frac{td(T(M)\otimes C)}{e(T(M))}$ in Eq.12.2.6. Next we need to find the contents of $\mathrm{ch}\left(\Lambda_g^0(M)-\Lambda_g^1(M)+\Lambda_{g-}^2(M)\right)$, that is,

$$\Lambda_g^0 = V_g \otimes \Lambda^0,\ \Lambda_g^1 = V_g \otimes \Lambda^1,\ \Lambda_{g-}^2 = V_g \otimes \Lambda_-^2. \tag{12.2.12}$$

where Λ^0, Λ^1 and Λ_-^2 are the sets of 0-forms, 1-forms, and anti-dual 2-forms on the manifold S^4, respectively. As a vector bundle on $M = S^4$, the instanton takes its values on the group space (G). Thus Λ_g^0, Λ_g^1 and Λ_{g-}^2 are all the twisted complexes which relate to the group G, with their specific forms given as follows:

1) Λ^0: The set of the 0-forms (which are 0-dimensional) on S^4.
$\Lambda_g^0 = V_g \otimes \Lambda^0$ is a trivial bundle of which the curvature is 0, and

$$\mathrm{ch}\left(\Lambda^0\right) = 1 \tag{12.2.13$_1$}$$

where "ch" is the Chern character (see definition in Eq.10.2.7).

2) Λ^1: The set of the 1-forms on S^4, where $\dim\left(S^4\right) = 4$. By the Decomposition Theorem of complex vector bundles (see Eq.10.2.8 and Eq.10.2.13), the complex vector bundle Λ^1 on S^4 can be decomposed into the sum of the complex line bundles as follows:

$$\Lambda^1 = L_a \oplus \overline{L}_a \oplus L_b \oplus \overline{L}_b \tag{12.2.13$_2$}$$

$$\begin{aligned}\mathrm{ch}\left(\Lambda^1\right) &= \mathrm{ch}\left(L_a \oplus \overline{L}_a \oplus L_b \oplus \overline{L}_b\right)\\ &= \mathrm{ch}\left(L_a\right) \oplus \mathrm{ch}\left(\overline{L}_a\right) \oplus \mathrm{ch}\left(L_b\right) \oplus \mathrm{ch}\left(\overline{L}_b\right)\\ &= e^{x_a} + e^{-x_a} + e^{x_b} + e^{-x_b}\end{aligned} \tag{12.2.14}$$

3) Λ_-^2: The set of the anti-dual 2-forms of which the $*$ eigenvalues are -1 on S^4.
In order to find $\mathrm{ch}\left(\Lambda_-^2\right)$, let us consider Λ^2, Λ_+^2, and Λ_-^2 on S^4.
Λ^1 has 4 base vectors on S^4: $\quad dx_1,\ dx_2,\ dx_3,\ dx_4.$ (12.2.15)
Λ^2 has 6 base vectors on S^4:

$$dx_1 \wedge dx_2,\ dx_1 \wedge dx_3,\ dx_1 \wedge dx_4,\ dx_2 \wedge dx_3,\ dx_2 \wedge dx_4,\ dx_3 \wedge dx_4. \tag{12.2.16}$$

The dual of Eq.12.2.16 is:

$$\begin{aligned}&*dx_1 \wedge dx_2 = dx_3 \wedge dx_4,\ *dx_1 \wedge dx_3 = -dx_2 \wedge dx_4,\ *dx_1 \wedge dx_4 = dx_2 \wedge dx_3,\\ &*dx_2 \wedge dx_3 = dx_1 \wedge dx_4,\ *dx_2 \wedge dx_4 = -dx_1 \wedge dx_3,\ *dx_3 \wedge dx_4 = dx_1 \wedge dx_2.\end{aligned} \tag{12.2.17}$$

The six base vectors of Λ_-^2 and Λ_+^2 with the $*$ eigenvalues ∓ 1 can be obtained by linear combinations:

(i) $*\left(dx_1 + i dx_2\right) \wedge \left(dx_3 - i dx_4\right)$

$$\begin{aligned}&= *\left(dx_1 \wedge dx_3 + dx_2 \wedge dx_4 + i dx_2 \wedge dx_3 - i dx_1 \wedge dx_4\right)\\ &= -dx_2 \wedge dx_4 - dx_1 \wedge dx_3 + i dx_1 \wedge dx_4 - i dx_2 \wedge dx_3\\ &= -\left(dx_1 + i dx_2\right) \wedge \left(dx_3 - i dx_4\right) \qquad \therefore\quad * = -\end{aligned} \tag{12.2.18}$$

(ii) $*\left(dx_1 - i dx_2\right) \wedge \left(dx_3 + i dx_4\right)$

$$=*\left(dx_1\wedge dx_3+dx_2\wedge dx_4-idx_2\wedge dx_3+idx_1\wedge dx_4\right)$$
$$=-dx_2\wedge dx_4-dx_1\wedge dx_3-idx_1\wedge dx_4+idx_2\wedge dx_3$$
$$=-\left(dx^1-idx^2\right)\wedge\left(dx^3+idx^4\right)\qquad\therefore\quad *=-\qquad(12.2.19)$$

(iii) $*\left(dx_1\wedge dx_2-dx_3\wedge dx_4\right)$
$$=\left(dx_3\wedge dx_4-dx_1\wedge dx_2\right)=-\left(dx_1\wedge dx_2-dx_3\wedge dx_4\right)$$
$$\therefore\quad *=-\qquad(12.2.20)$$

It is taken that $*=-$ in (i), (ii), and (iii). The basis belong to Λ_-^2.

(iv) $*\left(dx_1+idx_2\right)\wedge\left(dx_3+idx_4\right)$
$$=*\left(dx_1\wedge dx_3-dx_2\wedge dx_4+idx_1\wedge dx_4+idx_2\wedge dx_3\right)$$
$$=-dx_2\wedge dx_4+dx_1\wedge dx_3+idx_2\wedge dx_3+idx_1\wedge dx_4$$
$$=\left(dx_1+idx_2\right)\wedge\left(dx_3+idx_4\right)\qquad\therefore\quad *=+\qquad(12.2.21)$$

(v) $*\left(dx_1-idx_2\right)\wedge\left(dx_3-idx_4\right)$
$$=*\left(dx_1\wedge dx_3-dx_2\wedge dx_4-idx_1\wedge dx_4-idx_2\wedge dx_3\right)$$
$$=-dx_2\wedge dx_4+dx_1\wedge dx_3-idx_2\wedge dx_3-idx_1\wedge dx_4$$
$$=\left(dx_1-idx_2\right)\wedge\left(dx_3-idx_4\right)\qquad\therefore\quad *=+\qquad(12.2.22)$$

(vi) $*\left(dx_1\wedge dx_2+dx_3\wedge dx_4\right)$
$$=\left(dx_3\wedge dx_4+dx_1\wedge dx_2\right)=\left(dx_1\wedge dx_2+dx_3\wedge dx_4\right)$$
$$\therefore\quad *=+\qquad(12.2.23)$$

It is taken that $*=+$ in (iv), (v), and (vi). The basis belong to Λ_+^2.

Hence we know that Λ_-^2 has three base vectors (i), (ii) and (iii); Λ_+^2 has three base vectors (iv), (v) and (vi).

Now let us come back to Eq.12.2.6. Use the property of Chern character in Eq.10.2.8, we have

$$\mathrm{ch}\left(\Lambda_g^0(M)-\Lambda_g^1(M)+\Lambda_{g-}^2(M)\right)$$
$$=\mathrm{ch}\left(\Lambda_g^0(M)\right)-\mathrm{ch}\left(\Lambda_g^1(M)\right)+\mathrm{ch}\left(\Lambda_{g-}^2(M)\right)$$
$$=\mathrm{ch}\left[V_g\otimes\left(\Lambda^0(M)\ominus\Lambda^1(M)\oplus\Lambda_{g-}^2(M)\right)\right]\qquad\text{(by Eq.12.2.12)}$$
$$=\mathrm{ch}\left(V_g\right)\cdot\mathrm{ch}\left(\Lambda^0-\Lambda^1+\Lambda_-^2\right)(M)\qquad\text{(by Eq.10.2.8)}$$
$$=\mathrm{ch}\left(V_g\right)\cdot\left(\mathrm{ch}\,\Lambda^0-\mathrm{ch}\,\Lambda^1+\mathrm{ch}\,\Lambda_-^2\right)\qquad\text{(the symbol }(M)\text{ is omitted)}\qquad(12.2.24)$$

Λ^0 is the set of the 0-forms, and is equivalent to a trivial bundle with the curvature 0, that is, $\mathrm{ch}\,\Lambda^0=1$ (see Eq.12.2.13$_1$). Λ^1 is the set of the 1-forms with $\dim\left(S^4\right)=4$.

By Decomposition Theorem, $\mathrm{ch}(\Lambda^1)$ on S^4 can be written as

$$\mathrm{ch}(\Lambda^1) = \mathrm{ch}(L_a \oplus \overline{L}_a \oplus L_b \oplus \overline{L}_b) = e^{x_a} + e^{-x_a} + e^{x_b} + e^{-x_b}$$

where x_a and x_b are both purely imaginary numbers according to Eq.10.2.7.

Λ_-^2 is the set of the 2-forms of which the $*$ eigenvalues are (-1). According to Eq.10.2.13, Eq.10.2.14, and Eq.10.2.15 of Chapter 10, the complex vector bundle E can be decomposed into a direct sum of the complex line bundles (see Decomposition Theorem in Section 10.2). Therefore, by Eq.12.2.18 and Eq.12.2.19 we may define

the base of $\overline{L}_a$ to be $dx_1 + idx_2$, the base of L_a to be $dx_1 - idx_2$,

the base of $\overline{L}_b$ to be $dx_3 + idx_4$, and the base of L_b to be $dx_3 - idx_4$,

where L_a and L_b are both complex numbers. In contrast to this, $dx_1 \wedge dx_2 - dx_3 \wedge dx_4$ in Eq.12.2.20 is a real scalar which is invariant under rotation, and can be considered ≈ 1. Thus we have

$$\Lambda_-^2 = \overline{L}_a \otimes L_b + L_a \otimes \overline{L}_b + 1 \tag{12.2.25}$$

Therefore,

$$\begin{aligned}\mathrm{ch}(\Lambda_-^2) &= \mathrm{ch}\,\overline{L}_a \cdot \mathrm{ch}\,L_b + \mathrm{ch}\,L_a \cdot \mathrm{ch}\,\overline{L}_b + \mathrm{ch}\,1 \\ &= e^{-x_a} \cdot e^{x_b} + e^{x_a} \cdot e^{-x_b} + 1 \\ &= 1 + e^{x_a - x_b} + e^{-x_a + x_b}\end{aligned} \tag{12.2.26}$$

Combine Eq.12.2.13$_1$, Eq.12.2.14, and Eq.12.2.26, we have

$$\begin{aligned}&\mathrm{ch}(\Lambda^0 - \Lambda^1 + \Lambda_-^2) = \mathrm{ch}(\Lambda^0) - \mathrm{ch}(\Lambda^1) + \mathrm{ch}(\Lambda_-^2) \\ &= 1 - (e^{x_a} + e^{-x_a} + e^{x_b} + e^{-x_b}) + (1 + e^{x_a - x_b} + e^{-x_a + x_b}) \\ &= 2 - 4 + 2 - \frac{2x_a^2}{2!} - \frac{2x_b^2}{2!} + 2\frac{(x_a - x_b)^2}{2!} - 2\frac{x_a^4}{4!} - 2\frac{x_b^4}{4!} + 2\frac{(x_a - x_b)^4}{4!}\end{aligned}$$

(Since the manifold S^4 is 4-dimensional, then the exterior products on S^4 can not exceed 4-forms.)

$$\begin{aligned}&= -2x_a x_b - \frac{x_a^4 + x_b^4}{12} + \frac{(x_a - x_b)^4}{12} \\ &= -2x_a x_b - \frac{4x_a^3 x_b - 6x_a^2 x_b^2 + 4x_a x_b^3}{12} \\ &= x_a x_b \left(-2 - \frac{1}{3}x_a^2 + \frac{1}{2}x_a x_b - \frac{1}{3}x_b^2 \right)\end{aligned} \tag{12.2.27}$$

Then multiply Eq.12.2.27 and Eq.12.2.11, we obtain

$$\begin{aligned}&\mathrm{ch}(\Lambda^0 - \Lambda^1 + \Lambda_-^2) \cdot \frac{td(T(M) \otimes C)}{eT(M)} \\ &= x_a x_b \left(-2 - \frac{1}{3}x_a^2 + \frac{1}{2}x_a x_b - \frac{1}{3}x_b^2 \right) \frac{1}{x_a x_b} \left(1 - \frac{x_a^2}{12}\right)\left(1 - \frac{x_b^2}{12}\right)\end{aligned}$$

$$= x_a x_b\left(-2-\frac{1}{3}x_a^2+\frac{1}{2}x_a x_b-\frac{1}{3}x_b^2\right)\left(1-\frac{x_a^2}{12}-\frac{x_b^2}{12}\right)\frac{1}{x_a x_b}$$

$$= x_a x_b\left(-2+\frac{1}{6}x_a^2+\frac{1}{6}x_b^2-\frac{1}{3}x_a^2+\frac{1}{2}x_a x_b-\frac{1}{3}x_b^2\right)\frac{1}{x_a x_b}$$

$$=\left(-2-\frac{1}{6}\left(x_a^2+x_b^2\right)+\frac{1}{2}x_a x_b\right) \tag{12.2.28}$$

Since the manifold S^4 is 4-dimensional, no term higher than a 4-form will appear to the left of $\frac{1}{x_a x_b}$.

Recall Eq.12.2.6 and Eq.12.2.24, together with Eq.12.2.28, we obtain

$$\int_{M=S^4} \mathrm{ch}\left(\Lambda_g^0(M)-\Lambda_g^1(M)+\Lambda_{g-}^1(M)\right)\frac{td(T(M)\otimes C)}{e(T(M))}$$

$$=\int_{M=S^4} \mathrm{ch}\left(V_g\right)\cdot\left(\mathrm{ch}\,\Lambda^0-\mathrm{ch}\,\Lambda^1+\mathrm{ch}\,\Lambda_-^2\right)\frac{td(T(M)\otimes C)}{e(T(M))}$$

$$=\int_{M=S^4} \mathrm{ch}\left(V_g\right)\cdot\left(-2-\frac{1}{6}\left(x_a^2+x_b^2\right)+\frac{1}{2}x_a x_b\right) \tag{12.2.29}$$

where V_g is the adjoint representation of the $SU(N)$ group with $\left(N^2-1\right)$ generators. It forms a $\left(N^2-1\right)$-dimensional vector space. Take

$$\mathrm{ch}\left(V_g\right)=\left(N^2-1\right)+2Nk\eta \tag{12.2.30}$$

where η satisfies $\int_{M=S^4}\eta(M)=1$. Then we have

$$\text{Eq.12.2.29}=\int_{M=S^4}\left(\left(N^2-1\right)+2Nk\eta\right)\left(-2-\frac{1}{6}\left(x_a^2+x_b^2\right)+\frac{1}{2}x_a x_b\right)$$

$$=\int_{M=S^4}\left[\left(N^2-1\right)\left(-\frac{1}{6}\left(x_a^2+x_b^2\right)+\frac{1}{2}x_a x_b\right)-4Nk\eta\right] \tag{12.2.31}$$

(Since $-2\left(N^2-1\right)$ is a constant, and hence does not contribute to the integral $\int_{M=S^4}$.)

Let $\tau(M)$ be the symbolic difference:

$$\tau(M)=\int_{M=S^4}\frac{1}{3}\left(x_a^2+x_b^2\right). \tag{12.2.32}$$

Let $\chi(M)$ be the Euler's number:

$$\chi(M)=\int_{M=S^4}x_a x_b. \tag{12.2.33}$$

Plug $\tau(M)$ and $\chi(M)$ into Eq.12.2.31, we obtain

$$\text{Eq.12.2.29} = -\frac{1}{2}\tau(M)\left(N^2-1\right)+\frac{1}{2}\chi(M)\left(N^2-1\right)-4Nk \tag{12.2.34}$$

When $M = S^4$, we have $\tau\left(S^4\right)=0$ and $\chi\left(S^4\right)=2$. Thus Eq.12.2.34 can also be defined as

$$\text{Eq.12.2.29} = \left(N^2-1\right)-4Nk = -h^1 \tag{12.2.35}$$

For example, with $M = S^4$,

$$\left.\begin{array}{ll} SU(N): & h^1 = 4Nk-\left(N^2-1\right) \\ SU(2): & h^1 = 8k-3 \\ SU(3): & h^1 = 12k-8 \\ \cdots\cdots & \cdots\cdots \end{array}\right\} \tag{12.2.36}$$

§ 12.3 Topological Classification of Simply Connected 4-Manifolds

Consider a compact connected surface Σ without boundary. Let γ_1 and γ_2 be two closed curves on Σ. Then γ_2 can cross γ_1 (not tangent, not keep crossing through for infinitely many times) after a simple transformation. Take the intersection number and mod 2, we obtain a number which depends on the homotopy classes of γ_1 and γ_2 only.

The simplest non-orientable Σ is $RP(2)$ $(= P^2(R))$, of which the neighborhood of the closed curve γ_1 forms the Möbius strip. Let γ' be another closed curve on $RP(2)$, which is transformed into the Möbius strip. Then it is illustrated by the following figure that the intersection number of γ and γ' is 1 mod 2.

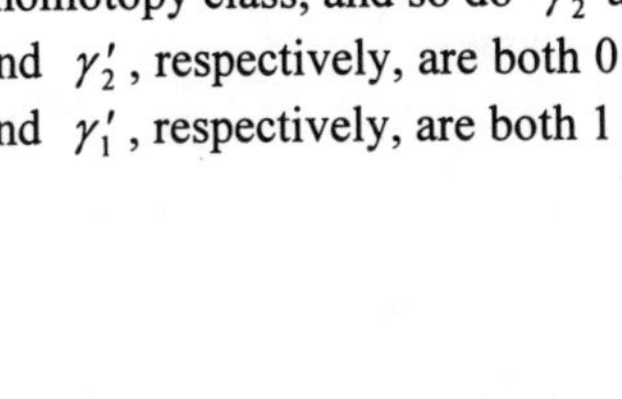

$$\mu = (1), \quad \text{a } 1\times 1 \text{ matrix} \tag{12.3.1}$$

The simplest oriented Σ is $S^1 \times S^1$ (torus). On $S^1 \times S^1$, γ_1 and γ_1' belong to the same homotopy class, and so do γ_2 and γ_2'. The intersection numbers of γ_1 and γ_1', γ_2 and γ_2', respectively, are both 0 mod 2. The intersection numbers of γ_1 and γ_2', γ_2 and γ_1', respectively, are both 1 mod 2.

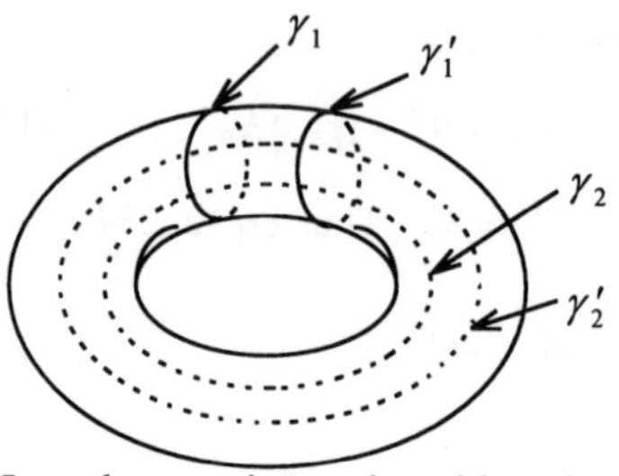

$$\mu = \begin{pmatrix} 0 & 1 \\ 1 & 0 \end{pmatrix}, \quad \text{a } 2\times 2 \text{ matrix} \tag{12.3.2}$$

Here the matrix μ is said to be an intersection form, of which the different columns and rows correspond to different homotopy classes. For example, γ_1 and γ_1' belong to the same homotopy class, γ_2 and γ_2' belong to another one, etc..

A matrix μ similar to those of Eq.12.3.1 and Eq.12.3.2 is called an *intersection form*, of which the different columns and rows correspond to different homotopy classes, such as that γ_1 and γ_1' belong to one homotopy class, γ_2 and γ_2' belong to another one, etc..

Since a closed curve on S^2 (sphere) can always be contracted into a single point, thus the smallest oriented Σ should be $S^1 \times S^1$, not S^2 (since the north and south poles of S^2 are not oriented).

To generalize the results to a more general Σ, we need the concept of the connected sum.

Connected Sum:

$\Sigma = A\#B$ is said to be a connected sum, which connects A and B in the following way:

$$\Sigma = (A-\sigma_1)\cup_h(B-\sigma_2); \quad h:\ \partial\sigma_1 \xrightarrow{\text{opposite orientation}} \partial\sigma_2 \text{ diffeomorphism} \tag{12.3.3}$$

where σ_1 and σ_2 are tiny regions which are taken out of A and B, respectively; $\partial\sigma_1$ and $\partial\sigma_2$ are the boundaries of σ_1 and σ_2, respectively; and h means to connect $\partial\sigma_1$ and $\partial\sigma_2$ according to the opposite orientation.

Therefore, after eliminating S^2 which is homotopically trivial, Σ can be divided into two classes: one is non-orientable, such as that in case of $W_1(T(M)) \neq 0$ (see Eq.10.5.6); the other one is oriented, such as that in case of $W_1(T(M)) = 0$ (see Eq.10.5.7).

Generalizing the previous discussion, we know that the intersection properties can be illustrated by μ matrix (or say the intersection form), of which the different columns and rows represent different homotopy classes.

I. Non-orientable Σ ($W_1 \neq 0$).

$$\text{Type I:} \qquad \mu = \begin{pmatrix} 1 & & & & \mathbf{0} \\ & 1 & & & \\ & & 1 & & \\ & \mathbf{0} & & \ddots & \\ & & & & 1 \end{pmatrix}$$

It is an odd form since the diagonal numbers are odd.

$$\Sigma = RP(2)\#\cdots\# RP(2). \tag{12.3.4}$$

II. Oriented Σ $(W_1 = 0)$.

$$\text{Type II:} \qquad \mu = \begin{pmatrix} 0 & 1 & & & \mathbf{0} \\ 1 & 0 & & & \\ & & \ddots & & \\ & \mathbf{0} & & 0 & 1 \\ & & & 1 & 0 \end{pmatrix}$$

It is an even form since the diagonal numbers are even (see Eq.12.3.2).

$$\Sigma = \left(S^1 \times S^1\right)\#\cdots\#\left(S^1 \times S^1\right). \tag{12.3.5}$$

Theorem 12.3.1: Two compact connected surfaces are diffeomorphic to each other iff their intersection forms are equivalent. (The proof is omitted.)

Hence we see that μ gives the classification of the diffeomorphic properties of non-trivial Σ.

Now we apply the above discussions to 4-manifolds.

The principal feature of a 4-dimensional compact simply connected manifold M^4 without boundary[1] is the simply connectivity, which makes the fundamental group $\Pi_1\left(M^4\right) = 0$ and hence the homology group $H_1\left(M^4\right) = 0$. Furthermore, according to Poincaré Duality Theorem and the dualities of $H_i\left(M^4\right)$ and $H^i\left(M^4\right)$ (see Eq.7.5.1 and the related corollaries), we should have the following formulae:

$$H_1\left(M^4\right) = H_3\left(M^4\right) = 0, \tag{12.3.6}$$

$$H_0\left(M^4, G\right) = H_4\left(M^4, G\right) = G. \tag{12.3.7}$$

Obviously no information for the classification of M^4 manifolds can be obtained from Eq.12.3.6 and Eq.12.3.7. Thus the homology group which provides information on M^4 must be $H_2\left(M^4\right)$.

Let F^k be a k-manifold, and F^l an l-manifold. $F^k, F^l \subset M^n$, where $n = k + l$. If F^k cross-intersects a closed submanifold G^l, then under the above restriction of the dimensions, they must intersect at finitely many points. Therefore, two cross-

1 A simply connected manifold M^4 is oriented.

intersecting 2-dimensional closed surfaces of M^4 intersect at finitely many points.

Similar to the case of 2-dimensional Σ, the intersection number of 2-dimensional surfaces F and G on 4-dimensional M^4 can be defined as well.

The following are two examples of M^4. Notice that the simply connected manifold M^4 is oriented.

(i) S^4: $H_2(S^4, Z) = 0$. The intersection number of all the circuits is zero, just as what has happened in Example 1 of Section 6.2. Therefore S^4 is a simply connected manifold.

(ii) $S^2 \times S^2$: $H_2(S^2 \times S^2, Z)$. The standard generators are the imbedded $S^2 \times \{\text{point}\}$ and $\{\text{point}\} \times S^2$, of which the self-intersection numbers are zero, and the cross-intersection number is 1.

$H_2(S^2 \times S^2, Z)$ on M^4 is similar to $H_1(S^1 \times S^1, Z)$ on M^2. For example, the standard generators of $H_1(S^1 \times S^1, Z)$ are the imbedded $S^1 \times \{\text{point}\}$ and $\{\text{point}\} \times S^1$, of which the self-intersection numbers are zero, and the cross-intersection number is 1. See the figure below, where $a = S^1 \times \{\text{point}\}$ and $b = \{\text{point}\} \times S^1$.

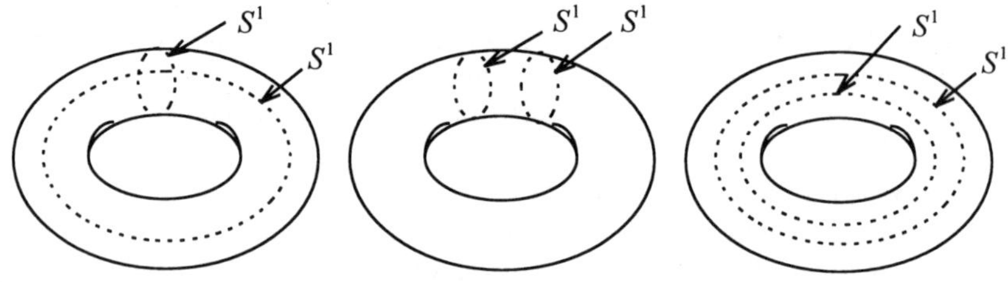

Intersection number of $a \cdot b = 1$ | Intersection number of $a \cdot a = 0$ | Intersection number of $b \cdot b = 0$

Figure 12.3.1

Similar to Eq.12.3.3, the *connected sum* $M^4 = P\#Q$ of 4-manifolds can be defined as follows:

$$\left.\begin{aligned} &M^4 = \left(P - \text{In } D_1^4\right) \underset{h}{\cup} \left(Q - \text{In } D_2^4\right); \\ &h:\ \partial \text{In } D_1^4 \xrightarrow{\text{opposite orientation}} \partial \text{In } D_2^4 \ \text{diffeomorphism}; \end{aligned}\right\} \tag{12.3.8}$$

where $\text{In } D_1^4$ and $\text{In } D_2^4$ are the small pieces taken from P and Q, respectively; $\partial \text{In } D_1^4$ and $\partial \text{In } D_2^4$ are the boundaries of $\text{In } D_1^4$ and $\text{In } D_2^4$, respectively; and h means to connect $\partial \text{In } D_1^4$ and $\partial \text{In } D_2^4$ according to the opposite orientation.

If P and Q are both simply connected, then so is the connected sum $P\#Q$.

After eliminating S^4 which is homotopically trivial, M^4 can be classified according to the following intersection forms (i.e. μ matrices). Notice that μ matrix is restricted to be *non-positive definite*, that is, the determinant of μ is not necessarily >0. The classification is as follows:

Intersection Form of Type I:

$$\mu \cong <1> \oplus \cdots \oplus <1> \oplus <-1> \oplus \cdots \oplus <-1>$$

$$= \begin{pmatrix} 1 & & & & & \\ & \ddots & & & \mathbf{0} & \\ & & 1 & & & \\ & & & -1 & & \\ & \mathbf{0} & & & \ddots & \\ & & & & & -1 \end{pmatrix} \tag{12.3.9}$$

It is an odd form (the diagonal numbers are odd) which satisfies

$$M^4 = CP^2 \# \cdots \# CP^2 \# \overline{CP}^2 \# \cdots \# \overline{CP}^2 \tag{12.3.10}$$

Since μ has to be non-positive definite, therefore we must have both CP^2 and $\overline{CP}^2$. That is, $<1>$ and $<-1>$ must both be in μ (see Eq.12.3.9).

Intersection Form of Type II:

$$\mu \cong H \oplus \cdots \oplus H \oplus E_8 \oplus \cdots \oplus E_8$$

$$= \begin{pmatrix} \boxed{\begin{matrix} 0 & 1 \\ 1 & 0 \end{matrix}} & & & & \mathbf{0} \\ & \ddots & & & \\ & & \boxed{\begin{matrix} 0 & 1 \\ 1 & 0 \end{matrix}} & & \\ & & & E_8 & \\ \mathbf{0} & & & & \ddots \\ & & & & & E_8 \end{pmatrix} \quad \text{where } H = \begin{pmatrix} 0 & 1 \\ 1 & 0 \end{pmatrix}. \tag{12.3.11}$$

It is an even form (the diagonal numbers are even) which satisfies

$$M^4 \cong (S^2 \times S^2) \# \cdots \# (S^2 \times S^2) \# E_8 \# \cdots \# E_8 . \tag{12.3.12}$$

Since μ has to be non-positive definite, therefore we must have at least one $(S^2 \times S^2)$. That is, μ must have at least one $\begin{pmatrix} 0 & 1 \\ 1 & 0 \end{pmatrix}$ (see the example of Eq.12.3.2 where the intersection number of the 2×2 matrix is 1).

CP^2 and $\overline{CP}^2$ in Eq.12.3.10, and $H = \begin{pmatrix} 0 & 1 \\ 1 & 0 \end{pmatrix} = S^2 \times S^2$ in Eq.12.3.12 are all well-defined (see Eq.12.3.2 and the example given by Figure 12.3.1 where the intersection number of $a \cdot b = 1$). E_8 in Eq.12.3.11 is defined by

$$E_8 \equiv \begin{pmatrix} 2 & 1 & 0 & 0 & 0 & 0 & 0 & 0 \\ 1 & 2 & 1 & 0 & 0 & 0 & 0 & 0 \\ 0 & 1 & 2 & 1 & 0 & 0 & 0 & 0 \\ 0 & 0 & 1 & 2 & 1 & 0 & 0 & 0 \\ 0 & 0 & 0 & 1 & 2 & 1 & 1 & 0 \\ 0 & 0 & 0 & 0 & 1 & 2 & 0 & 0 \\ 0 & 0 & 0 & 0 & 1 & 0 & 2 & 1 \\ 0 & 0 & 0 & 0 & 0 & 0 & 1 & 2 \end{pmatrix} \tag{12.3.13}$$

Notice that the non-zero off-diagonal elements of E_8 happen to be consistent with the following Dynkin figure of E_8 in Cartan classification:

E_8 : 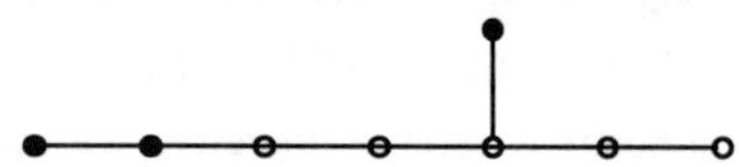 (12.3.14)

Permute the rows and the columns of E_8 in (12.3.14), we obtain:

$$E_8 = \begin{pmatrix} 2 & 1 & 0 & 0 & 0 & 0 & 0 & 0 \\ 1 & 2 & 1 & 0 & 0 & 0 & 0 & 0 \\ 0 & 1 & 2 & 1 & 0 & 0 & 0 & 0 \\ 0 & 0 & 1 & 2 & 1 & 0 & 0 & 0 \\ 0 & 0 & 0 & 1 & 2 & 1 & 1 & 0 \\ 0 & 0 & 0 & 0 & 1 & 2 & 0 & 0 \\ 0 & 0 & 0 & 0 & 1 & 0 & 2 & 1 \\ 0 & 0 & 0 & 0 & 0 & 0 & 1 & 2 \end{pmatrix} = (-1) \begin{pmatrix} 2 & 1 & 0 & 0 & 0 & 0 & 0 & 0 \\ 1 & 2 & 1 & 0 & 0 & 0 & 0 & 0 \\ 0 & 1 & 2 & 1 & 0 & 0 & 0 & 0 \\ 0 & 0 & 1 & 2 & 1 & 0 & 0 & 0 \\ 0 & 0 & 0 & 1 & 2 & 1 & 1 & 0 \\ 0 & 0 & 0 & 0 & 1 & 0 & 2 & 0 \\ 0 & 0 & 0 & 0 & 1 & 2 & 0 & 1 \\ 0 & 0 & 0 & 0 & 0 & 0 & 1 & 2 \end{pmatrix}$$

$$= (-1)^2 \begin{pmatrix} 2 & 1 & 0 & 0 & 0 & 0 & 0 & 0 \\ 1 & 2 & 1 & 0 & 0 & 0 & 0 & 0 \\ 0 & 1 & 2 & 1 & 0 & 0 & 0 & 0 \\ 0 & 0 & 1 & 2 & 1 & 0 & 0 & 0 \\ 0 & 0 & 0 & 1 & 2 & 1 & 1 & 0 \\ 0 & 0 & 0 & 0 & 1 & 2 & 0 & 1 \\ 0 & 0 & 0 & 0 & 1 & 0 & 2 & 0 \\ 0 & 0 & 0 & 0 & 0 & 1 & 0 & 2 \end{pmatrix} = (-1)^3 \begin{pmatrix} 2 & 1 & 0 & 0 & 0 & 0 & 0 & 0 \\ 1 & 2 & 1 & 0 & 0 & 0 & 0 & 0 \\ 0 & 1 & 2 & 1 & 0 & 0 & 0 & 0 \\ 0 & 0 & 1 & 2 & 1 & 0 & 0 & 0 \\ 0 & 0 & 0 & 1 & 2 & 1 & 1 & 0 \\ 0 & 0 & 0 & 0 & 1 & 2 & 0 & 1 \\ 0 & 0 & 0 & 0 & 0 & 1 & 0 & 2 \\ 0 & 0 & 0 & 0 & 1 & 0 & 2 & 0 \end{pmatrix}$$

$$= (-1)^4 \begin{pmatrix} 2 & 1 & 0 & 0 & 0 & 0 & 0 & 0 \\ 1 & 2 & 1 & 0 & 0 & 0 & 0 & 0 \\ 0 & 1 & 2 & 1 & 0 & 0 & 0 & 0 \\ 0 & 0 & 1 & 2 & 1 & 0 & 0 & 0 \\ 0 & 0 & 0 & 1 & 2 & 1 & 0 & 1 \\ 0 & 0 & 0 & 0 & 1 & 2 & 1 & 0 \\ 0 & 0 & 0 & 0 & 0 & 1 & 2 & 0 \\ 0 & 0 & 0 & 0 & 1 & 0 & 0 & 2 \end{pmatrix} = \begin{pmatrix} 2 & 1 & 0 & 0 & 0 & 0 & 0 & 0 \\ 1 & 2 & 1 & 0 & 0 & 0 & 0 & 0 \\ 0 & 1 & 2 & 1 & 0 & 0 & 0 & 0 \\ 0 & 0 & 1 & 2 & 1 & 0 & 0 & 0 \\ 0 & 0 & 0 & 1 & 2 & 1 & 0 & 1 \\ 0 & 0 & 0 & 0 & 1 & 2 & 1 & 0 \\ 0 & 0 & 0 & 0 & 0 & 1 & 2 & 0 \\ 0 & 0 & 0 & 0 & 1 & 0 & 0 & 2 \end{pmatrix} \tag{12.3.15}$$

Notice that the first, third, and fifth equivalent forms of E_8 matrix in Eq.12.3.15 are all diagonally symmetric.

§ 12.4 Donaldson's Theorem

Let M^4 be a compact simply connected 4-manifold which is oriented and smooth. The intersection form μ_M of this manifold is positive definite. Then μ_M must be equivalent to a diagonal form, that is,

$$\mu_M \cong <1> \oplus \cdots \oplus <1> = \begin{pmatrix} 1 & & 0 \\ & \ddots & \\ 0 & & 1 \end{pmatrix} \tag{12.4.1}$$

which gives Donaldson's Theorem. Before proving this theorem, let us consider the instanton of $k=1$ (i.e. the Yang-Mills self-dual solution) for $SU(2)$ group on this 4-manifold M^4. The moduli space of this instanton is the 5-manifold $\mathcal{M}$ (see Eq.12.2.36 for $SU(2)$: when $k=1$, the dimension of the moduli space $=8-3=5$). The instanton also has the following properties:

a. There are n singular points on $\mathcal{M}$, that is, $(p_1, p_2, \cdots, p_n) \in \mathcal{M}$. Furthermore, $\mathcal{M} - (p_1, p_2, \cdots, p_n)$ is a smooth (i.e. differentiable) 5-manifold.
b. The neighborhood of each singular point p_i is a CP^2 cone, of which the vertex is the singular point p_i, and the base (of 4-dimensional) is a CP^2.
c. $\mathcal{M}$ is oriented. See Figure 12.4.2.
d. $\mathcal{M} - (p_1, p_2, \cdots, p_n) \neq \varnothing$. That is, the 5-manifold $\mathcal{M}$ does exist. It contains a "collar" $(0, \lambda] \times M^4 \subset \mathcal{M}$ which is open below and is closed above (see Figure 12.4.2).
e. $\overline{\mathcal{M}} = \mathcal{M} \cup M^4$ is compact. Since M^4 is attached to the lower border of $\overline{\mathcal{M}}$, thus the "collar" of $\overline{\mathcal{M}}$ is $[0, \lambda] \times M^4 \subset \overline{\mathcal{M}}$.

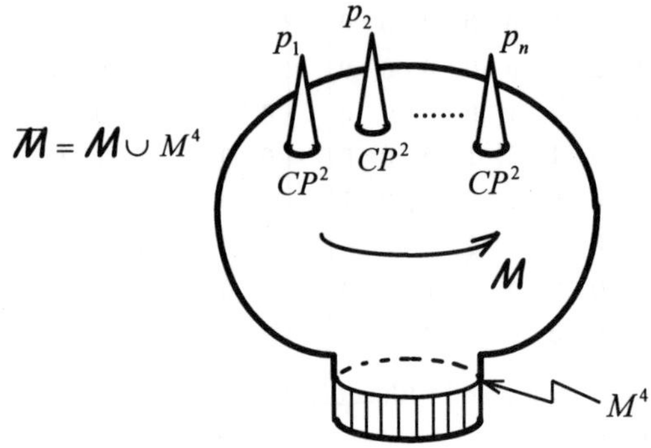

Figure 12.4.2

The thick black curves represent the compactness of $\mathcal{M}$ (see part **e**). The arrow above $\mathcal{M}$ represents the orientation (see part **c**).

Cut off the surface C_{p_i} of the cone at the neighborhood of each singular point p_i ($i = 1, 2, \cdots, n$), we obtain the following figure:

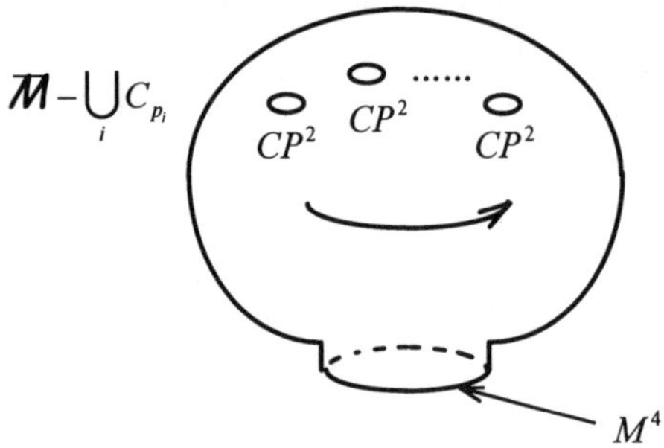

Figure 12.4.3

$\mathcal{M} - \bigcup_i C_{p_i}$ in Figure 12.4.3 is an oriented 5-manifold, of which the borders are M^4 and $\underbrace{\pm CP^2 \pm CP^2 \cdots \pm CP^2}_{\text{(There are } n \text{ terms of } (\pm CP^2))}$. The reason we have $\pm$ signs in front of CP^2 is that they might have different orientations.

Cobordism Relation:

Example 1. See Figure 12.4.4, where the arrow $\Longrightarrow$ represents the orientation of $\mathcal{M}^5$. Therefore we can see that

i) M_1^4 has the same orientation as that of $\mathcal{M}^5$, and hence M_1^4 takes the $(+)$ sign;

ii) M_2^4 has the opposite orientation as that of $\mathcal{M}^5$, and hence M_2^4 takes the $(-)$ sign.

Thus the boundary $\partial \mathcal{M}^5$ of $\mathcal{M}^5$ should be

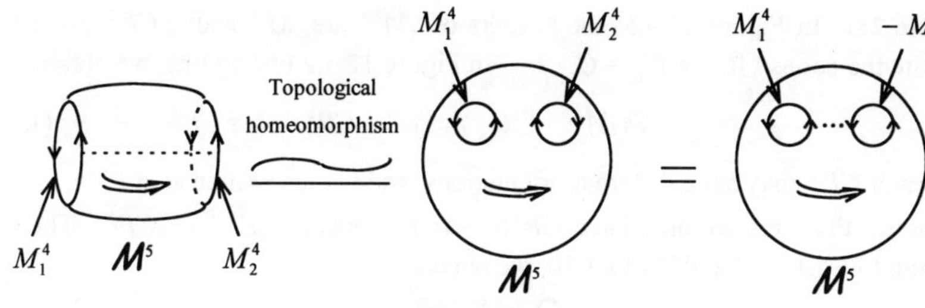

Figure 12.4.4

$$\partial M^5 = M_1^4 \cup (-M_2^4) \tag{12.4.2}$$

The above formula is said to be the cobordism relation between M_1^4 and $-M_2^4$, where ∂M^5 represents the boundary of M^5.

Example 2.

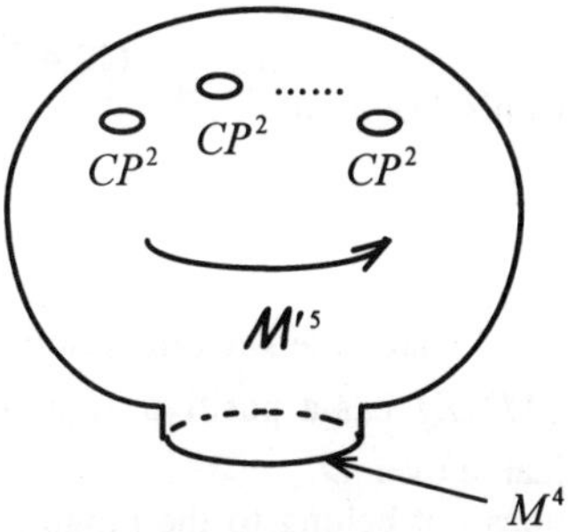

Figure 12.4.5

According to Figure 12.4.3, the orientation of CP^2 could be either positive or negative. Thus we have

$$\partial(M'^5 - \bigcup_i C_{p_i}) = M^4 \pm CP^2 \pm \cdots \pm CP^2 \tag{12.4.3}$$

which gives the cobordism relation between M^4 and $\pm CP^2 \pm \cdots \pm CP^2$. Notice that there are finitely many terms of $\pm CP^2$ since the number of the singular points is finite. We have $\pm$ signs since the orientation of CP^2 could be either positive or negative.

Cobordism Theorem:

The sum of the signatures of all the borders which possess the cobordism relation is zero.

This property can be demonstrated by the cobordism relations of the previous two examples. Let Q represent the border, and $\sigma(Q)$ be the signature of Q. Then we have

Example 1a: In Figure 12.4.4, M_1^4 and M_2^4 have opposite directions of rotation. Hence we have

$$\sigma(Q_{M^5}) = \sigma(Q_{M_1^4}) - \sigma(Q_{M_2^4}) = 0 \tag{12.4.4}$$

where

$$-\sigma(Q_{M_2^4}) = \sigma(Q_{-M_2^4}). \tag{12.4.5}$$

Example 2a: In Figure 12.4.5, the borders of M'^5 are M^4 and $\pm CP^2 \pm \cdots \pm CP^2$. Eliminate the cones $\bigcup_i C_{p_i} = C_{p_1} + C_{p_2} + \cdots$ in Figure 12.4.2 one by one, we obtain

$$\partial(M'^5 - \bigcup_i C_{p_i}) = M^4 \pm CP^2 \pm \cdots \pm CP^2 \qquad (12.4.6)$$

where each CP^2 may have different orientations, and is thus written as $\pm CP^2$.

Notice that here we only have $CP^2 \# \cdots \# CP^2$, but no $\overline{CP}^2 \# \cdots \# \overline{CP}^2$. Therefore according to Eq.12.3.9 and Eq.12.3.10, we obtain

$$\mu_M \cong \begin{pmatrix} 1 & & & & 0 \\ & 1 & & & \\ & & 1 & & \\ & & & \ddots & \\ 0 & & & & \ddots \\ & & & & & 1 \end{pmatrix} = <1> \oplus <1> \oplus \cdots \oplus <1> \qquad (12.4.7)$$

which is consistent with Eq.12.4.1. That proves Donaldson's Theorem.

12.5 Taubes' Theorem

Let M^4 be a smooth simply connected 4-manifold, of which the smooth periodic end is topologically homeomorphic to $S^3 \times R$. If $H_2(M^4, Z)$ has a positive definite unimodular intersection form, then this form is diagonalizable over Z.

Since Z is the set of integers, the manifold Z does not belong to the range of smooth manifolds. Therefore, R_r^4 and R_s^4 are not diffeomorphic to each other for all $0 < r < s < 1$. These R_r^4, R_s^4, which are not diffeomorphic are usually called Taubes R^4's.

In case that more details of Taubes' Theorem are concerned, there are three references [8], [9], [10] listed for the readers.

References

1. T. Eguchi, P. B. Gilkey, and A. C. Hanson, *Phys. Rep.* **66** (1980), 213.
2. S. S. Chern and W. H. Chen, *Text Book on Differential Geometry* (Beijing University Publication, 1983. Chinese).
3. Michael H. Freedman and Feng Luo, *Selected Applications of Geometry to Low-Dimensional Topology* (University Lecture Series, 1047-3998; 1; 1989).
4. F. Hirzebruch, *Topological Methods in Algebraic Geometry* (1966).
5. R. Bott and L. W. Tu, *Differential Forms in Algebraic Topology* (1982).
6. Patrick Shanahan, *The Atiyah-Singer Index Theorem, An Introduction* (Lecture Notes in Mathematics 638, Springer-Verlag, 1978).
7. M. F. Atiyah and I. M. Singer, *Bull. Amer. Math. Soc.* **69** (1963), 422-433.
8. Clifford Henry Taubes, *Self-Duel Yang-Mills Connections on Non-Self-Duel 4-Manifolds*, J. Differential Geometry **17** (1982), 139-170.
9. Clifford Henry Taubes, *Gauge Theory on Asymptotically Periodic 4-Manifolds*, J. Differential Geometry **25** (1987), 363-430.
10. Clifford Henry Taubes, *A Framework for Morse Theory for the Yang-Mills Functional*, Invent. Math. **94**, 327-402 (1988).

Index